Natural Fibres: Advances in Science and Technology Towards Industrial Applications

RILEM BOOKSERIES

Volume 12

RILEM, The International Union of Laboratories and Experts in Construction Materials, Systems and Structures, founded in 1947, is a non-governmental scientific association whose goal is to contribute to progress in the construction sciences, techniques and industries, essentially by means of the communication it fosters between research and practice. RILEM's focus is on construction materials and their use in building and civil engineering structures, covering all phases of the building process from manufacture to use and recycling of materials. More information on RILEM and its previous publications can be found on www.RILEM.net. Indexed by SCOPUS and Springerlink.

More information about this series at http://www.springer.com/series/8781

Raul Fangueiro · Sohel Rana
Editors

Natural Fibres: Advances in Science and Technology Towards Industrial Applications

From Science to Market

Springer

Editors
Raul Fangueiro
School of Engineering
University of Minho
Guimaraes
Portugal

Sohel Rana
University of Minho
Guimaraes
Portugal

ISSN 2211-0844 ISSN 2211-0852 (electronic)
RILEM Bookseries
ISBN 978-94-017-7513-7 ISBN 978-94-017-7515-1 (eBook)
DOI 10.1007/978-94-017-7515-1

Library of Congress Control Number: 2015957095

Printed on acid-free paper

This Springer imprint is published by SpringerNature
The registered company is Springer Science+Business Media B.V. Dordrecht

Preface

In current times, natural fibres are coming up as a promising material for various industrial applications. One can realize the importance of natural fibre-based products from the numerous published articles and filed patents during the last few years. A number of natural fibre-based technical products are already available in the market. Nevertheless, in spite of several attractive properties, coupled with low cost and environmental benefits, the growth of natural fibre market is restricted due to a number of technical and political issues. A great deal of efforts has been directed towards analysing these various factors and developing new approaches to overcome these problems. Moreover, research and innovations on extracting or exploring new fibres, new products and applications are being highly encouraged to improve the market growth of natural fibres. The present book discusses all these issues related to natural fibre in a single volume, providing up-to-date information and current state of research and developments in the field of natural fibres.

The present book is based on the collection of selected high-quality research papers submitted to ICNF 2015—2nd International Conference on Natural Fibers. Wide range of topics related to various aspects of natural fibres such as agriculture, extraction and processing, surface modification and functionalization, advanced structures, nanofibres, composites and nanocomposites, design and product development, applications, market potential, environmental impacts are covered in this book. Divided into different sections on these various topics, the book presents the latest and high-quality research work addressing different approaches and techniques to improve processing, performance, functionalities and cost-effectiveness of natural fibre- and natural-based products, in order to promote their applications in different advanced technical sectors. This book will be a useful reference for materials scientists, teachers and students from various disciplines as well as for R&D staff from different industries using natural fibre-based materials.

The editors would like to thank all the authors who have contributed in this book. Sincere thanks are also due to the FIBRENAMICS research group members for their great efforts in organizing ICNF 2015.

Raul Fangueiro
Sohel Rana

Contents

About the Editors

Prof. Raul Fangueiro is currently professor and senior researcher in the School of Engineering at the University of Minho, Portugal. He is the head of the Fibrous Materials Research Group of the same university with expertise in advanced materials (nano, smart, composites) and structures (3D, auxetic, multiscale) with 25 researchers. He is the mentor and the coordinator of the FIBRENAMICS International Platform (www.fibrenamics.com) including 200 partners developing promotion, dissemination, technology transfer and research activities on fibre-based advanced materials. He has more than 110 published papers in international reputed scientific journals, 320 conference publications, 36 books and 14 patents. He is the scientific coordinator of several national and international research projects on advanced fibrous and composite materials, mainly for building, architectural and healthcare applications. He supervised various Ph.D. and Pos-Doc scientific works and is an expert of the European Technological Textile Platform and member of the editorial board of several leading international scientific journals on composite and fibrous materials.

Dr. Sohel Rana is currently a Senior Scientist at Fibrous Materials Research Group, University of Minho, Portugal. He obtained his bachelor's degree in Textile Technology from University of Calcutta, India, and master's degree and Ph.D. in Fiber Science and Technology from Indian Institute of Technology (IIT), Delhi, India. His current research areas are advanced fibrous and composite materials, natural fibres, nanocomposites, electrospinning, multifunctional and biocomposite materials, and so on. He is author of one book, edited 4 books and contributed in 14 book chapters, 6 keynote and invited papers, and about 100 publications in various refereed journals and international conferences. He is also participating in the editorial board of several scientific journals and is a potential reviewer for numerous scientific journals including Composite Science and Technology, Composites Part A, Composite Interfaces, Journal of Composite Materials, Journal of Reinforced Plastics and Composites, Materials, Powder Technology, Journal of Nanomaterials and Journal of Applied Polymer Science.

Part I
Properties and Functionalities of Natural Fibres and Structures

Fibre Science: Understanding How It Works and Speculating on Its Future

Mário de Araújo

Abstract Fibers are very versatile materials used in many technical applications as well as apparel and fashion. Their mechanical properties are very interesting as they may exhibit very high tensile properties but very low bending and torsional rigidities. This, together with their high specific surface area and low density enables them to have superior performance in many high tech applications. This paper attempts to explain why strong materials have to be slender and fibre like and how fibre assemblies can be designed with controlled anisotropy. Advances in biotechnology, nanotechnology and computer science may enable the design and manufacture (growth) of very advanced fibres, fibre assemblies and fibrous products just by mimicking nature's natural selection process.

Keywords Fibres · Computer and information technology · Biotechnology · Nanotechnology · Mimicking

Introduction

Evolution is the change in the inherited characteristics of biological populations over successive generations. Evolutionary processes give rise to diversity at every level of biological organization, including species, individual organisms and molecules such as DNA and proteins (Hall et al. 2008).

In nature, the evolution of life started in the sea. It is interesting to note that many higher order animals use some kind of scaly material for protection. This is true of fish, reptiles, birds and mammals.

In the case of mammals their body is covered with hairs and each hair is protected by a scaly like cuticle. This scaly structure provides toughness whilst

M. de Araújo (✉)
University of Minho, Fibrous Materials Research Group, Campus de Azurém, 4800-058 Guimarães, Portugal
e-mail: mario.dearaujo@gmail.com

R. Fangueiro and S. Rana (eds.), *Natural Fibres: Advances in Science and Technology Towards Industrial Applications*, RILEM Bookseries 12,
DOI 10.1007/978-94-017-7515-1_1

keeping flexibility. These hairs protect both from mechanical aggression as well as the environment by entrapping air and providing thermal insulation and by waterproofing with the aid of oily body secretions.

By mimicking and intuition mankind has also been using scaly structures for protection, be it for body armor in the form of plaques or the roof of buildings in the form of roof tiles and slabs.

We've also been using the hairs of other animals for protection when we lost most of our own and we call these animal fibres.

Natural selection is the key mechanism of evolution and may be described as the gradual process by which heritable biological traits become either more or less common in a population as a function of the effect of inherited traits on the differential reproductive success of organisms interacting with their environment. The term "natural selection" was popularized by Darwin (1859), who intended it to be compared with artificial selection, now more commonly referred to as selective breeding, i.e. the process of modifying organisms by selection in breeding controlled by humans.

Humans have been domesticating plants and animals since around 12,000 BC, using selective breeding or artificial selection, as contrasted with natural selection (Kingsbury 2009).

The process of selective breeding, in which organisms with desired traits (and thus with the desired genes) are used to breed the next generation and organisms lacking the trait are not bred, is the oldest form of genetic modification by humans (Root 2007; Zohary et al. 2012). Genetic engineering, the direct manipulation of genes using biotechnology, was first accomplished by Herbert Boyer and Stanley Cohen (Anon 1973). Whereas selective breeding is based on naturally occurring genetic variation within a population or species, genetic engineering can involve intentional introduction of genes from different species. Advances have allowed scientists to manipulate, remove, and add genes to a variety of different organisms to induce a range of different traits. Since 1976 the technology has been commercialized, with companies producing and selling genetically modified food and medicine. When genetic material from a different species is added, the resulting DNA is called recombinant DNA and the organism is called a transgenic organism. The first recombinant DNA molecules were produced by Paul Berg (Jackson et al. 1972).

Natural selection, selective breeding and genetic engineering have played important roles in developing today's fibres and no doubt that especially the latter will play an extremely important role in future developments.

The Way Fibres and Fibrous Structures Work

A fibre is a unit of matter characterized by flexibility, fineness and a high ratio of length to thickness (Anon 1963). Because fibres have a high surface to volume ratio, they can be extremely strong materials (Griffith 1921).

Fibres are normally constituted by long and chain-like molecules known as macromolecules or polymers, which may be of organic or inorganic nature. These molecules are able to pack together closely to each other resulting in regions of crystallinity. The degree of orientation of these regions is an important factor in determining the usefulness of a fibre for a particular application. There are other regions however, where the molecules do not hold together and form random arrangements or amorphous regions. The crystalline regions provide strength and rigidity to the fibres while the amorphous regions are responsible for flexibility and reactivity. The ratio of crystalline to amorphous material has an important influence on the properties of the fibres.

Due to the higher stiffness-to-weight and strength-to-weight ratios of some fibres as compared to other materials, such as metals, they are favourite candidates in many technical applications (de Araujo 2011) (Table 1).

Fibres are anisotropic materials and they normally are very strong and stiff along the fibre length but may bend under their own weight. Fibres have low bending and torsional rigidities and they buckle easily.

The structure and properties of fibres: tenacity, modulus, toughness, thermal resistance, chemical resistance and so on must be transferred (translated) into yarn and fabric structures in order to produce products with the desired properties for an application. In the case of the mechanical properties the translation efficiency will very much depend on the degree of the fibres orientation in the yarn and fabric structures developed. It should be noted that the structure and geometry of fabrics also play a major role in achieving particular properties (i.e. thermal resistance and porosity).

Figure 1 illustrates how fibre properties are translated or transferred to a final application (i.e. a composite material).

Textile fabrics designed with load-bearing functions for many technical applications, are often required to exhibit functional elements in multiple directions. In this context a biaxial woven fabric being orthotropic exhibits high strength and low compliance in the axial and radial directions and the opposite in the bias directions. The same is true for a Raschel-knitted fabric incorporating warp and weft inlays. A braided fabric, on the other hand, exhibits low strength and high compliance in

Table 1 Specific properties of some high-performance fibres in comparison with some metals

Specific properties	Carbon HM	Para-aramid HM	E-glass	Hemp	Aluminium	Steel
Specific modulus: E/ρ (N m/kg)[a]	256	80	28	21.7	26	27
Specific strength: σ/ρ (N m/kg)[b]	1.2	2	0.775	0.47	0.05–0.23	0.04–0.27

ρ = Density
σ = Tensile strength
E = Young's modulus
[a]Also known as stiffness-to-weight ratio
[b]Also known as strength-to-weight ratio

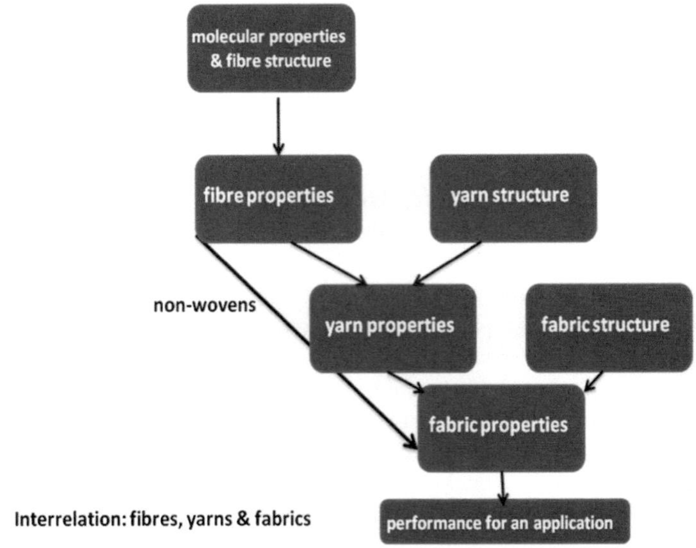

Fig. 1 Transference of fibre properties to a final application

the axial and radial directions and the opposite in the bias directions. Many non-woven fabrics made up of randomly oriented constituent elements enjoy a degree of isotropy. However, conventional products of all four fabric-formation systems are invariably planar sheets, having no functional element in the thickness direction. Even if these planar sheets are joined one on top of the other, as in laminates, the functional elements in the third direction do not exist (Fig. 2).

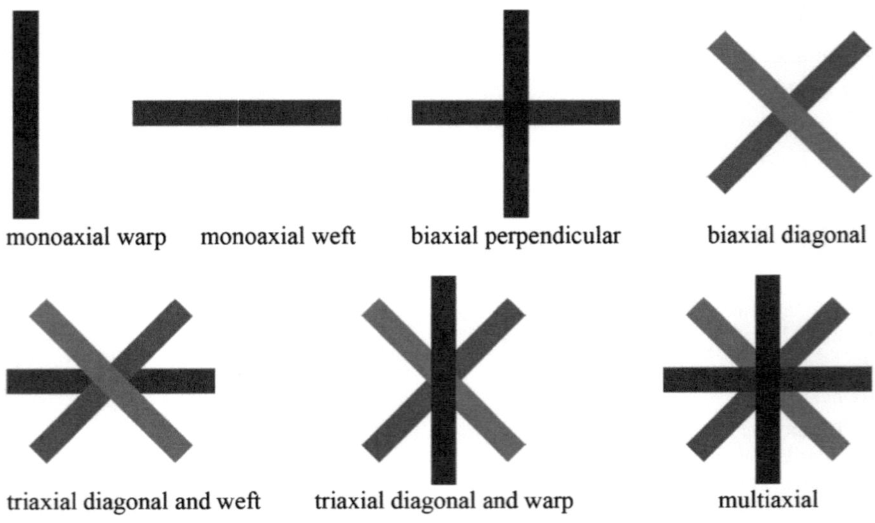

Fig. 2 The range of directionally oriented structures

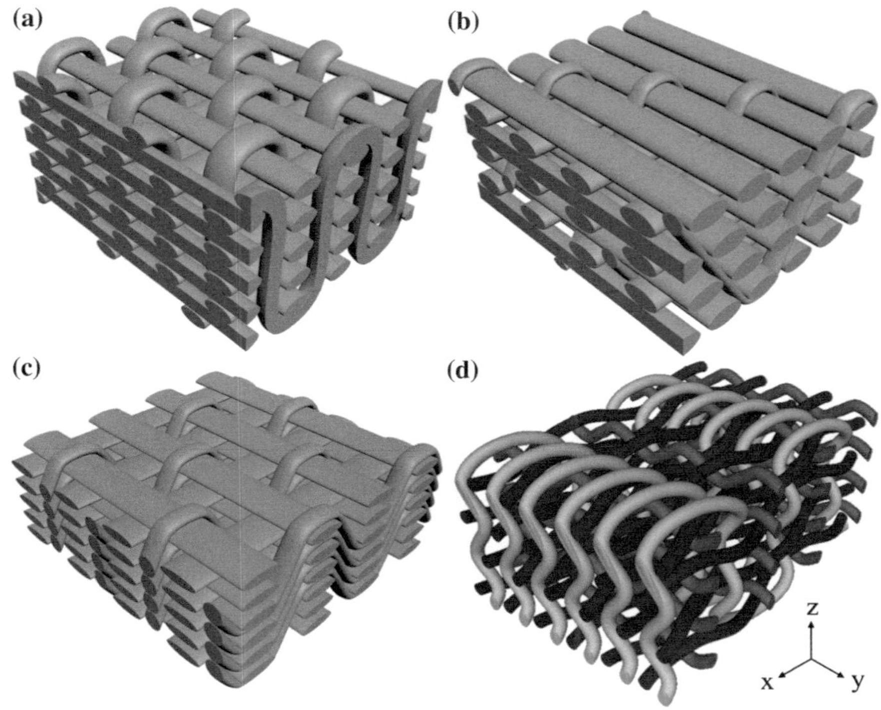

Fig. 3 3D woven structures: **a** orthogonal; **b** through-thickness angle interlock; **c** layer-to-layer angle interlock; **d** fully interlaced

In order to achieve an improved planar isotropy in woven, braided and knitted fabrics, yarns may be introduced in different directions. This approach results in triaxial woven and triaxial braided fabrics as well as multiaxial woven and knitted fabrics. The introduction of load-bearing elements in the third direction in woven, knitted and braided structures has led to the three-dimensional or 3-D fabric-formation systems (Banerjee 2014) (Fig. 3).

It should also be noted that for apparel and fashion fabrics, handle and comfort are very important features and these are related to the initial part of the stress-strain curve where very small loads produce very large deformations as is the case of knitted structures (Fig. 4).

Handle is a complex concept and it depends on the mechanical properties of the fabric under small loads (tensile, compression, bending and sheer), its thermal properties such as heat flow at the initial stage of contact (Q_{max}), which is related to the warm/cool feeling of the material, the fabric surface properties, such as surface friction and surface roughness and some other parameters like fabric mass and thickness.

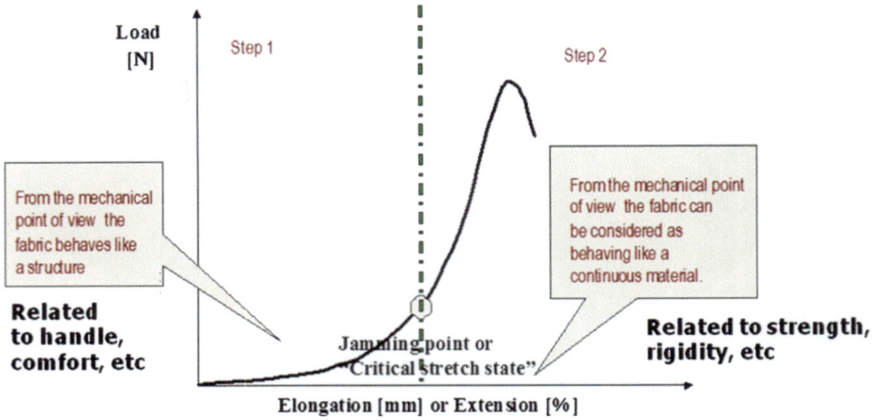

Fig. 4 Typical load-extension curve characteristic of a weft knitted fabric

Fine fibres and softly twisted yarns are essential in the development of fabrics with a soft handle as both the flexural and torsional rigidities of fibres are very much related to the 4th power of fibre diameter as can be seen in Eqs. 1 and 2.

$$G_o = B \times R = \pi d^4 E/64 \tag{1}$$

where

G_o flexural rigidity of a fibre is the couple required to produce unit curvature
B bending moment (couple)
R radius of curvature in bending
l/R curvature
E Young's modulus of the fibre
d fibre diameter

$$\Gamma = M/\tau = \pi G d^4/32 \tag{2}$$

where

Γ torsional rigidity of a fibre is the couple required to produce a rotation of 360° on a fibre length equal to 1 cm
M torque (couple)
d diameter of fibre
τ torsion (angle of rotation in radians/unit length)
G shear modulus

Structures designed for thermal insulation should be light and thick which implies fine fibres, possibly hollow and crimpy and a lot of air pockets or air layers. Equation 3 shows in a very simplified form the principle of heat flow through a fabric (Fourier's law: the rate of heat flow through a fabric is proportional to the negative temperature gradient across the fabric).

$$\mathbf{q} = -k\frac{\partial u}{\partial x} \tag{3}$$

where
q heat flow
k thermal conductivity of the fabric
u temperature

As concerns moisture management, this is understood to be the ability of a textile to absorb gaseous or liquid humidity from the skin, to transport it from the inside of a textile to the outer surface and to release it into the surrounding air. In a simplified form it involves such physical phenomena as diffusion and sorption. Normally the gas part (water vapor) is transferred by diffusion from an area of high concentration to an area of low concentration through pores. This can be illustrated in a very simplified form by Eq. 4 (Fick's 1st law).

$$J = -D\frac{\partial \phi}{\partial x} \tag{4}$$

where
J is the "diffusion flux" which is the amount of water vapour that will flow through unit area in unit time (mol m^{-2} s^{-1})
D is the diffusion coefficient or diffusivity (m^2 s^{-1})
ϕ is the concentration which is the amount of water vapour per unit volume (mol/m^3)
x is the position or length (m)

The liquid part is a**b**sorbed into the amorphous part of the fibre or a**d**sorbed on the fibre surface. This adsorbed part of the liquid is done through the capillaries formed by the fibre assemblies and may be termed wicking. If the diameter of the capillaries formed by the fibres are sufficiently small, then the combination of surface tension (which is caused by cohesion within the liquid) and adhesive forces between the liquid and the capillaries act to lift the liquid. In short, the capillary action is due to the pressure of cohesion and adhesion which cause the liquid to work against gravity. This can be illustrated in a very simplified form by Eq. 5 (Laplace equation).

$$\Delta P = \frac{2\gamma\cos\theta}{Rc} \tag{5}$$

where
ΔP difference in pressure due to liquid flow (column) (Pa)
γ surface tension (N/m);

θ contact angle (rad);
R_c capillary radius (m)

Wetting is the ability of a liquid to maintain contact with a solid surface, resulting from intermolecular interactions when the two are brought together. The degree of wetting (wettability) is determined by a force balance between adhesive and cohesive forces. Wetting deals with the three phases of materials: gas, liquid and solid.

For comfort we need fabrics that dry very quickly (especially for sports) so as to keep the body dry and so, absorption by the fibre is not required as the fabric takes a long time to dry. Normally, hydrophobic fibres with a large specific surface area are preferred as moisture transfer is done very quickly by diffusion and wicking. However, for everyday life, especially in hot climates, some degree of humidity is required for comfort as slower drying gives a cooling effect that is appreciated and in this case hydrophilic fibres are an interesting option.

In any case, garment fit and size are very important parameters to consider in thermo-physiological comfort. Tight garments worn close to the skin are essential to provide wicking and loose garments favor the formation of air layers that are important for thermal insulation.

Impact of Emerging Sciences and Technologies in Fibre Science

The emerging sciences and technologies such as the ones that are shaping the intelligence revolution, the biotechnology revolution and the quantum revolution (Kaku 2007) are the fundamental areas that we must watch in order to find better solutions both for old and new problems. It is there, at the frontier of science that we must be in order to be able to develop the next generation of products that will lead to a sustainable development that is truly knowledge based (Fig. 5).

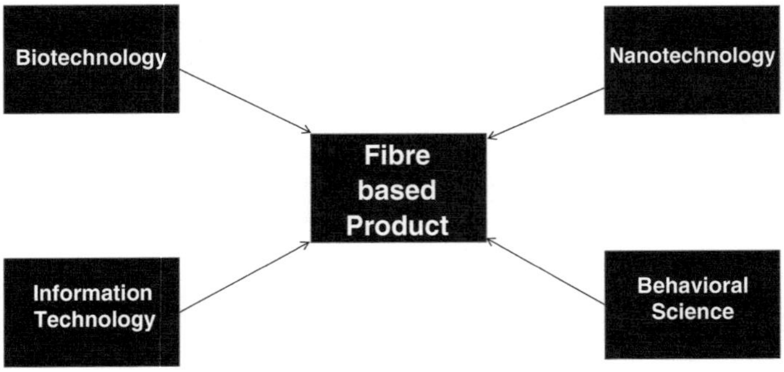

Fig. 5 Emerging sciences and technologies impacting on fibre based products

New generations of fibre based products will integrate micro and nano computers (processing by digital and quantum computing) driven by applications that will perform a variety of tasks (de Araujo 2010a). The common PC will soon disappear into the day to day products used by consumers, as we move into the post-desktop model of human-computer interaction in which information processing is thoroughly integrated into everyday objects and activities. In the course of ordinary activities, someone "using" *ubiquitous computing* will engage many computational devices and systems simultaneously, and may not necessarily even be aware that they are doing so.

The effects of the biotechnology revolution will be far reaching. Even in its nascent stages, this new paradigm is positively affecting the environment, agriculture, criminal investigations and most importantly—the quality of human life. In the environment, biotechnology applications are replacing or enhancing the cleanup of hazardous waste sites, oil spills and abandoned industrial sites. Biotechnology is also being used to prevent pollution by replacing chemicals in laundry detergents with enzymes that are friendlier to the environment.

The use of biomaterials is of great interest for medical textile applications.

Biomimetics is the process of understanding and applying biological principles to human design. Innovation can often be inspired by nature and scientists and engineers are increasingly turning to biomimetics, as the solutions arrived at by natural selection are often a good starting point in the search for answers to scientific and technical problems.

Examples of biomimetic design in textiles include:

- Blue morphus butterfly mimicking for color without pigments
- Pine cones mimicking for ventilation of dampness from sweat
- Shark skin mimicking for fast swimming
- Super waterproof spider mimicking
- Lotus effect for water repellent and self-cleaning
- Sandfish lizard skin mimicking for extremely low friction
- Polar bear coat mimicking for super thermal insulation
- Spider silk mimicking for super high energy absorption (super high strength and super high elastic deformation).

Of these, the latter three have not yet been well understood or mimicked (Vendl et al. 2006).

Examples of biomimetic fabrication in textiles include:

- Mimicking the silkworm for producing man-made fibres
- Mimicking snake skin growth for producing hosiery
- Mimicking insect exoskeleton growth for producing clothing and armor
- Mimicking cotton fibre growth for producing clothing.

Of these, the latter three have not yet been well understood or mimicked (de Araujo 2010b).

Generally, nanotechnology deals with structures of the size 100 nm or smaller in at least one dimension and involves developing materials or devices within that size. It is concerned with controlling matter at an atomic and molecular scale dealing with approaches based upon molecular self-assembly and the development of new materials. Some of the new materials have amazing specific properties and can be extremely light and strong, super absorbing and even invisible, such as the metamaterials which are artificial materials engineered to provide properties which may not be readily available in nature.

With all these sources of inspiration and knowledge in mind for the development of the future generations of fiber based products, there is no doubt that, as these will be ever more sophisticated and complex, product design will have to be accomplished by multidisciplinary design teams and excellence in design management (de Araujo 2010a).

Speculating on Future Product Development

The most advanced product of nature is mankind and it took millions of years of natural selection to arrive at the modern human. Therefore, nature is excellent at designing and fabricating many advanced products, including fibres and so we should learn, get inspired and imitate nature in the development and fabrication of fibrous materials.

In principle, by combining biotechnology, nanotechnology and information technology at the atomic level one could build up a model to design, develop and fabricate of almost anything we wish.

Nature is the prime example of this. By the process of molecular self-assembly of the basic atomic elements using bottom-up fabrication in a multitude of ways, it produces all natural products.

Some examples of man mimicking nature's fabrication are as follows:

- Mimicking the silkworm for producing man-made fibres (achieved many years ago)
- Mimicking snake skin growth for producing hosiery (not yet realized)
- Mimicking insect exoskeleton growth for producing clothing and armor (not yet realized).

Snakes have a skin covered in scales. These scales are made of keratin, the same fibrous structural material that wool is made of (…and hair, nails, skin epidermis and so on). The complete outer layer of skin is shed in one layer, similar to a sock being turned inside out. Can we copy this technology for the production of hosiery? (Fig. 6).

The arrangement of the chitin fibres in the external skeleton of insects is self-assembled. When the right components come together in the right way the exoskeleton just builds itself up and eventually is shed. Can we copy this technology for the production of clothing and armor? (Fig. 7).

If nature can do it we can do it too. It is doable; it is just a matter of time.

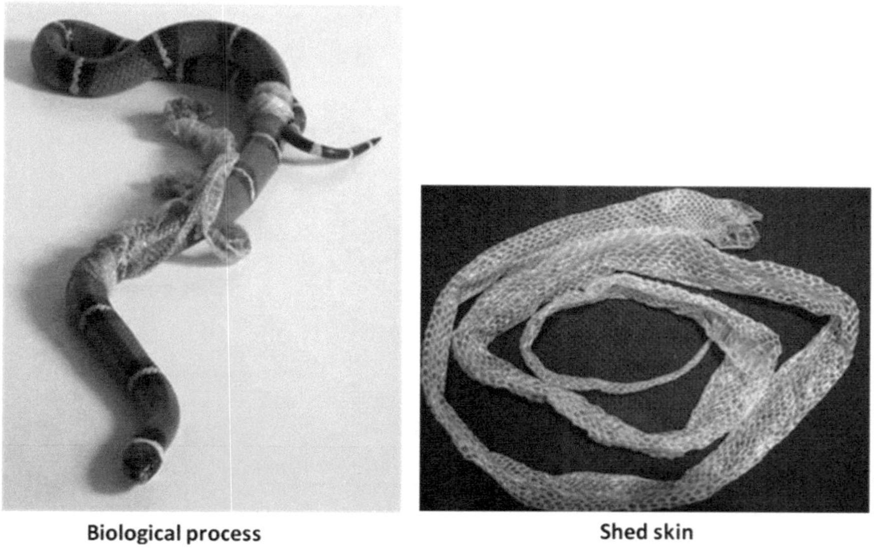

Fig. 6 Mimicking snake skin growth for producing hosiery

Fig. 7 Mimicking insect exoskeleton growth for producing clothing and armor

The building blocks of such a model are atoms, with which molecules are built up and with molecules, cells are built up. Many fibres are made of cells and with fibres, textiles can be manufactured. So, in principle, what is required is to design fibrous materials and use bottom-up fabrication for their growth.

Therefore, we should move from the current assisted-assembly which is mechanical to biological self-assembly of fibrous materials.

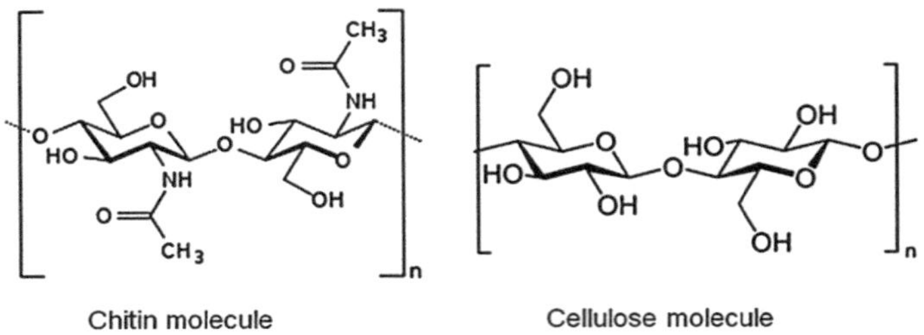

Fig. 8 Comparison of the molecules of chitin and cellulose

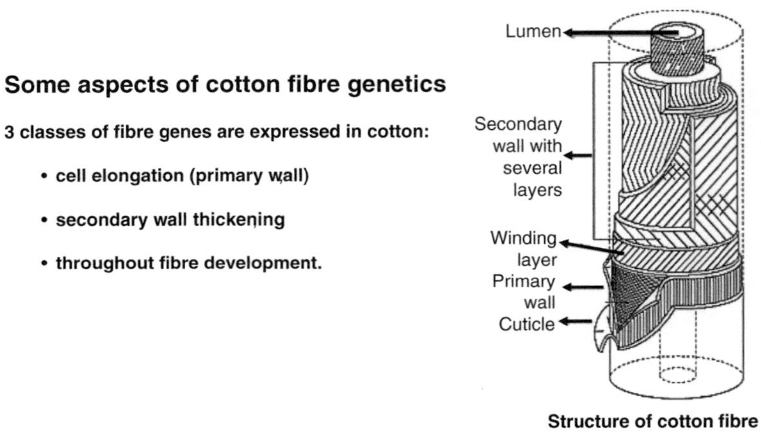

Fig. 9 Mimicking cotton fibre growth for producing fibrous products

Which types of fibres are most suited for a self-assembly model? Cell based fibres such as cotton could be the inspiration. After all, cotton's basic composition is cellulose and the molecule of cellulose is similar to that of chitin (Fig. 8).

Three classes of genes are expressed in cotton growth. One for the cell elongation and the creation of the primary wall. A second one is responsible for the thickening of the secondary wall and a third is expressed throughout the development of the fibre (Fig. 9).

If we want to fabricate 3D clothing, modeled on cotton growth, then, by DNA programming, four stages of development may be considered:

1. "*fibre*" initiation
2. cell growth (primary wall): 3D shape and size (scaffold)
3. secondary wall deposition: color, pattern and physical properties (microfibrils synthesis and deposition)
4. maturation.

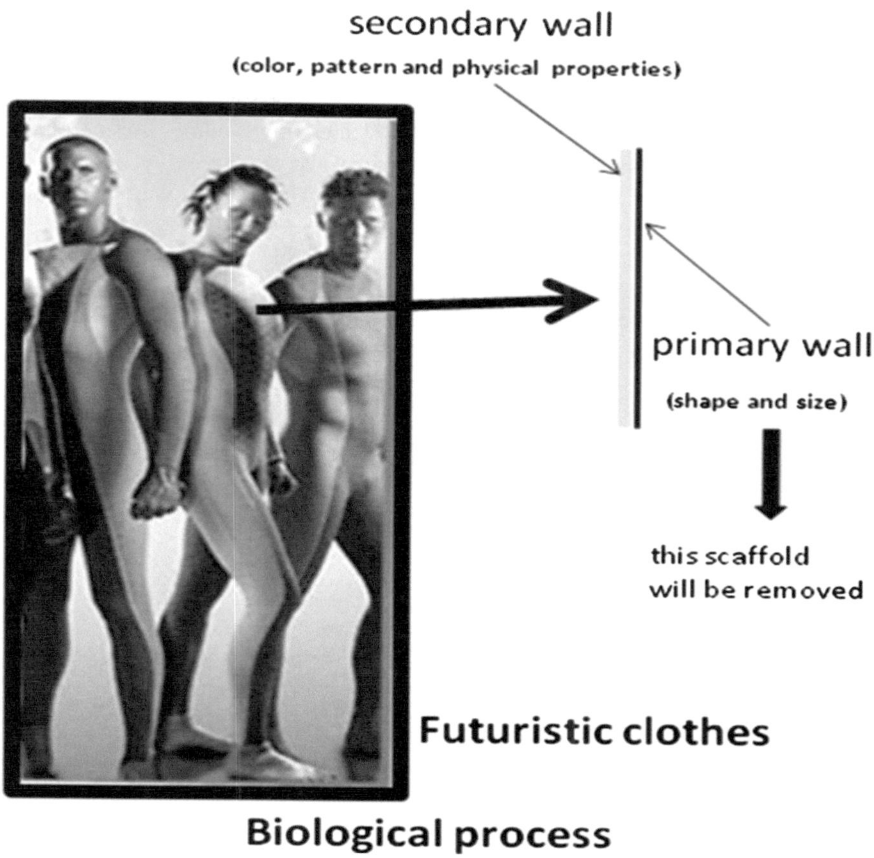

Fig. 10 Hypothetical example of mimicking cotton fibre growth for producing clothing

This would be followed by dehydration and the resulting exocellular matrix would be the designer textile product.

Figure 10 shows a conceptual idea of how clothes may look like in the future.

The benefits of choosing the cotton fibre as a model system for plant development, is that a culture method for cotton ovules was perfected a few decades ago (Kim and Triplett 2001).

The plant-like fibrous assemblies and clothes envisaged for the future may be grown from specially designed seeds, in specially designed growth chambers and growth rooms with control and monitoring systems for temperature, humidity, light intensity (including day/night programmable cycle with dawn/dusk effect), air-flow and so on.

Conclusions

Society is undergoing great changes and markets are saturated with similar products from different suppliers and not really with many alternative solutions to a problem. Many of these products have very short life cycles and are not really designed for sustainability. Bearing this in mind, companies must use innovation and best practices in order to gain competitive advantage.

A move towards a more personalized offer is inevitable and inspiration from the emerging science and technology will lead to increased performance.

Engineering design should concentrate on fibres and fabrication of 3D-products should concentrate on going from assisted-assembly into self-assembly.

Biological processes will be increasingly used in manufacturing.

A multi and even interdisciplinary approach to the development and fabrication of new products will be of paramount importance.

References

Anon, (1963) Textile terms and definitions, Manchester, The Textile Institute, 63.

Anon,(1973) "Genome and genetics timeline - 1973". Genome news network.

Banerjee, P.K. (2014) Principles of fabric formation, Boca Raton - FL 33487 ,CRC Press-Taylor & Francis Group,425

Clive Root (2007) Domestication. Greenwood Publishing Groups.

Daniel Zohary, Maria Hopf, Ehud Weiss (2012). Domestication of Plants in the Old World: The origin and spread of plants in the old world. Oxford University Press.

Darwin, Charles (1859) On the Origin of Species by Means of Natural Selection, or the Preservation of Favoured Races in the Struggle for Life (1st ed.). London: John Murray. LCCN 06017473. OCLC 741260650.

de Araujo, M. (2010a) New Product Development in the Age of Radical Change, 10th AUTEX Conference, University of Kaunas, Vilnius, Lithuania, 21-23 June, (invited plenary paper).

de Araujo, M. (2010b) Radical Innovation In Textile Product Design And Fabrication, XIIth International Izmir Textile and Apparel Symposium, Çesme, Turkey, Oct 28 – 30, (invited paper).

de Araujo, M. (2011) Chapter 1: Natural and Man-made Fibers: Physical Properties, in F. Fangueiro (Ed) – Fibrous and composite materials for civil engineering applications, Woodhead Publishing Ltd. , Cambridge, UK, P.3-28

Griffith, A. A. (1921), "The phenomena of rupture and flow in solids", Philosophical Transactions of the Royal Society of London, A 221: 180–183,

Hall,B.K.,Hallgrimsson, B.(2008), Strickberger's Evolution (4th ed), Sudbury, Jones and Bartlett Publishers, ISBN 978-0-7637-0066-9

Kaku, M.,(2007) Visions of The Future, BBC4 series.

Kim, H.J. and Triplett, B.A. (2001) Cotton Fiber Growth in Planta and in Vitro. Models for Plant Cell Elongation and Cell Wall Biogenesis, Plant Physiol. Vol. 127

Kingsbury, N. (2009) Hybrid: The History and Science of Plant Breeding, University of Chicago Press, Oct 15.

Jackson, DA; Symons, RH; Berg, P (1 October 1972). "Biochemical Method for Inserting New Genetic Information into DNA of Simian Virus 40: Circular SV40 DNA Molecules Containing Lambda Phage Genes and the Galactose Operon of Escherichia coli". PNAS69 (10): 2904 2909. Bibcode:1972PNAS…69.2904J. doi:10.1073/pnas.69.10.2904. PMC 389671. PMID 4342968.

Vendl, A. and Nicholls, S., (2006) Nature Tech, *MR-FILM production for ORF series,* 2006.

Wool in Human Health and Well-Being

Raechel Laing and Paul Swan

Abstract This paper reviews published and unpublished literature on the role of wool in human health and well-being. Human-based investigations, or those involving human simulations (manikins) were the focus. The principal parameters in the review were skin health, physical contact between textiles/garments and human skin (tactile acceptability—prickle, friction, allergies), thermal and moisture properties, human body odour, and sleep (bed clothes/sleepwear, bedding).

Keywords Wool fibre · Skin health · Perceptual characteristics

Introduction

The objective of the review was to identify and critically review published and unpublished literature on the role of wool in human health and well-being, where possible, accounting for inconsistencies in evidence in that literature. Several exclusions were applied i.e. wool in very specialised applications such as high-level human performance, and medical interventions; burning behaviour/flammability; wool in the built environment; and wool in animal health. Findings from investigations on most of these topics have been well published.

Evidence of the role of wool in human health and well-being was obtained by reviewing published peer-reviewed literature, unpublished reports, and personal communications. More than 240 documents were examined. The focus was evidence based on human studies (or those from human manikins), rather than laboratory reports of fabric properties, although the latter were sometimes also considered. The principal parameters identified for the review were (i) skin health,

R. Laing (✉)
University of Otago, Dunedin, New Zealand
e-mail: raechel.laing@otago.ac.nz

P. Swan
Australian Wool Innovation, Sydney, Australia

R. Fangueiro and S. Rana (eds.), *Natural Fibres: Advances in Science and Technology Towards Industrial Applications*, RILEM Bookseries 12,
DOI 10.1007/978-94-017-7515-1_2

(ii) physical contact between textiles/garments and human skin (including tactile acceptability—prickle, friction, allergies), (iii) thermal and moisture properties, (iv) human body odour, and (v) sleep—bed clothes, sleepwear, bedding. Some methodological and data issues also warranted comment with regard to interpretation of findings.

Skin Health

Skin is the human body barrier to the external environment, attributable to its physical structure and properties. Indicators of skin health include pH, water content of the stratum corneum (SC), transepidermal water loss (TEWL), skin elasticity. Skin pH is typically acid, ranging from 4.0 to 7.0 pH: skin with <5.0 pH is considered more desirable than that with >5.0 pH (Lambers et al. 2006); superficial layers of skin are typically 4.0–4.5 pH) (Darlenski et al. 2009; Schmid-Wendtner and Korting 2006). Water content of the SC (SC hydration) reflects the skin response to changing environmental conditions, and when not functioning correctly, SC is typically 'dry' (Rawlings and Harding 2004). Transepidermal water loss also reflects the skin response to changing conditions and is another effective indicator of skin barrier function (Elkeeb et al. 2010), with small changes detectable (Fluhr et al. 2006). Water content of the skin is related to elasticity and linked to age (Potts et al. 1984).

Because the skin may be in direct contact with wool/wool products, evidence of skin health and changes in this after such contact is of interest. There is a dearth of publications on skin health and its relationship with fabrics manufactured from either wool or other fibre types. Effects of dry and wet fabric samples ($n = 16$ different fibres and structures, including wool) on the forearm of females ($n = 35$) for 75 min showed little effect on SC hydration when dry, but an increase when wet with both wool and cotton fabrics (Schneider et al. 1996). Effects of fibre content of socks on health of the skin on human feet has also been investigated (Laing et al. 2015). Dress socks in different fibre/yarn types (100 % acrylic, 100 % wool 24.5 μm, 100 % wool 20.5 μm, 100 % cotton) were worn (with standard boots) for a minimum of 8 h per day over an 8-week period. Each participant ($n = 16$ males) was his own control with % change in indicators analysed. Differences in the effect of sock types, wool and cotton, at the heel were observed: although effects on SC were not significant, the order of mean values was wool 20.5 μm, wool 24.5 μm, both positive; cotton, acrylic, both negative. Improved barrier function, indicated by a decrease in TEWL and an elevated SC, was evident at the heel (Fig. 1).

No other completed investigation of skin health related to fibre content of skin coverings made from wool has been identified. Preliminary findings from an investigation on effects of wool fabrics against the skin of those predisposed to atopic dermatitis seem promising (Swan 2014, personal communication).

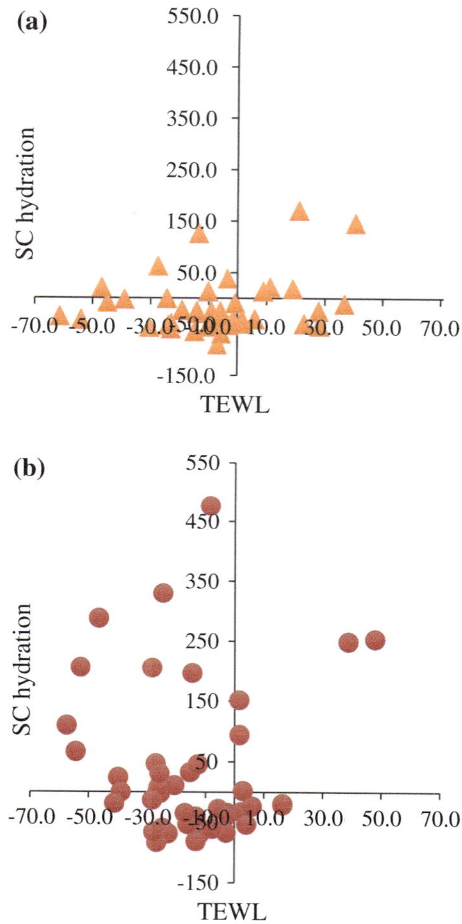

Fig. 1 Percentage change from baseline TEWL and from baseline SC hydration at the heel **a** cotton sock, **b** wool sock 20.5 μm (reproduced from Textile Research Journal 2015 85(17) 1849–1863 doi: 10.1177/0040517515573413 with permission)

Physical Contact Between Textiles/Garments and Human Skin

An aversion to next-to-skin wool garments has been attributed to discomfort or a sensation of prickle, and beliefs and experiences with wool in childhood can influence future use (Sneddon et al. 2012a, b). Prickle is understood to be mechanical irritation of the skin by coarse fibre ends indenting the skin and activating nerve endings (Garnsworthy et al. 1988; Naylor 1992; Naylor et al. 1997). Fibre, top, yarn, and fabric factors can influence the tendency of a fabric to exhibit prickle (Naylor 1997; Naylor et al. 1997) e.g. length of fibre protruding from the fabric surface, presence of coarse fibre ends (Naebe et al. 2015). Methods for overcoming prickle effects include enzyme treatments (Bishop et al. 1998; Das and

Ramaswamy 2006), and yarn structural modification (wrap yarn reductions of 1, 3 and 3 μm for 25, 29 and 31 μm diameter wool fibre respectively (Miao et al. 2005)).

Perceptions of wearers have been linked to prickle (McGregor et al. 2013; Stanton et al. 2014). A British-based wear trial on next-to-skin/thermal underwear in which 21 μm diameter wool was used, reported approximately 30 % of the male participants indicated garments were 'scratchy' and that same percentage of participants indicated the garments were thermally unacceptable (Harnett 1984a). In a subsequent wear trial using 19 μm diameter wool, no participant perceived the garments as itchy or scratchy (Harnett 1984b). Fabric 'pleasantness' has been reported to decrease as temperature and relative humidity increase (Gwosdow et al. 1986), so it is likely participants who perceived the garments too warm were those who perceived the garments as scratchy. Two Australian studies are pertinent. In one, wool fabrics manufactured from 19 μm fibres were reported to have no greater prickle value than cotton fabrics when assessed in a controlled, non-standard atmosphere (22 ± 1 °C, 65 ± 5 %RH) (n = 60 volunteers) (Naylor et al. 1992). In a second, fabrics manufactured in a next-to-skin long-sleeved, fitted garment composed of fine wool (16.5 μm) were ranked as most preferred by adult females (n = 39 Australian, n = 47 USA) over comparable garments manufactured from 18.5 μm and 20.5 μm wool. Garments were handled as if purchasing, so both tactile and visual cues were present (Sneddon et al. 2012b). What is not always clear in these studies is whether all manufacturing parameters (yarn, fabric structure, fabric mass per unit area) other than fibre type, were identical, and whether test conditions were also identical. Skin temperature and relative humidity are known to affect the sensation of prickle of products worn against the skin (Gwosdow et al. 1986), and when sweating begins, wool garments may become less comfortable (Wang et al. 2003). The perception of skin wetness in turn is influenced by interactions between thermal and mechanical stimuli (Filingeri et al. 2014), with warm temperatures suppressing the perception of skin wetness and coldness seeming to dominate (Filingeri et al. 2015).

How the human skin responds to movement of fabric across its surface (friction between fabrics and skin) is of interest in applications such as next-to-skin garments. A first response may be redness of the skin, an indicator of irritation. Several studies on fabric type and blood flow and/or skin temperature have been undertaken (Gan et al. 2010; Hatch et al. 1990), although most do not include wool fabrics. The study by Gan et al. (2010) measured fabrics for just five minutes on the forearm of one participant. Hatch et al. (1990) observed skin redness in some participants, suggesting increased blood flow to that area was caused by contact between fabric and skin, although blood flow at the site was not measured.

In relation to foot coverings, the three important parameters are how textiles respond in the presence of moisture, friction between the skin surface and the sock fabric, and compression and compressional resistance/recovery of the sock fabric. Reports of the superiority in sock performance of one fibre type over another, particularly through the 1980s and 1990s, included comparisons of acrylic with cotton or wool socks (Brooks et al. 1990; Euler 1985; Herring and Richie 1990, 1993; Morris et al. 1984). Some years later, sock fabrics manufactured from mid-micron and fine wool were reported to have a lower coefficient of static friction than fabrics composed of acrylic in both dry and damp conditions when measured against a synthetic skin (van Amber et al. 2015b). Fibre type, yarn structure, and fabric structure are all controllable manufacturing variables, thus their relative effects on frictional properties also controllable. Fabrics in this study were prepared for a factorial experimental design (100 % mid-micron wool (26 μm), 100 % fine wool (19 μm), 100 % acrylic (19 μm); yarn high twist, low twist, single; single jersey, full terry, half terry). The lowest coefficient of static and dynamic friction was evident with the single jersey. With respect to damp fabrics, that made from fine wool exhibited the lowest coefficient of static friction and acrylic the highest.

Compression/compressional resilience of socks, including wool socks has been reported. Wool cushioned socks have been shown to have a greater shock-attenuating effect than walking barefoot (Howarth and Rome 1996), and wool socks were also associated with an increased time to peak force, and decreased propulsive force than when the participant(s) was walking bare footed (Blackmore et al. 2011), unsurprising in both cases. Differences detected were attributed to differences in sock thickness (i.e. fabric structure and sock construction had a more important shock-attenuating effect than the type of fibre (Blackmore et al. 2011). Similarly, fibre type was not a significant factor in the percentage of thickness retained under compression (van Amber et al. 2015a), but was relevant in compression: recovery ratios, the acrylic fabrics being superior to the two wool fabrics when fabrics were damp. The authors did caution against potential misinterpretation, as the fabrics held different amounts of moisture (van Amber et al. 2015a).

Claims of allergic reactions to wool arise from three potential sources: chemical (e.g. lanolin, residues of chemical substances used in processing), physical (irritation from prickle, largely resolved), and allergens (e.g. insects/mites typically associated with carpets/furnishings). In the 1980s, wool fibres had been reported as causing acute and cumulative irritant dermatitis and as aggravating atopic dermatitis (Hatch and Maibach 1985), but by the 1990s, wool as a skin irritant caused by prickle, was better understood (Hatch and Maibach 1995). Prickle or itch is still sometimes misconstrued as a 'wool allergy'. For example, 35–40 % of interviewees who would not consider buying wool garments nominated prickle as the reason, and 7–10 % nominated allergy (Starick 2013). Although some residual misunderstanding remains (Fujimura et al. 2011), atopic dermatitis is not wool allergy.

One possible chemical explanation for wool allergy is related to lanolin on wool rather than the fibre itself. A review of 24,449 patients in Britain from 1982 to 1996 reported a 1.7 % annual rate of sensitivity to lanolin or wool alcohol, a relatively low sensitivity in this sample (Wakelin et al. 2001). During the 1980s Hatch and others sought to identify dermatological problems related to fibre content and dyes (Scheurell et al. 1985), with little published since that time. Whether a causal link exists between chemical substances in textile products and allergic reactions was investigated in a cooperative study in Italy (Associazione Tessile e Salute-Health and Textile Association 2013). The report identified contact dermatitis to be caused mostly by fabric dyes, but also finishing resins and adhesive resins, particularly when these release formaldehyde. Prevalence and sensitisation to formaldehyde was reportedly decreasing. 'Emerging allergens' were noted, attributable to non-European manufacturers (Associazione Tessile e Salute-Health and Textile Association 2013).

An epidemiological study (n = 401 patients, aged 5–84 years) showed fabrics were the cause of contact dermatitis in approximately 70 % of cases, metallic and garment accessories in approximately 17 % of cases, and shoes in approximately 14 % of cases (Associazione Tessile e Salute-Health and Textile Association 2013). Dyes and intermediary agents accounted for 44 % of cases: garments most commonly involved were nylon stockings (8.9 %), underwear (13.4 %), shirts (13.0 %), trousers and skirts (13.8 %), and sportswear (8.1 %) (Associazione Tessile e Salute-Health and Textile Association 2013). Other than nylon (stockings), no mention was made of fibre content. That sweat on the skin under fabric is likely to affect permeability of those fabrics to chemical transfer (i.e. from dyes, finishes) has been long recognised (Raheel 1991), although no wool-related evidence has been identified. Companies manufacturing pure new wool textiles are required to ensure the safety of their products through declaring compliance with Restricted Substance Lists. These restrictions appear specific to members of the European Union.

Thermal and Moisture Properties

Many advantages of including wool fibres in textiles for human health and well-being derive from the chemical composition and structure of the fibre, particularly those related to thermal and moisture properties. Molecules in the fibre are able to create hydrogen bonds with water, immobilising the water and incorporating it into the fibre, with a small amount of heat released. This is detectable in fabric form (Laing et al. 2007), in garments in use (Laing et al. 2008), and in bedding (Naylor 2014, personal communication). Physical structure of the wool fibre (e.g. crimp, scaled surface) is a major determinant of yarn properties irrespective of whether processed by the woollen or worsted system, and largely irrespective of yarn twist. That wool (i.e. wool products) is considered by end-users to be 'warm' is due primarily to fibre bulk and crimp, which lead to bulkier yarns and yield thicker, more thermally resistant fabrics/end products (Harnett 1984a; Leeder 1984), and many papers published during the 20th century show this.

Thermal properties of a fabric are derived from one or more of four parameters: thickness, thermal conductivity, absorptivity, and heat of absorption. Wool fabrics are regarded as providing superior thermal properties under both damp and wet conditions. The small amount of heat released with absorption of water, and wool fibres/fabrics reported as having a lower thermal conductivity than cotton, polypropylene or acrylic, underscore these advantages (Schneider et al. 1992). Fabrics composed of wool have been reported as absorbing more moisture than matched fabrics composed of synthetic fibres such as nylon, polyester, and acrylic (Collie 2002; Laing et al. 2007; van Amber 2013). Absorption of liquid from the skin results in perception of drier and therefore warmer skin (Laing 2009). Further evidence confirmed perception of warmth is affected by perception of wetness (Filingeri et al. 2014), with warm temperatures suppressing the perception of wetness (Filingeri et al. 2015). Surface wetness of fabrics in contact with the skin is known to influence more general perceptions of comfort (Hatch et al. 1987).

Garments composed of wool when worn have been reported to lower the relative humidity at the skin surface compared to effects of garments composed of acrylic fibres (Li 2005). Wearing wool garments has also been reported to delay the on-set of sweating and result in smaller changes in core temperature than when wearing matched polyester and wool/polyester plated garments (Laing et al. 2008). Buffering is the probable explanation for these smaller changes in core temperature and smaller increases in heat content in both cold and hot conditions than identical polyester garments. Participants wearing the wool garment had a lower heart rate during all test conditions (Laing et al. 2008). Buffering effects of wool fabrics/garments compared to polyester have also been reported by Li et al. (1992), and wool blankets compared to an acrylic/cotton blend reported by Umbach (1986).

Smooth, lightweight wool fabrics have been shown to be perceived as cooler to the touch (forearm test, n = 20 participants) than comparable woven fabrics of cotton, polyester, and a wool/polyester blend, with an immediate drop of 0.4–0.8 °C observed in skin temperature on the forearm (Cameron et al. 1997). These and other similar findings on lightweight fabrics of the latter part of the 20th century led to new commercial markets for apparel.

The thermal resistance of wool fabrics has also been attributed to improvement in specific aspects of user health. Use of wool undergarments (and bedding) over a six-week period was reported to reduce symptoms and drug use of patients suffering from fibromyalgia (Kiyak 2009). Details on construction, mass per unit area, thickness, and laundering practices were not reported for either the undergarments or the bedding however.

Human Body Odour—Textiles, Clothing, Footwear

Basic understanding of body odour and several links to clothing have been known since the 1950s (Shelley et al. 1953): human apocrine sweat, a sterile fluid, is acted on by bacteria and other organisms residing on the skin [e.g. *Staphylococcus*, *Micrococcus*, *Bacillus*, *Acinetobacter*, *Klebsiella*, *Enterobacter* and *Streptomyces* (Kloos and Musselwhite 1975)], releasing volatile acids and indoles (e.g. isovaleric, butyric, carbonic). Major sources of body sweat odour are those emitted by the apocrine glands (i.e. fatty acids, steroids, amines). Axillary hair collects sweat and other debris, and in relation to foot odour, similar patterns occur (Ara et al. 2006; Marshall et al. 1987, 1988). Strong foot odour has been linked with greater total population densities of microflora, staphylococci and aerobic coryneforms, rather than an increase in any single type of bacterium (Marshall et al. 1988). Minimising this odour is another measure of ensuring well-being.

In the 1970s, isovaleric acid (the 'sweat-like' odour) was identified as one of six main primary odours (Amoore 1977). A study of ten males (aged 25–30 years) (n = 5 strong foot odour, n = 5 low or no odour) showed isovaleric acid identifiable in all participants with foot odour, but not in those without (Kanda et al. 1990). While isovaleric acid is responsible for foot odour (Kanda et al. 1990), propionic acid, isobutyric acid and butyric acid are also contributors (Ara et al. 2006).

Clothing adjacent to the axilla or textiles covering the foot (socks, shoes) are implicated in human body odour, as organisms are able to reside in various fabrics (Shelley et al. 1953) and in footwear. Understanding the interactions among axillary odours, bacterial count and fibre type improved during the early part of the 21st century (McQueen et al. 2007a, b, 2008). McQueen et al. assessed the odour of fabric samples (wool, polyester, cotton; interlock, single jersey, 1 × 1 rib, matched constructions) which had been worn by five males. Polyester fabrics exhibited the strongest and wool fabrics the least intense odour (Fig. 2). No inherent antibacterial properties were evident, with bacterial counts at day 1 similar on all fabrics.

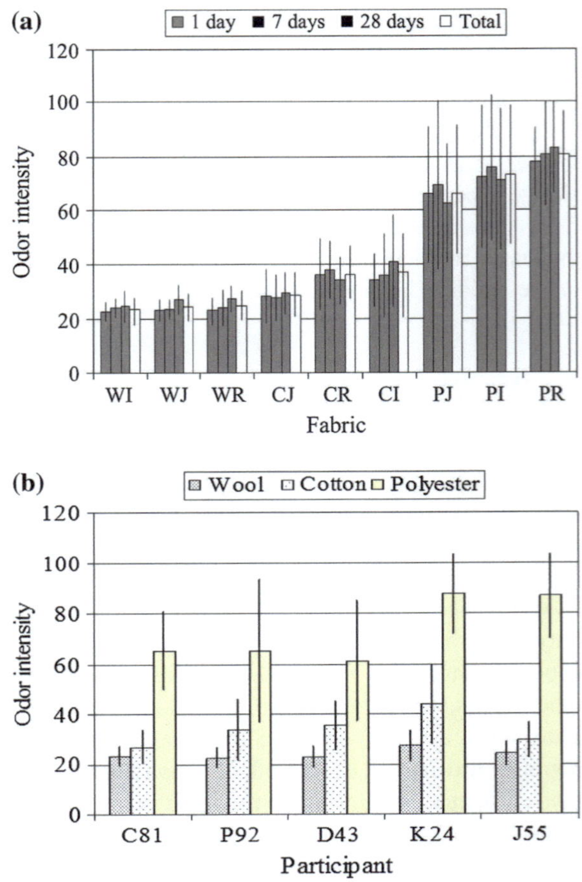

Fig. 2 Odour intensity ratings of wool, cotton, polyester fabrics (mean ± s.e. m.) **a** stored for different times, **b** five participants, (reproduced from Textile Research Journal 2007 77, 449 doi: 10.1177/ 0040517507074816)

Bacterial populations were present on all fabrics for up to 28 days, with numbers remaining relatively stable in wool fabrics, and declining in polyester fabrics (McQueen et al. 2007a). Whether the bacterial count was related to the method of extraction is unknown but is considered unlikely to explain the observed differences in odour intensity.

Intermittent claims are made about wool being naturally antibacterial, but no evidence to support these claims has been identified. McQueen's work (2007) did not show a link between the number of bacteria and odour. Nor has any evidence of anti-fungal properties related to wool been identified. The notion of natural antibacterial/antimicrobial properties may arise from the fact that wool products such as garments do not retain/release odour volatiles during and following wear as do comparable products in other fibre types. Exposure of a matrix of known odour volatiles to wool, cotton, polyester fibres/yarns under controlled conditions provided some explanation of the mechanisms involved (Yao et al. 2015), and this work is continuing.

Treatments to confer antimicrobial properties on textiles have been developed and many reviewed (Gao and Cranston 2008). One method applied to wool and wool products has been inclusion of silver (Li et al. 2010; Tang et al. 2011) although concerns have been expressed about development of bacterial resistance to silver (Percival et al. 2005). Effectiveness of other wool-metal complexes/salts has also been examined (Freddi et al. 2001; Zhao and Sun 2006; Zhu and Sun 2004), as has chitosan as an antimicrobial treatment for woollen fabrics (Hseih et al. 2004).

The effect of footwear on temperature and humidity, and related microbial populations on skin of feet was investigated in the 1960s, but with little information on the footwear, including socks. A study of the socks and shoes of hospital patients with symptoms of *tinea pedis* but not undergoing treatment (n = 30) focussed on these items of clothing as carriers of fungal spores (Brown and McLarnon 2007). Although non-significant, there was a trend for participants with *tinea pedis* infections to also have infected footwear (Brown and McLarnon 2007). That fungal spores remain in socks, shoes, and on various surfaces has been assumed given high levels of transmittance and re-infection (Ajello and Getz 1954). Surprisingly, wool socks appear not to have been included in these studies.

Sleep—Bed Clothes, Sleepwear, Bedding

Clothing worn during sleep (pyjamas, nightdresses, 'stretch-and-grows'/'onesies') has the potential to influence sleep quality, with several variables indicating sleep quality (e.g. time to the on-set of sleep, duration of sleep, duration of wakefulness). Sleep in adult humans becomes disrupted once thermoregulatory mechanisms are elicited (Bach et al. 2002), and information on many aspects of temperature and sleep have been published (REM (Muzet et al. 1983; Okamoto-Mizuno et al. 1999); room temperature, bed temperature, and various body temperature indicators (Muzet et al. 1983; Okamoto-Mizuno et al. 1999); effects of elevated skin wettedness (Okamoto-Mizuno et al. 1999)). Temperature of the sleep environment is even more important for infants, as temperature has been suggested as one cause of Sudden Infant Death (Muzet et al. 1983; Stanton 1984).

There is some evidence that the quality of sleep can be affected by the type of fabrics in both bedding and clothing worn in bed. Most investigations have not included wool (e.g. pyjamas of cotton/elastane compared with those of polyester/elastane (Yao et al. 2007)). Patterns of sleep of pre-school children (n = 101, aged 2–5 years) was the focus of an Australian study on effects of bedding, bedroom environment, sleep hygiene. Sleepwear was typically cotton in winter and summer, with cotton/synthetic sleepwear reportedly worn by about half the children. Few children wore sleep clothing made from wool fabrics, just 7 % of children in winter (Richdale 2013). A sleep problem was associated with synthetic bed wear in winter and in summer, and feeling 'too cold' during sleep (Richdale 2013).

Effects of sleeping apparel (wool, cotton) and bedding (wool, synthetic) on sleep of 17 participants (adults, aged 24.6 ± 6.9 years; n = 10 males, n = 7 females) over nine nights of sleep have been examined (Shin et al. 2014). Participants were randomly allocated to one of two ambient conditions (17, 22 °C), to the sleeping apparel, and to the bedding types. Wool bed apparel was suggested as promoting sleep through a shorter time to its on-set, and also the total sleep time was of longer duration. Differences were attributed to the general hygroscopicity, high moisture absorption rate, and thermal resistance of the wool textiles. Effects of bedding were less clear.

A pilot study in Britain in the early 1980s reported low birth weight babies gained more weight when nursed on lambswool (either lambswool in an artificial fibre backing or natural lambskins) rather than cotton sheets (Scott et al. 1983). The authors hypothesised that the wool provided a more thermally neutral, 'less stressful' environment, although these findings were based on absolute weight gain rather than a percentage of initial body weight (Roberts et al. 1986). The study was repeated with monitored energy intake, and no significance detected (Roberts et al. 1986). During the latter part of the 20th century and early 21st century Wilson and colleagues examined infant sleep arrangements and the links with Sudden Infant Death (Wilson et al. 1994), particularly thermal parameters (Wilson et al. 2000, 2002). Use of a wool under-blanket with a 'waterproof' covering had previously been reported as contributing to a lower likelihood of the incidence of Sudden Infant Death (Wilson et al. 1994). These investigations demonstrate the complexity of bedding and its effects.

Since the classic 1980s investigation by Umbach (1986), understanding the role of bedding is slowly being enhanced. Umbach compared performance of a wool blanket with another from acrylic (matched for thickness, mass per unit area), a comparison based on a skin model, instrumented human manikin (Charlie) dressed in pyjamas, and also humans (n = 4 males in a climate chamber). The wool blanket performed better in terms of thermal insulation, moisture absorption, and moisture buffering. With respect to underlays, investigations during the latter part of the 20th century focussed largely on sheepskins or pile-type constructions. Participants (n = 10) sleeping on a wool ‘fleecy’ underblanket were observed to have 20 % more periods of immobile sleep and more participants self-reported feeling better in the morning and having improved sleep quality (Dickson 1984). Better understanding of potential effects of fibre/fabric/structure would be possible with detail on both mattress pads (e.g. fibre content, construction, mass, thickness).

Bed covers used on Australian pre-school children (n = 101) (Richdale 2013) in winter were either duvets or blankets (approximately equal numbers of each), and wool (fill) in 26 % of duvets and 28 % of blankets. In summer, the most common cover was a cotton blanket (44 %) and wool just 12 %. Sheets were common in both summer and winter. In terms of underlays, these were cotton for approximately one third of the sample irrespective of season: wool underlays were much less common, with 12 % in winter and 10 % in summer (Richdale 2013). This study involved multiple factors: differences in fabrics/structures, fibre types self-reported, and some categories of variables not mutually exclusive (e.g. the underlay mattress protector (waterproof) may have been used with either a cotton or a wool underlay). However, children who slept in any synthetic bed wear in summer or winter were more likely to be reported as having a sleep problem.

Conclusions

This review provides evidence of the contribution of wool to human health and well-being. What is known includes

1. Indicators of skin health are measurable, and textiles in close contact with the skin can affect these. Wool fabrics in close contact with the skin can maintain and/or enhance normal moisture levels of the skin.
2. The mechanisms underlying prickle resulting from skin contact with wool fabrics are well understood, and can be ameliorated by manufacturing processes at fibre, yarn, and fabric stages. Perception of wool products remains a barrier to use for some consumers.
3. Effects of fabric structure dominate many performance properties when all variables other than fibre type are controlled. This relates particularly to thermal and moisture relationships.
4. There is no evidence to support claims that wool fibre causes allergic reactions. Allergic reactions may occur as sweat from the human body reacts to dyes and/or finishes on wool fabrics, or fabrics of any other fibre type, which are in contact with the skin surface.
5. Wool in garment form provides benefits during exercise by slowing thermo-physiological responses, thereby allowing adaptation to the changed environment.
6. There is physical and perceptual evidence of smooth, lightweight, woven wool fabrics being cooler to the touch than comparable fabrics in several other fibres/blends.

7. Thermal and moisture transmission through a layered assembly is not the sum of properties of the individual layers and air spaces. Effective cooling of the human body through evaporative heat loss is decrementally affected by layers, up to 80 % reduction in effectiveness.
8. No evidence has been identified which shows wool is intrinsically antibacterial or antimicrobial (and thus wool is biodegradable, an advantage). Wool yarns/fabrics are treated to confer antimicrobial properties, and while of some benefit, consumer concerns have been raised.
9. Wool fabrics (garments) after wear, have less intense odour than matched cotton or polyester fabrics: intensity of odour seems not to correspond with the number of bacteria extracted, and adsorption/release of odour volatiles from wool differs to that from other fibres.
10. There is some evidence that wool blankets perform better than equivalent acrylic/blend blankets from a human thermophysiological perspective. Other than for babies and infants, little is known about effects of underlays on sleep.

Understanding the role of wool in human health and well-being is not straight forward as both investigative approaches and level of detail provided in reports/papers differ widely. Although standard test methods may be followed (e.g. ISO, ASTM, BS), different standard methods purportedly measuring the same parameter may not do so. Description of methods followed and materials used are not always complete. Fibre-based comparisons are often confounded with other manufacturing variables by insufficient control of yarns/fabrics/finishes. Caution in interpretation of findings is therefore required.

Acknowledgments This paper is based on a review prepared for and funded by the International Wool Textile Organisation, Belgium, and Australian Wool Innovation, Sydney, Australia. Contributions to sections of that review are acknowledged and included: Mr. David Crowe, AWTA, Melbourne, Australia; Mr. Mauro Rossetti, General manager, Associazione Tessile e Salute, Biella, Italy; Mr. Pier Giorgio Minazio, European technical service manager, Woolmark Italy, Biella, Italy.

References

Ajello, L. and Getz, M. E. 1954. Recovery of dermatophytes from shoes and shower stalls. Journal of Investigative Dermatology 22 (1): 17-24.

Amoore, J. E. 1977. Specific anosmia and the concept of primary odors. Chemical Senses and Flavor 2: 267-281.

Ara, K., Hama, M., Akiba, S., Koike, K., Okisaka, K., Hagura, T., Kamiya, T. and Tomita, F. 2006. Foot odor due to microbial metabolism and its control. Canadian Journal of Microbiology 52 (4): 357-364.

Associazione Tessile e Salute-Health and Textile Association 2013. Chemical substances in textile products and allergic reactions. Biella, Italy: Associazione Tessile e Salute-Health and Textile Association.

Bach, V., Telliez, F. and Libert, J.-P. 2002. The interaction between sleep and thermoregulation in adults and neonates. Sleep Medicine Reviews 6 (6): 481-492.

Bishop, D. P., Shen, J., Heine, E. and Hollfelder, B. 1998. The use of proteolytic enzymes to reduce wool-fibre stiffness and prickle. Journal of The Textile Institute (Special Issue) 89 (3): 546-553.

Blackmore, T., Ball, N. and Scurr, J. 2011. The effect of socks on vertical and anteroposterior ground reaction forces in walking and running. The Foot 21 (1): 1-5.

Brooks, J. H., Capablanca, J. S., Watt, I. C. and Holcombe, B. V. 1990. Comparative comfort of socks. In 8th International Wool Textile Research Conference, V Vol. 256-265. Christchurch, New Zealand: Wool Research Organisation of New Zealand (Inc).

Brown, L. and McLarnon, N. A. 2007. Do patients with untreated tinea pedis have concomitant fungal contamination within their footwear and hosiery? British Journal of Podiatry 10 (4): 134-138.

Cameron, B. A., Brown, D. M., Dallas, M. J. and Brandt, B. 1997. Effect of natural and synthetic fibers and film and moisture content on stratum corneum hydration in an occlusive system. Textile Research Journal 67 (8): 585-592.

Collie, S. R. 2002. Comfort and blister prevention properties of wool and synthetic socks. Christchurch, New Zealand: Wool Research Organisation of New Zealand (Inc.), Christchurch, New Zealand. Unpublished report.

Darlenski, R., Sassning, S., Tsankov, N. and Fluhr, J. W. 2009. Non-invasive in vivo methods for investigation of the skin barrier physical properties. European Journal of Pharmaceutics and Biopharmaceutics 72 (2): 295-303.

Das, T. and Ramaswamy, G. N. 2006. Enzyme treatment of wool and specialty hair fibers. Textile Research Journal 76 (2): 126-133.

Dickson, P. R. 1984. Effect of a fleecy woollen underlay on sleep. The Medical Journal of Australia 140: 87-89.

Elkeeb, R., Hui, X., Chan, H., Tian, L. and Maibach, H. I. 2010. Correlation of transepidermal water loss with skin barrier properties in vitro: comparison of three evaporimeters. Skin Research and Technology 16 (1): 9-15.

Euler, R. D. 1985. Creating "comfort" socks for the U.S. consumer. Knitting Times (May): 47-50.

Filingeri, D., Redortier, B., Hodder, S. and Havenith, G. 2014. Thermal and tactile interactions in the perception of local skin wetness at rest and during exercise in thermo-neutral and warm environments. Neuroscience 258: 121-130.

Filingeri, D., Redortier, B., Hodder, S. and Havenith, G. 2015. Warm temperature stimulus suppresses the perception of skin wetness during initial contact with a wet surface. Skin Research and Technology 21 (1): 9-14.

Fluhr, J. W., Feingold, K. R. and Elias, P. M. 2006. Transepidermal water loss reflects permeability barrier status: validation in human and rodent in vivo and ex vivo models. Experimental Dermatology 15 (7): 483-492.

Freddi, I., Arai, T., Colonna, G. M., Boschi, A. and Tsukada, M. 2001. Binding of metal cations to chemically modified wool and antimicrobial properties of the wool-metal complexes. Journal of Applied Polymer Science 82 (14): 3513-3519.

Fujimura, T., Takagi, Y., Sugano, I., Sano, Y., Yamaguchi, N., Kitahara, T., Takema, Y. and Rizer, R. L. 2011. Real-life use of underwear treated with fabric softeners improves skin dryness by decreasing the friction of fabrics against skin. International Journal of Cosmetic Science 33: 566-571.

Gan, Y., Cheng, L., Ding, X. and Pan, N. 2010. Blood flow fluctuation underneath human forearm skin caused by local thermal stimuli of different fabrics. Journal of Thermal Biology 35 (7): 372-377.

Gao, Y. and Cranston, R. 2008. Recent advances in antimicrobial treatments of textiles. Textile Research Journal 78 (1): 60-72.

Garnsworthy, R. K., Gully, R. L., Kandiah, R. P., Kenins, P., Mayfield, R. J. and Westerman, R. A. 1988. Understanding the causes of prickle and itch from the skin contact of fabrics. Australasian Textiles 4 (88): 26-29.

Gwosdow, A. R., Stevens, J. C., Berglund, L. G. and Stolwijk, J. A. J. 1986. Skin friction and fabric sensations in neutral and warm environments. Textile Research Journal 56 (9): 574-580.

Harnett, P. 1984a. Functions and properties of 'thermal' underwear. Wool Science Review 60: 3-11.
Harnett, P. 1984b. Wool underwear for warmth and comfort. Wool Science Review 60: 47-60.
Hatch, K. L. and Maibach, H. I. 1985. Textile fiber dermatitis. Contact Dermatitis 12 (1): 1-11.
Hatch, K. L. and Maibach, H. I. 1995. Textile dermatitis: an update. Contact Dermatitis 32 (6): 319-326.
Hatch, K. L., Markee, N. L., Maibach, H. I., Barker, R. L., Woo, S. S. and Radhakrishnaiah, P. 1990. In vivo cutaneous and perceived comfort response to fabric Part III: Water content and blood flow in human skin under garments worn by exercising subjects in a hot, humid environment. Textile Research Journal 60 (9): 510-519.
Hatch, K. L., Wilson, D. R. and Maibach, H. I. 1987. Fabric-caused changes in human skin: in vivo stratum corneum water content and water evaporation. Textile Research Journal 57 (10): 583-591.
Herring, K. M. and Richie, D. H. 1990. Friction blisters and sock fiber composition: a double-blind study. Journal of the American Podiatry Association 80 (2): 63-71.
Herring, K. M. and Richie, D. H. 1993. Comparison of cotton and acrylic socks using a generic cushion sole design for runners. Journal of the American Podiatric Medical Association 83 (9): 515-522.
Howarth, S. J. and Rome, K. 1996. A short-term study of shock-attenuation in different sock types. The Foot 6 (1): 5-9.
Hseih, S. H., Huang, Z. K., Huang, Z. Z. and Tseng, Z. S. 2004. Antimicrobial and physical properties of woolen fabrics cured with citric acid and chitosan. Journal of Applied Polymer Science 94 (5): 1999-2007.
Kanda, F., Yagi, E., Fukuda, M., Nakajima, K., Ohta, T. and Nakata, O. 1990. Elucidation of chemical compounds responsible for foot malodour. British Journal of Dermatology 122 (6): 771-776.
Kiyak, E. K. 2009. A new nonpharmacological method in fibromyalgia: the use of wool. The Journal of Alternative and Complementary Medicine 15 (4): 399-405.
Kloos, W. E. and Musselwhite, M. S. 1975. Distribution and persistance of Staphylococcus and Micrococcus species and other aerobic bacteria on human skin. Applied Microbiology 30 (3): 381-395.
Laing, R., Wilson, C., Dunn, L. and Niven, B. 2015. Detection of fiber effects on the skin health of the human foot. Textile Research Journal 85 (17): 1849-1863.
Laing, R. M. 2009. Assessing fabrics for cold weather apparel: the case of wool. In Textiles for Cold Weather Apparel, edited by Williams, J. T., 33-55. Woodhead Publishing Limited.
Laing, R. M., Niven, B. E., Barker, R. L. and Porter, J. 2007. Response of wool knit apparel fabrics to water vapor and water. Textile Research Journa*l* 77 (3): 165-171.
Laing, R. M., Sims, S. T., Wilson, C. A., Niven, B. E. and Cruthers, N. M. 2008. Differences in wearer response to garments for outdoor activity. Ergonomics 51 (4): 492-510.
Lambers, H., Piessens, S., Bloem, A., Pronk, H. and Finkel, P. 2006. Natural skin surface pH is on average below 5, which is beneficial for its resident flora. International Journal of Cosmetic Science 28 (5): 359-370.
Leeder, J. D. 1984. Structure of the wool fibre. In Wool - Nature's Wonder Fibre, 8-12. Ocean Grove, Victoria: Australasian Textiles Publishers and J.D. Leeder.
Li, W.-R., Xie, X.-B., Shi, Q.-S., Zeng, H.-Y., Ou-Yang, Y.-S. and Chen, Y.-B. 2010. Antibacterial activity and mechanism of silver nanoparticles on Escherichia coli. Applied Microbiology and Biotechnology 85 (4): 1115-1122.
Li, Y. 2005. Perceptions of temperature, moisture and comfort in clothing during environmental transients. Ergonomics 48 (3): 234-248.
Li, Y., Holcombe, B. V. and Apcar, F. 1992. Moisture buffering behavior of hygroscopic fabric during wear. Textile Research Journal 62 (11): 619-627.
Marshall, J., Holland, K. and Gribbon, E. 1988. A comparative study of the cutaneous microflora of normal feet with low and high levels of odour. Journal of Applied Bacteriology 65 (1): 61-68.

Marshall, J., Leeming, J. P. and Holland, K. T. 1987. The cutaneous microbiology of normal human feet. Journal of Applied Bacteriology 62 (2): 139-146.

McGregor, B. A., Naebe, M., Stanton, J., Speijers, J., Beilby, J., Peruzzini, S. and Tester, D. 2013. Relationship between wearer prickle response with fiber and garment properties and Wool ComfortMeter assessment. Journal of The Textile Institute 104 (6): 618-627.

McQueen, R. H., Laing, R. M., Brooks, H. J. L. and Niven, B. E. 2007a. Odor intensity in apparel fabrics and the link with bacterial populations. Textile Research Journal 77 (7): 449-456.

McQueen, R. H., Laing, R. M., Delahunty, C. M., Brooks, H. J. L. and Niven, B. E. 2008. Retention of axillary odour on apparel fabrics. Journal of The Textile Institute 99 (6): 515-523.

McQueen, R. H., Laing, R. M., Wilson, C. A., Niven, B. E. and Delahunty, C. M. 2007b. Odor retention on apparel fabrics: development of test methods for sensory detection. Textile Research Journal 77 (9): 645-652.

Miao, M., Collie, S. R., Watt, J. D. and Glassey, H. E. 2005. Prickle and pilling reduction by modified yarn structures. In11th International Wool Textile Research Conference. Leeds, England.

Morris, M. A., Prato, H. H. and White, N. L. 1984. Relationship of fiber content and fabric properties to comfort of socks. Clothing and Textiles Research Journal 3 (1): 14-19.

Muzet, A., Ehrhart, J., Candas, V., Libert, J. P. and Vogt, J. J. 1983. REM sleep and ambient temperature in man. International Journal of Neuroscience 18 (1-2): 117-125.

Naebe, M., McGregor, B. A., Swan, P. and Tester, D. 2015. Associations between the physiological basis of fabric-evoked prickle, fiber and yarn characteristics and Wool ComfortMeter value. Textile Research Journal 85 (11): 1122-1130.

Naylor, G. R. S. 1992. The role of coarse fibres in fabric prickle using blended acrylic fibres of different diameters. Wool Technology and Sheep Breeding 40 (1): 14-18.

Naylor, G. R. S. 1997. Fabric-evoked prickle in worsted spun single jersey fabrics Part II: the role of fiber length, yarn count, and fabric cover factors. Textile Research Journal 67 (5): 354-358.

Naylor, G. R. S., Phillips, D. G., Veitch, C. J., Dolling, M. and Marland, D. J. 1997. Fabric-evoked prickle in worsted spun single jersey fabrics Part I: the role of fiber end diameter characteristics. Textile Research Journal 67 (4): 288-295.

Naylor, G. R. S., Veitch, C. J., Mayfield, R. J. and Kettlewell, R. 1992. Fabric-evoked prickle. Textile Research Journal 62 (8): 487-493.

Okamoto-Mizuno, K., Mizuno, K., Michie, S., Maeda, A. and Lizuka, S. 1999. Effects of humid heat exposure on human sleep stages and body temperature. Sleep 22 (6): 767-773.

Percival, S. L., Bowler, P. G. and Russell, D. 2005. Bacterial resistance to silver in wound care. Journal of Hospital Infection 60 (1): 1-7.

Potts, R. O., Buras, E. M. and Chrisman, D. A. 1984. Changes with age in the moisture content of human skin. Journal of Investigative Dermatology 82 (1): 97-100.

Raheel, M. 1991. Pesticide transmission in fabrics: effect of perspiration. Bulletin of Environmental Contamination and Toxicology 46 (6): 837-844.

Rawlings, A. V. and Harding, C. R. 2004. Moisturization and skin barrier function. Dermatologic Therapy 17 (s1): 43-48.

Richdale, A. 2013. Bedding, diet and sleep in children aged 2 to 5 years. Melbourne, Australia: La Trobe.

Roberts, S. B., Savage, J. and Lucas, A. 1986. Does lambswool promote growth in preterm infants? The Lancet 327 (8486): 921-922.

Scheurell, D. M., Spivak, S. M. and Hollies, N. R. S. 1985. Dynamic surface wetness of fabrics in relation to clothing comfort. Textile Research Journal 55 (7): 394-399.

Schmid-Wendtner, M.-H. and Korting, H. C. 2006. The pH of the skin surface and its impact on the barrier function. Skin Pharmacology and Physiology 19 (6): 296-302.

Schneider, A. M., Holcombe, B. V. and Stephens, L. G. 1996. Enhancement of coolness to the touch by hygroscopic fibers: part 1 subjective trials. Textile Research Journal 66 (8): 515-520.

Schneider, A. M., Hoschke, B. N. and Goldsmid, H. J. 1992. Heat transfer through moist fabrics. Textile Research Journal 62 (2): 61-66.

Scott, S., Cole, T., Lucas, P. and Richards, M. 1983. Weight gain and movement patterns of very low birthweight babies nursed on lambswool. The Lancet 322 (8357): 1014-1016.
Shelley, W. B., Hurley, H. J. and Nichols, A. C. 1953. Axillary odor - experimental study of the role of bacteria, apocrine sweat, and deodorants. Archives of Dermatology 68 (4): 430-446.
Shin, M., Swan, P. and Chow, C. M. 2014 The effects of textile fabrics for sleeping apparel and bedding on sleep quality at ambient conditions of 17°C and 22°C. personal communication.
Sneddon, J. N., Lee, J. A. and Soutar, G. N. 2012a. Exploring consumer beliefs about wool apparel in the USA and Australia. Journal of The Textile Institute 103 (1): 40-47.
Sneddon, J. N., Lee, J. A. and Soutar, G. N. 2012b. Making sense of consumers' wool apparel preferences. Journal of The Textile Institute 103 (4): 405-415.
Stanton, A. N. 1984. Overheating and cot death. The Lancet 324 (8413): 1199-1201.
Stanton, J. H., Speijers, J., Naylor, G. R. S., Pieruzzini, S., Beilby, J., Barsden, E. and Clarke, A. 2014. Skin comfort of base layer knitted garments. Part 1 Description and evaluation of wearer test protocol. Textile Research Journal 84 (13): 1385-1399.
Starick, L. 2013. Global fabric tracker study. Sydney, Australia: Australian Wool Innovation in association with Nielsen.
Tang, B., Wang, J., Xu, S., Afrin, T., Xu, W., Sun, L. and Wang, X. 2011. Application of anisotropic silver nanoparticles: multifunctionalization of wool fabrics. Journal *of* Colloid and Interface Science 356 (2): 513-518.
Umbach, K. H. 1986. Comparative thermophysiological test on blankets made from wool and acrylic-fibre-cotton blends. Journal of The Textile Institute 77 (3): 212-222.
van Amber, R. R. 2013. Sock fabrics: the effect of fibre type, yarn type and fabric structure on selected properties. In Dunedin, New Zealand: University of Otago.
van Amber, R. R., Lowe, B. E., Niven, B. E., Laing, R. M. and Wilson, C. A. 2015a. Sock fabrics: relevance of fiber type, yarn, fabric structure and moisture on cyclic compression. Textile Research Journal 85 (1): 26-35.
van Amber, R. R., Lowe, B. E., Niven, B. E., Laing, R. M., Wilson, C. A. and Collie, S. R. 2015b. The effect of fiber type, yarn structure and fabric structure on the frictional characteristics of sock fabrics. Textile Research Journal 85 (2): 115-127.
Wakelin, S. H., Smith, H., White, I. R., Rycroft, R. J. G. and McFadden, J. P. 2001. A retrospective analysis of contact allergy to lanolin. British Journal of Dermatology *(1951)* 145 (1): 28-31.
Wang, G., Zhang, W., Postle, R. and Phillips, D. 2003. Evaluating wool shirt comfort with wear trials and the forearm test. Textile Research Journal 73 (2): 113-119.
Wilson, C., Laing, R. and Niven, B. 2000. Multi-layer bedding materials and the effect of air spaces on 'wet' thermal resistance of dry materials. Journal of the Human-Environment System 4 (1): 23-32.
Wilson, C. A., Laing, R. M. and Carr, D. J. 2002. Air and air spaces - the invisible addition to thermal resistance. Journal of the Human-Environment System 5 (2): 69-77.
Wilson, C. A., Taylor, B. J., Laing, R. M., Williams, S. M., Mitchell, E. A. and New Zealand Cot Death Study Group 1994. Clothing and bedding and its relevance to sudden infant death syndrome: further results from the New Zealand Cot Death Study. Journal of Paediatrics and Child Health 30 (6): 506-512.
Yao, L., Laing, R. M., Bremer, P. J., Silcock, P. J. and Leus, M. J. 2015. Measuring textile adsorption of body odor compounds using proton-transfer-reaction mass spectrometry. Textile Research Journal 85 (17): 1817-1826.
Yao, L., Tokura, H., Li, Y., Newton, E., Gohel, M. D. I. and Chung, W. J. 2007. Mechanism of pajama material on stratum corneum water content under mild cold conditions: explored by hierarchical linear regression. Skin Research and Technology 13 (4): 412-416.
Zhao, T. and Sun, G. 2006. Antimicrobial finishing of wool fabrics with quaternary aminopyridinium salts. Journal of Applied Polymer Science 103 (1): 482-486.
Zhu, P. and Sun, G. 2004. Antimicrobial finishing of wool fabrics using quaternary ammonium salts. Journal of Applied Polymer Science 93 (3): 1037-1041.

Correlations Between the Physiochemical Characteristics of Plant Fibres and Their Mechanical Properties

Marie-Joo Le Guen, Roger H. Newman, Alan Fernyhough, Stefan J. Hill and Mark P. Staiger

Abstract Lignocellulosic fibres harvested from different plant types exhibit variations in mechanical properties that are associated with their chemical composition and physical features. This diversity indicates that plant fibres could be selected based on their physio-chemical properties for tailored applications such as enhanced vibration damping. In this study, bast, leaf, and mesocarp fibre bundles were investigated to understand correlations between their physiochemical characteristics and their mechanical properties with a particular focus on their vibrational damping ability. Due to the interrelations between the investigated variables such as cellulose content and microfibril angle, a multivariate analysis (principal component analysis) was applied to elucidate trends. The stiffness and strength of the fibre bundles were found to be positively correlated to high cellulose content and low microfibril angle while high toughness was correlated with fibre bundles of high lignin content and high microfibril angle. Conversely, the damping coefficient was found to be positively correlated to fibres containing high level of hemicelluloses, such as those extracted from leafy plants.

Keywords Plant fibers · Chemical composition · Physical characteristics · Mechanical properties

M.-J. Le Guen (✉) · R.H. Newman · A. Fernyhough · S.J. Hill
Scion, Private Bag 3020, Rotorua 3010, New Zealand
e-mail: mariejoo.leguen@scionresearch.com

A. Fernyhough
e-mail: AFernyhough@ashland.com

S.J. Hill
e-mail: stefan.hill@scionresearch.com

M.P. Staiger
Department of Mechanical Engineering, MacDiarmid Institute for Advanced Materials and Nanotechnology, University of Canterbury, Private Bag 4800, Christchurch 8140, New Zealand
e-mail: mark.staiger@canterbury.ac.nz

R. Fangueiro and S. Rana (eds.), *Natural Fibres: Advances in Science and Technology Towards Industrial Applications*, RILEM Bookseries 12,
DOI 10.1007/978-94-017-7515-1_3

Introduction

Lignocellulosic fibres extracted from different parts of the plant including the bast, leaf, or seed may exhibit significant differences in physical and chemical properties. These differences in properties have been tentatively correlated with their biological role in the plant (Newman 2008; Eder and Burgert 2010). High tensile strength and stiffness are associated with fibres containing a high proportion of cellulose such as that found in bast fibres. Conversely, fibres of high cellulose microfibril angle (MFA) are associated with low strength and stiffness but high elongation and toughness (e.g. stipe, mesocarp fibres) (Chakravarty and Hearle 1967; McLaughlin and Tait 1980; Satyanarayana et al. 1982).

However, in terms of vibration damping of plant fibres, limited knowledge is available. The viscous behaviour of European flax fibres has been attributed to the amorphous matrix components of the fibre that include pectin and hemicellulose (Keryvin et al. 2015). Yet, in contrast, Gassan and Bledzki attributed the higher damping capacity of European flax composites compared to jute composites to higher cellulose content in flax fibre indicating that damping could originate from the reinforcing component of the fibres (i.e. the microfibrils) (Gassan and Bledzki 1999).

In wood research, high damping coefficients were observed for specific MFA values of 27° (Reiterer et al. 2001) and 30° (Chauhan et al. 2006). However, in another study, the variation of the damping coefficient with MFA ranging between 15° and 45°, was described with a power law relationship indicating that there was no optimal value of MFA and, the higher the MFA, the higher the damping coefficient. For instance, wood fibre with an MFA of $\sim$45° exhibited a damping coefficient that was 50 % higher compared to an MFA of 15° (Bremaud 2012). Beside MFA, vibration damping in wood was also found to be influenced by its chemical composition and particularly to its hemicelluloses content and their hydration state (Entwistle 2005). Clearly, further work is necessary to fully identify the structural parameters that most influence the damping capacity of plant and wood fibres.

In the present study, it is hypothesised that lignocellulosic fibres of a higher MFA (or a specific value of MFA) will exhibit a maximum in vibration damping capacity indicating that certain fibre types maybe more suited than others in vibration damping applications. However, as indicated previously, the MFA may not be the only parameter of importance and others, such as the moisture content or hemicelluloses content could also influence the damping coefficient. The difficulties with proving such a hypothesis arise due to the number of parameters involved (moisture content, MFA, carbohydrate content etc.) and their interrelations. For example, high cellulose content has been associated in previous research with low MFA and low lignin content (Eder and Burgert 2010). Hence, the different parameters cannot be investigated separately without potentially masking some synergetic or underlaying effects. In this work, a statistical approach is proposed using principal component analysis to further elucidate the mechanisms of damping in plant fibres.

Experimental Procedures

Materials

Fibres were purchased and/or harvested from garden centre plants in Rotorua, New Zealand (38.1378° South, 176.2514° East). European flax (*Linum usitatissimum*) and ramie fibres (*Boehmeria nivea*) were purchased as slivers from Tongling Worldbest Linen and Ramie Textile Co Ltd. The flax fibres were retted, scutched and hackled into textile grade slivers. No additional information was provided concerning the origins or the processing of the ramie fibres. Jute fabric (*Corchorus*) was purchased from a domestic supplier (Spotlight), while sisal twine (*Agave sisalana*) was obtained from Donaghys Ltd. and loose mesocarp coir fibres (*Cocos nucifera*) from Horticom (using two suppliers and labelled coir 1 and 2). Fibre bundles of jute were obtained from the weft and warp directions of a plain weave fabric, while the sisal fibres were extracted from the twine. All of the fibre bundles were straightened in water to remove any waviness caused by the weave or twine, dried and equilibrated at room temperature. Windmill palm fibres (*Trachycarpus fortunei*) were gathered from the sheath of the leaves of a palm tree growing in Rotorua. Harakeke fibre bundles (*Phormium tenax*) were collected from a private garden in Rotorua and were hand-extracted from mature leaves harvested from the top, middle and bottom of leaves in July 2013. All of the fibre specimens were weighed using a Mettler AT261 Delta range scale (Metler Toledo, Switzerland) with five decimal precision to calculate the linear mass in tex after conditioning (>48 h at 23 °C and 50 % relative humidity). The equilibrium moisture content (EMC) of the fibres was also measured by weighing following conditioning at 23 °C and 50 % RH for a period of 1 week, and then weighing specimens once a constant weight was achieved following drying at 105 °C.

Chemical Composition of Fibres

The chemical composition of the fibres was carried out by hydrolysis and anion exchange liquid chromatography on a Dionex ICS 3000 (Thermo Fisher Scientific Inc., Waltham, MA, USA). 2 g of fibres were pre-ground to 40 mesh through a Wiley mill (Thomas Scientific Inc., Swedesboro, NJ, USA). Ground samples (including routine *Pinus radiata* for quality control) were extracted with dicholoromethane in a Soxhlet apparatus (Tecator Soxtec System Model HT 1043) using a boiling time of 1 h and a rinsing time of 1 h. The total lignin content was determined in duplicate as the sum of Klason plus acid-soluble lignins following Tappi standard T222 om-88 1988 and Tappi standard UM-250 1991. In the cases of the windmill palm, the analysis was scaled down to analyse 0.25 g fibre sample due to the limited amount available. Monomeric sugars in the filtrates from Klason lignin determinations were analysed in duplicate by ion chromatography (Petterson 1991).

The results were expressed as anhydrosugar units. Lignin and carbohydrate results were validated when the duplicates were within 2 % of each other and sample total sugar and lignin values within the range 90–100 %.

Measurement of the Microfibril Angle

Wide angle X-ray scattering (WAXS) patterns were collected on the SAXS/WAXS beamline at the Australian Synchrotron (Melbourne, Victoria). The radiation wavelength was 0.8266 Å. Diffraction patterns were collected with a two-dimensional MAR-165 CCD detector (Rayonix LLC, Evanston, IL, USA), located 120.714 mm from the specimen. The beam size was approximately 150 μm in diameter and the detector pixel size was 79 × 79 μm. 15 scans of 1 s exposure at different positions were recorded and average values were taken using 5 replicate specimens. The data was processed using the Australian Synchrotron software Scatterbrain Version 1.235. The windmill palm fibres were not characterised by X-ray diffraction as they were not available at the time of the experiments.

Tensile Properties of Fibres

Tensile testing was performed on a RSA-G2 DMTA instrument (TA instruments, New Castle, DE, USA) using a 35 N load cell. The gauge length was 10 mm for all tests. 20–30 fibre bundles per type of fibre were tested to obtain average values. Quasi-static tests were carried out using a crosshead speed of 2 mm s^{-1}. Load-displacement curves were recorded, providing values of tenacity in Newton/tex and strain to failure (ε_f). The stiffness was determined from the initial slope of the load-displacement curve (between strains of 0–0.1 %). The toughness was derived from the tenacity-strain curves by numerical integration of the area under the curve. The damping coefficient (tan δ) was determined during dynamic tensile tests at a temperature of 23 °C and 50 % R.H. Consecutive frequency sweeps from 0.1 to 100 Hz were performed at 0.05 % strain along with a preload of ~10 % of the failure load ensuring that the test was carried out in the linear visco-elastic range. Tan δ was measured at a frequency of 1 Hz and averaged over the 10th and 20th cycles to avoid potential damping artefacts caused by the handling of the fibres during clamping.

Principal Component Analysis

Principal component analysis (PCA) was performed on the average values of 10 variables using a correlation matrix. The variables included the cellulose content, lignin content, hemicellulose content, MFA, tenacity, stiffness, strain, toughness, damping and water content at 50 % R.H. The PCA was interpreted using R version 3.1.0 (R Development Core Team), and R studio version 0.98.945 (RStudio, Inc.).

Results and Discussion

Carbohydrate and Lignin Content of the Plant Fibres

Cellulose, lignin and hemicellulose were the main components of the chemical composition of the fibres (Table 1). "Other" components included waxes, minerals and pectin. Only chemicals that were common to all of the fibres were taken into account in the analysis, although it must be noted that pectin is a significant component of European flax and as such may influence damping behaviour. The chemical composition of the fibres was similar to previous work, highlighting the significant variation in chemical composition across different natural fibres (Table 1) (Müssig et al. 2010).

Differences in the cellulose to lignin ratio were observed for the harakeke fibres sampled at three different locations in the leaf (Table 2). Harakeke fibres extracted from the bottom of the leaf were richer in lignin compared to that of the top part of the leaf. Richter et al. made similar observations in the lignin distribution through

Table 1 Chemical composition of plant fibres (wt%)

	Cellulose	Hemicelluloses	Lignin (Klason)	Others
Pinus radiata	44.1	19.8	27.6	7.9
European flax	72.6	8.7	3.2	14.7
Ramie	96.0	0.5	0.4	3.1
Jute	46.2	12.5	15.4	23.3
Harakeke	49.2	21.2	6.8	18.7
Sisal	45.3	20.4	15.1	15.5
Windmill palm	36.2	17.8	35.6	18.6
Coir 1	32.9	25.2	35.0	5.5
Coir 2	34.1	20.9	33.6	9.6

Table 2 Chemical composition of Harakeke fibres as a function of the location within the leaf (wt%)

	Cellulose	Hemicelluloses	Lignin (Klason)	Others
Top	59.0	22.0	5.1	10.0
Middle	57.5	22.4	5.7	10.1
Bottom	55.6	22.6	7.6	9.5

harakeke leaf using Raman spectroscopy. It was hypothesised that increased lignin at the base of the harakeke plant provides the additional required stiffness in this part of the leaf (Richter et al. 2011). In contrast, little variation in the hemicellulose content was observed across different locations in the harakeke leaf (Table 2).

Measurement of the Plant Fibres Microfibril Angle

The MFA determined by X-ray diffraction were consistent with values given in the literature with the exception of the ramie fibres which was lower than previously reported (Table 3) (Eder and Burgert 2010). Trends in the MFA were clearly observed with bast fibres having an MFA of ~4–10°, while that of leaf fibres was ~10–20° and that of mesocarp/stipe fibres was >20°. The MFA value of 39.5° for windmill fibres was used from literature (Zhai et al. 2013).

Tensile Properties of Plant Fibres

The stiffnesses of the fibres from European flax, ramie and jute were found to be of the same order of magnitude (Table 4). European flax fibres exhibited the highest ultimate tensile strength, while coir fibres had the highest strain to failure and toughness as expected from previous research (Müssig et al. 2010). In terms of intermediate properties, harakeke fibres exhibited the highest combined performance in terms of the ultimate tensile strength, strain to failure and stiffness.

The tensile properties of harakeke fibres extracted from the tip and mid sections of the leaf were similar, while the fibres extracted from the lower section were stiffer although not as strong or as tough as the fibres from the other sections (Table 5). These findings are also supported by the work of Richter et al. that reported a higher lignin content within the lower part of the harakeke leaf (Richter

Table 3 Experimental and literature values of the MFA of the various plant fibres studied in this work

Plant species	Fibre type	MFA (Exp.)	MFA (literature)	Reference
European flax	Bast	4.5 (0.1)	4.4	Müller et al. (1998)
Ramie	Bast	4.2 (0.1)	7–8	Bledzki and Gassan (1999)
Jute	Bast	6.6 (0.1)	8	Bledzki and Gassan (1999)
Harakeke	Leaf	11.4 (0.3)	16	Richter et al. (2011)
Sisal	Leaf	15.2 (0.2)	20	Pavithran et al. (1987)
Windmill palm	Leaf/Stipe	N/A	37.8	Zhai et al. (2013)
Coir 1	Mesocarp	46.8 (0.8)	45	Martinschitz et al. (2008)
Coir 2	Mesocarp	N/A	45	Martinschitz et al. (2008)

Values in parentheses represent 1 standard deviation

Table 4 Tensile properties of the selected plant fibres

	Ultimate stress (N/tex)	Ultimate strain (%)	E-modulus (N/tex)	Toughness (N/tex)
European flax	0.52 (0.08)	1.89 (0.21)	20.8 (3.7)	0.0051 (0.0014)
Ramie	0.31 (0.07)	1.84 (0.16)	26.8 (4.4)	0.0057 (0.0008)
Jute	0.39 (0.09)	1.53 (0.21)	24.6 (5.6)	0.0032 (0.0009)
Harakeke	0.48 (0.07)	2.77 (0.31)	17.2 (2.5)	0.0095 (0.0024)
Sisal	0.40 (0.05)	2.65 (0.30)	18.7 (3.0)	0.0058 (0.0012)
Windmill palm	0.12 (0.01)	17.94 (2.71)	4.0 (0.2)	0.0154 (0.0029)
Coir 1	0.12 (0.02)	24.44 (4.21)	3.8 (0.6)	0.0210 (0.0051)
Coir 2	0.14 (0.02)	38.50 (5.48)	3.1 (0.3)	0.0365 (0.0069)

The value in parentheses is the 95 % confidence interval

Table 5 Tensile properties of the harakeke fibres extracted from the tip, middle, and bottom of the leaves

	Ultimate stress (N/tex)	Ultimate strain (%)	E-modulus (N/tex)	Toughness (N/tex)
Top	0.49 (0.06)	3.66 (0.24)	17.0 (2.7)	0.0099 (0.0014)
Middle	0.48 (0.10)	3.47 (0.38)	17.2 (2.5)	0.0095 (0.0024)
Bottom	0.37 (0.10)	3.02 (0.32)	18.7 (2.0)	0.0071 (0.0015)

The value in parentheses is the 95 % confidence interval

et al. 2011). From a biological perspective, the lower part of the leaves requires greater stiffness that supports the plant structure above. Conversely, the top of the leaves requires greater flexibility to reduce mechanical stresses at the base of the plant (e.g. during windy weather) and also maximise the orientation of the leaves surface towards sunlight.

During tensile loading, the force acting on the cellulose microfibrils is a function of the MFA. The larger the MFA, the less the cellulose microfibrils contribute to the stiffness at the beginning of a tensile test explaining the low stiffness of mesocarp fibres compared to bast fibres (Table 4).

In the case of the bast fibres, the microfibrils realignment was minimal due to their initial low MFA (≤10°). Hence, the stress-strain deformation was quasi-linear. In the case of leaf fibres the MFA was ranging between 10° and 20°. The initial deformation was elastic, similarly to bast fibres but then gradually deviated from linear behaviour indicating irreversible plastic deformation.

When the MFA was greater than 20°, the deformation was bi-phasic and the transition was clearly indicated by a yield point. The initial elastic region of the curve was matrix-dominated. Past the yield point, the load gradually transferred to the cellulose microfibrils exhibiting a second linear behaviour with a steeper slope than the beginning of the test, caused by the realignment of the cellulose microfibrils (Martinschitz et al. 2008).

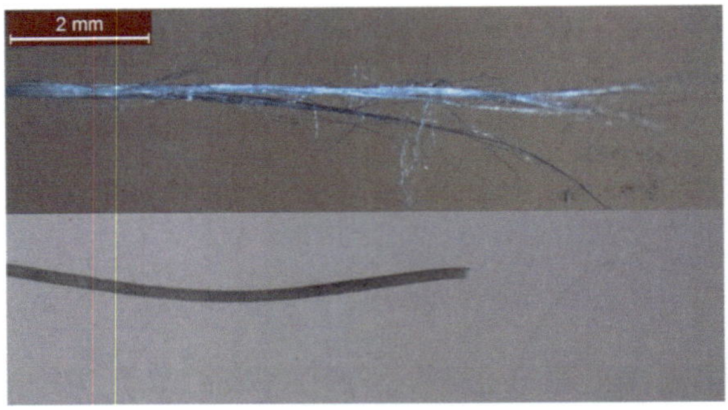

Fig. 1 Tensile fracture of coir fibre (*bottom*) and harakeke fibre (*top*) indicating a cohesive in-plane fracture of the coir fibre and defibrillation of the harakeke fibre

In terms of fracture, coir and windmill fibres failed abruptly after extensive plastic deformation. Coir fibre had a smooth fracture surface after tensile testing where all of the cells within a single fibre fractured on the same plane (Fig. 1). In contrast, the cells within the harakeke, sisal and European flax fibres failed independently of each other, causing the fibres to slide past each other and leading to extensive defibrillation (Fig. 1).

Damping Coefficient of Plant Fibres

The range of damping coefficient was measured from 0.0125 to 0.0198 (Table 6). Berthelot et al. reported damping coefficients of 0.012 for glass fibres, 0.0062 for carbon fibres and 0.0165 for aramid fibres measured by acoustic method. Their results indicated that plant fibres have a significant advantage in damping applications compared to glass and carbon fibres (Berthelot and Sefrani 2007). Differences in the damping coefficient between the fibre types were observed;

Table 6 Damping coefficient for selected plant fibres

	Damping coefficient	Moisture content (wt%)
European flax	0.0140 (0.0018)	10.6
Ramie	0.0142 (0.0020)	10.2
Jute	0.0128 (0.0010)	9.4
Harakeke	0.0176 (0.0027)	10.2
Sisal	0.0198 (0.0023)	9.5
Windmill palm	0.0194 (0.0018)	10.2
Coir 1	0.0180 (0.0012)	9.6
Coir 2	0.0165 (0.0026)	8.1

The value in parentheses is the 95 % confidence interval

Table 7 Damping coefficient of harakeke fibres extracted from the tip, middle, and bottom of the leaves

	Damping coefficient	Moisture content (wt%)
Top	0.0167 (0.0018)	10.1
Middle	0.0188 (0.0012)	10.2
Bottom	0.0187 (0.0026)	10.1

The value in parentheses is the 95 % confidence interval

however, no obvious trend emerged with MFA, moisture content or location in the leaf (Table 7). As mentioned previously, the damping coefficient was believed to be influenced by a combination of several variables (e.g. chemical composition, MFA). Hence, the influence of these variables on damping were investigated using multivariate analysis.

Determination of the Correlation Between Variables Using Principal Component Analysis

The principal components analysis was limited to the first two components as they accounted for 89 % of the variance. A bi-plot of the data in the space of the first two principal components was produced with the fibres' scores and the variables (loadings) (Fig. 2). The analysis could highlight 3 clusters of variables:

- Group 1: MFA, lignin, strain and toughness
- Group 2: cellulose, stiffness, water content and tenacity
- Group 3: hemicellulose and damping

Group 1 and 2 were oppositely related. This observation indicated that when the variables from Group 1 (MFA, lignin, strain, toughness) were high, the variables from Group 2 were low (Fig. 2). The fibre toughness was strongly correlated with high strain, high MFA and high lignin content (e.g. windmill palm and coir fibre). During a tensile test, the cellulose microfibrils are known to realign towards the fibre axis, enabling extensive plastic deformation (Martinschitz et al. 2008). Hence, while bast fibre such as European flax exhibited a tenacity 4 times higher than coir fibres, their toughness was up to 7 times lower (Le Guen 2014). Intermediate toughness was found for fibres with MFA between 10° and 20°, for which the deformation curves indicated a bi-phasic behaviour and the absence of a distinctive yield point (Le Guen 2014). The microfibrils could reorient in the axis direction, causing enough shear stresses in the hemicellulose/lignin matrix to provoke a gradual transition from elastic to plastic deformation as exhibited by harakeke and sisal fibres. In contrast to studies on wood, the damping coefficient did not correlate with the MFA (Chauhan et al. 2006). The second cluster of variables indicated correlation between cellulose content, stiffness, tenacity and water content (Fig. 2). A high tenacity was correlated with a high cellulose content (i.e. low lignin content) similar to previous studies (Eder and Burgert 2010). The larger is the MFA, the

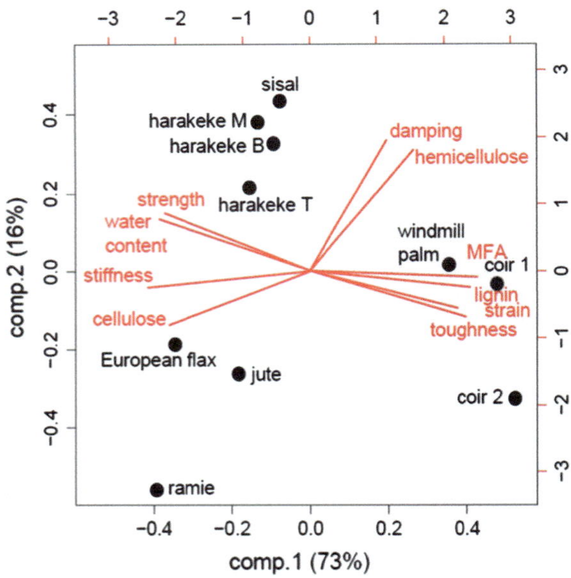

Fig. 2 Bi-plot of the variables (loading and scores) using principal component 1 and 2

lesser is the cellulose microfibrils' contribution to the stiffness at the beginning of the test explaining the low stiffness of mesocarp fibres compared to bast fibres. Water content was not associated with the damping coefficient. This observation indicated that the source of damping was not directly linked to the motion of the water molecules but the plasticising of the hemicelluloses (Salmen 2004). However, no correlation was found between hemicellulose and water content. Jarvis hypothesised that the cellulose chains at the surface of a microfibril are loosely packed compared to the cellulose within the crystal, enabling more hydrogen bonding between the cellulose and the hemicellulose or water (Jarvis 2005). It was also believed that a small amount of hemicellulose was present between the cellulose microfibrils enabling access for a limited amount of water (Jarvis 2005). These observations are in line with the correlation between water and cellulose content. The last cluster of variables included the damping coefficient and hemicellulose content (Fig. 2). Fibres that contain a higher proportion of hemicelluloses (e.g. harakeke, sisal) exhibited a higher damping coefficient. In wood research, energy losses are explained by the hemicellulose molecular hydrogen bonding structure, especially, in the presence of water. Vibrational energy loss in wood was reported to increase almost linearly with the hydration of hemicellulose (Entwistle 2005). Bucur reported that the damping coefficient typically increased from 5.5×10^{-3} to 18.5×10^{-3} as the moisture content increased from 5 to 35 % (Bucur 2006). The hydration of hemicelluloses increases the energy dissipation by the (i) motion of water molecules and (ii) hydrogen bonding shifts between hydrated hemicellulose and cellulose (Bremaud et al. 2010; Ebrahimzadeh and Bertilsson 1998; Newman 2005). The first mechanism relies on the water molecules capacity to move freely due to their small size and the second mechanism on the plasticisation of the hemicelluloses by water (Entwistle 2005). It was shown previously that there was no direct correlation between damping and water content; hence, the second damping mechanism is favoured. Hydrated hemicellulose molecules do not have the same mobility due to their size, molecular structure and conformation (Åkerholm and Salmén 2001; Newman 2005). However, no obvious trend emerged with the different types of hemicelluloses in this study (results not presented here) (Le Guen 2014).

Conclusions

Selected plant fibres were characterised for their mechanical properties (strength, stiffness, strain, toughness and damping), physical property (morphology and MFA) and chemical composition (hemicellulose, cellulose and lignin content). Correlations between variables were analysed by PCA. The highest mechanical properties in terms of stiffness and strength were observed for bast fibres, while leaf and mesocarp fibres exhibited the higher toughness.

In terms of energy absorption properties, the toughness of the technical fibres increased with increasing MFA and was correlated with fibres containing high level of lignin and low levels of cellulose. The damping coefficient was only related to the hemicellulose content. The literature states that damping increases with moisture due to plasticisation of the hemicelluloses. However, the present findings indicate clearly that the amount of absorbed water was only correlated with cellulose content. Harakeke and sisal fibre exhibited intermediate mechanical properties to bast and mesocarp fibres (e.g. tenacity, toughness) and the highest damping coefficient. Hence, natural fibres that contain a higher proportion of hemicelluloses would be preferable for maximising vibration damping applications if used in a hydrated-like state.

Acknowledgements The authors thank the Biopolymer Network Ltd for funding under the New Zealand Ministry of Business, Innovation and Employment contract number BPLY0801, Mrs. Sunita Jeram for carrying out the chemical analysis and Dr. Nigel Kirby for his help during the X-ray diffraction experiments.

References

Åkerholm M, Salmén L (2001) Interactions between wood polymers studied by dynamic FT-IR spectroscopy. Polym 42(3):963-969.

Berthelot J-M, Sefrani Y (2007) Longitudinal and transverse damping of unidirectional fibre composites. Compos Sci Technol 79(3):423-431.

Bledzki AK, Gassan J (1999) Composites reinforced with cellulose based fibres. Prog Polym Sci 24(2):221-274.

Bremaud I (2012) What do we know on resonance wood properties? In: Acoustics 2012 Nantes Conference, 23-27 April 2012, Nantes, France.

Bremaud I, Minato K, Langbour P, Thibault B (2010) Physico-chemical indicators of inter-specific variability in vibration damping of wood. Ann For Sci 67:707.

Bucur V (ed) (2006) Acoustics of wood. 2nd edn. Springer-Verlag, Berlin Heidelberg.

Chakravarty AC, Hearle JW (1967) Observations on the tensile properties of ultimate cells of some plant fibres. J Text Inst 58(12):651-656.

Chauhan S, Donnelly R, Huang C-L, Nakada R, Yafang Y, Walker JCF (2006) Wood quality: multifaceted opportunities. In: Walker JCF (ed) Primary Wood Processing: Principles and Practice 2nd edn. Springer, Dordrecht, p 159-202.

Ebrahimzadeh P, Bertilsson H (1998) Effect of impregnation on mechanosorption in wood and paper studied by dynamic mechanical analysis. Wood Sci Technol 32(2):101-118.

Eder M, Burgert I (2010) Natural fibres – Function in nature. In: Müssig J (ed) Industrial Applications of Natural Fibres Structure, Properties and Technical Applications. Wiley, Chichester, p 23-39.

Entwistle KM (2005) The reaction of hemicelluloses to applied stresses with emphasis on the effect of changes of water concentration. In: Entwistle KM and Walker JCF (eds) The hemicelluloses workshop. Wood Technology Research Centre, University of Canterbury, Christchurch, p 137-146.

Gassan J, Bledzki AK (1999) Possibilities for improving the mechanical properties of jute/epoxy composites by alkali treatment of fibres. Compo Sci Technol 59(9):1303-1309.

Jarvis MC (2005) Cellulose structure and hemicellulose-cellulose association. In: Entwistle KM and Walker JCF (eds) The hemicelluloses workshop. Wood Technology Research Centre, University of Canterbury, Christchurch, p 87-102.

Keryvin V, Lan M, Bourmaud A, Parenteau T, Charleux L, Baley C (2015) Analysis of flax fibres viscoelastic behaviour at micro and nano scales. Composites Part A 68(0):219-225.

Le Guen M-J (2014) Damping behaviour of plant-fibre composite materials. Dissertation, University of Canterbury.

Martinschitz K, Boesecke P, Garvey C, Gindl W, Keckes J (2008) Changes in microfibril angle in cyclically deformed dry coir fibers studied by in-situ synchrotron x-ray diffraction. J Mater Sci 43(1):350-356.

McLaughlin E, Tait R (1980) Fracture mechanism of plant fibres. J Mater Sci 15(1):89-95.

Müller M, Czihak C, Vogl G, Fratzl P, Schober H, Riekel C (1998) Direct observation of microfibril arrangement in a single native cellulose fiber by microbeam small-angle X-ray scattering. Macromol 31(12):3953-3957.

Müssig J, Fischer H, Graupner N, Drieling A (2010) Testing methods for measuring physical and mechanical fibre properties (Plant and animal fibres). In: Müssig J (ed) Industrial Applications of Natural Fibres Structure, Properties and Technical Applications. Wiley, Chichester, p 269-310.

Newman RH (2005) Solid-state NMR as a tool for studying dancing molecules. In: Entwistle KM and Walker JCF (eds) The hemicelluloses workshop. Wood Technology Research Centre, University of Canterbury, Christchurch, p 77-86.

Newman RH (2008) Development of non-wood natural-fibre composites. In: Pickering KL (ed) Properties and performance of natural-fibre composites, Woodhead Publishing, Cambridge and CRC Press, Boca Raton, p193-204.

Pavithran C, Mukherjee P, Brahmakumar M, Damodaran A (1987) Impact properties of natural fibre composites. J Mater Sci Lett 6(8): 882-884.

Pettersen RC (1991) Wood sugar analysis by anion chromatography. J Wood Chem Technol 11(4):495-501.

Reiterer A, Lichtenegger H, Fratzl P, Stanzl-Tschegg S (2001) Deformation and energy absorption of wood cell walls with different nanostructure under tensile loading. J Mater Sci 36 (19):4681-4686.

Richter S, Müssig J, Gierlinger N (2011) Functional plant cell wall design revealed by the Raman imaging approach. Planta 233(4):763-772.

Salmén L (2004) Micromechanical understanding of the cell-wall structure. Comptes Rendus Biologies 327:873-880.
Satyanarayana K, Pillai C, Sukumaran K, Pillai S, Rohatgi P, Vijayan K (1982) Structure property studies of fibres from various parts of the coconut tree. J Mater Sci 17(8):2453-2462.
Zhai S, Horikawa Y, Imai T, Sugiyama J (2013) Cell wall characterization of windmill palm (*Trachycarpus Fortunei*) fibers and its functional implications. IAWA J 34(1):20-33.

Influence of Stem Morphology and Fibres Stiffness on the Loading Stability of Flax

Alain Bourmaud, Marianne Gibaud, Anaële Lefeuvre, Claudine Morvan and Christophe Baley

Abstract Flax fibers (*Linum usitatissimun* L.) with good mechanical properties are required to reinforce polymers. This work provides an original method to determine the stiffness of dried or green fibers contained inside the plant. We studied three recent flax varieties (Marylin, Eden and TDL25) with distinct lodging resistances. After mechanical characterization on elementary fibers and morphological analysis on stems, we showed, by flexural tests on green and dried stems, that it was possible to correlate their stiffness with the Young modulus of the elementary fibers as well as their internal organization. This result confirmed the crucial role of fibers in supporting the plant and in particular in the lodging behaviour.

Keywords Flax fibers · Stem morphology · Fiber stiffness · Lodging stability

Introduction

The use of flax fibers as composite reinforcements is justified by the renewable character of these fibers as well as by their good specific mechanical properties (Lefeuvre et al. 2014). Moreover, their mechanical performances are yearly reproducible, for a same variety (Lefeuvre et al. 2013) or for numerous batches (Baley and Bourmaud 2014). In order to ensure a reliable culture of flax, the lodging stability of the plant could be considered as a key point. Thus, an important varietal selection work is performed in order to develop new varieties with high fibers yield and good disease or lodging resistance, to ensure a sufficient income for the farmer.

A. Bourmaud (✉) · M. Gibaud · A. Lefeuvre · C. Baley
LIMATB, University of South Brittany, Lorient, France
e-mail: alain.bourmaud@univ-ubs.fr

C. Morvan
PBS Laboratory, University of Rouen, UMR 6270 CNRS FR 3038, Mont Saint Aignan, France

R. Fangueiro and S. Rana (eds.), *Natural Fibres: Advances in Science and Technology Towards Industrial Applications*, RILEM Bookseries 12,
DOI 10.1007/978-94-017-7515-1_4

The lodging originates because of buckling of the stems due to abundant rain, strong winds or numerous visiting animals (Menoux et al. 1982); in case of severe lodging, the plants might be lying down on the ground. The lodging of flax stems is highly influenced by the stem stiffness, the environmental solicitations and especially the additional water weight. Consequently, the distribution of the water drops on the plant increases its mass and, with wind, the risk of instability becomes even more significant. Moreover, plants could react to environmental stimuli by different ways; this phenomenon, called thigmomorphogenesis, generally induces a decrease into the plant height conjugated to an increase of their diameter and ramification (Telewski and Jaffe 1986). In addition, mechanically induced stress could cause an improvement of the cell wall stiffness (Biddington 1986).

In this article, we study the correlation between the stem stiffness and the elementary fibers modulus based on their distribution. Green and dried stems were both analysed. We selected three varieties of flax (*Linum usitatissimum*), Marylin, Eden and TDL25 having various lodging resistances. Firstly, we analysed the fiber yield for each variety as well as the stems morphology. Secondly, we studied the elementary fibers' mechanical properties after dew retting and scutching according to the fibers' location along the stem. The third part of the paper is dedicated to the measurement of the dried and green stem bending stiffness, giving an estimation of the modulus of elementary fibers in a green plant. The last part concerns an estimation of the risk of stem buckling.

Materials and Methods

Marylin, Eden and TDL 25 textile flax varieties were used for this study. Flax samples were obtained from the agricultural cooperative Terre De Lin (France). Eden is a variety registered since 2009 and is known for its lodging resistance, whereas TDL25 isn't registered yet (recently selected) and presents a high lodging sensitivity. Flax stems were grown in France (Saint Pierre le Viger, France) in 2009, 2010, 2011 and 2012 (Marylin) or 2013 (Eden and TDL25) with almost same densities (around 1500 plants/m^2) which corresponds to the conventional density for flax culture (Bert 2013). An original climatic scenario characterized each year. The year 2009 corresponded to a base/average year with accumulated sunshine hours and a pluviometry close to the mean values calculated over 30 years. The year 2010 exhibited a drought during the vegetative phase with accumulated sunshine hours slightly higher than the average value and with a water deficit in April and May. In 2011 there was a drastic drought during the vegetative phase with accumulated sunshine hours largely higher than the normal/than the average value; this drought was increased by a lack of water in March, April and May with water reserves already affected by the drought in the year 2010. The droughts during the vegetative phase had repercussions on the growth of the stem and generated short flax stems. Year 2012 exhibited an opposite climatic scenario compared to the previous years. The accumulated sunshine hours were close to the normal; however there was one important excess of water in April and June inducing lodging for some fields. Lastly, in 2013, weather conditions were favourable and close to normal ones; there was no lodging in the cultivated fields for this work.

Mechanical properties (Young's modulus, ultimate strength and failure strain) of single flax fibers were obtained from tensile tests. Fibers are hand-extracted and bonded onto a paper frame and clamped on a universal MTS type tensile testing machine equipped with a 2 N capacity load cell. Tensile tests were carried out according to the AFNOR XP T-501-2 standard (AFNOR 2009) taking into account the compliance of the cell-load. The gauge length was 10 mm and fibers were tested at a crosshead displacement speed of 1 mm min^{-1} and at controlled temperature (23 °C) and humidity (50 %). The mean diameter used for calculations is the average of 6 values measured along the fiber (Lefeuvre et al. 2013). For each sampling area (bottom and middle of the stem), at least 50 elementary fibers were tested.

External dimensions (height and external diameter) of the stems were assessed using a tape measure and a calliper. The bottom of the stem was defined as the beginning of the aerial part of the plant. Internal morphology of the stems and characterization of the fibers were performed by using histological cuts. The flax stem was embedded in elder marrow and cut with a Ranvier's microtome and a razor blade. Cuts on green stems were performed about 30 min after being removed from the freezer to avoid drying. Cuts on dried stems were made after immersion for 20 min in a 50 % ethanol-50 % water solution.

Sections were observed by using an optical microscope and pictures were taken. To enable the detection of fibers, these were first manually encircled using the Gimp® software. Images were then analysed with image J®, an image analyser program including a cell-counter. Thus, the following information was obtained: fiber number, areas, and filling rate, as well as the area of transverse section and stem diameter. The analysis was first performed on green stems and then on dried stems in order to measure the variation of these previous parameters according to the stem water content.

For 3 points bending tests, flax stems were cut into 15 cm long segments. Each section was weighed and submitted to a three-point bending test. Tests were performed by using a MTS universal testing machine with a 50 N load cell with a span length of 120 mm. Given the diameter of the stem (around 2 mm) and the selected span length, the shear effect can be ignored here. The crosshead speed is set at 20 mm min^{-1}. The force displacement curve was registered and used to determine the bending stiffness EI of the stem. Considering the stem as a uniform beam with circular cross section, Eq. 1 gives displacement in the middle of the section:

$$y = \frac{FL^3}{48 \cdot EI} \tag{1}$$

where F is the applied force, L the span length and EI the bending stiffness.

Assuming that fibers are the strengthening tissues responsible for the lodging resistance, the flax stem is considered as a hollow circular beam. Cross-sections obtained by using a microtome were observed and the axial second moment of the area was determined using image analysis, as given by Eq. 2:

$$I = \Pi \cdot \frac{(D^4 - d^4)}{64} \tag{2}$$

where D and d are the external and internal fiber crown, respectively.

Thus, Young's modulus E of the stem is given by the formula:

$$E = \frac{dF}{dY} \cdot \frac{64 \cdot L^3}{48 \cdot \Pi \cdot (D^4 - d^4)} \tag{3}$$

where dF/dY is the initial slope of the force displacement curve.

The tests were first performed on green stems 30 min after being removed from the freezer. The tests were stopped before the rupture of the stem to stay in the elastic zone. The stems were then stored at a controlled temperature and humidity for drying. Once drying was completed, similar tests were performed on these same stems to measure the influence of water content on the bending rigidity of the stem.

Results

In first time, we selected Marylin flax stem cultivated in the same geographical area over 4 successive years. Various parameters influencing the stem stiffness were analysed: the height and the diameter of the stems, their internal organization including the proportion of fiber bundles, the diameter and mechanical properties of elementary fibers.

Stem Height and Fibre Yield

Figure 1 presents the relationship between the technical length of long fibres and the production of scutched fibres by hectare. It clearly evidences that the more the stem height increased the more important the percentage of collected fibres. Today in France, according to the professional data (Bert 2013), the average mass of fibres (mass of the long fibres and tows after scutching) collected by hectare is around 2500 kg. The average yield for the 11 batches was 2059 ± 941 kg/ha. The fibre yields were significantly lower than the average for the samples submitted in drastic drought periods in 2011 and significantly higher for one sample, which received the highest amount of rainfall in 2012. Thus, fibre production was strongly correlated with the climatic conditions, decreasing with the drought intensity and increasing with rainfall. Thus, in normal rain-fall and sunshine conditions, the shorter plants

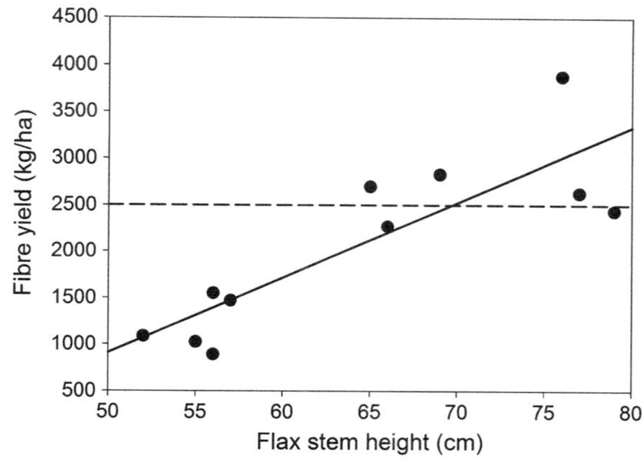

Fig. 1 Relationship between the length of technical long fibres and the fibre yield. The *dashed line* is the usual yield from the professions' data

could exhibit a significant scutched fibre yield. This phenomenon could be due to a better fibre thickening or to differences in the fibre number.

Regarding the average fibre yield after retting and scutching, the average was estimated to be 36.5 ± 3.5 %; this value is well correlated with the usual yield published by the profession (37 %) (Bert 2013). Globally the fibre rate increased when the technical length increased. This effect was the most remarkable within the 2011 year due to the drastic drought occurring this year. For each year, 2009–2011, there were shorter stems producing larger fibre rates, suggesting that the relationship was complex. This might happen, especially because the fibre rate depended not only on the fibre biosynthesis but also on the retting extent as well as the degree of humidity in the scutched plant. On a general way, the data pointed out the strong impact of an extreme drought climate on the fibre yield and rate of fibres in the biomass. On the other hand, there was no major effect of an excess rainfall compared with normal growth conditions.

Elementary Fibres Mechanical Properties

Mechanical properties of the fibers were investigated. The studied fibres, whatever their batch, exhibited a Young modulus of 57.5 ± 0.3 GPa, a strength at rupture of 1034 ± 6 MPa and an elongation of 2.0 ± 0.1 %. Baley and Bourmaud (2014) analyzed the tensile mechanical properties of 50 batches of flax fibres (*Linum usitatissimum*) cultivated in France between 1993 and 2011 submitted to dew retting and scutching. They were characterized under the same conditions as those of the present study in accordance with the XPT 25-501-2 (AFNOR 2009) standard (constant T° and hygrometry). The average Young modulus of these 50 batches was 52.6 ± 8.6 GPa, the strength at rupture 945 ± 200 MPa and the elongation at rupture 2.07 ± 0.45 %. Interestingly, most of our data exhibited higher values for E and σ than the mean estimated for the 50 reference samples (Fig. 2).

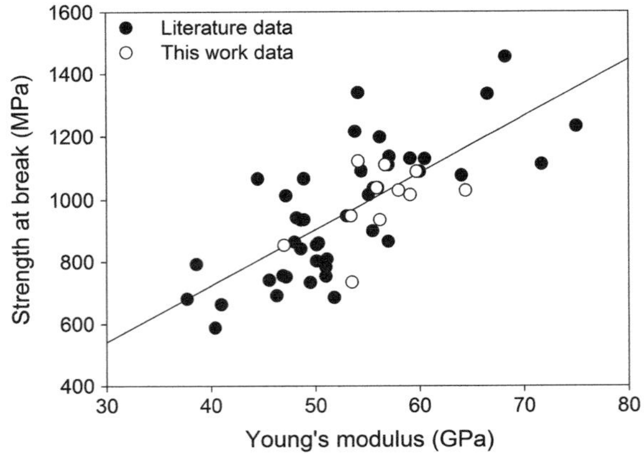

Fig. 2 Relationship between the strength at break and the Young modulus of flax fibres (Baley and Bourmaud 2014)

Due to their low density (1.53 vs. 2.54 g/cm^3 for the glass fibres), the Marylin fibres average specific stiffness and stress at rupture were estimated to be 36.9 GPa cm^3/g and 646 MPa cm^3/g, respectively. These values are close to, and sometimes, for the Young's modulus, better than those of glass fibres (28.3 GPa cm^3/g and 695 MPa cm^3/g) (Coroller et al. 2013). Moreover, although the flax fibre tensile properties are widely scattered, this phenomenon also occurred with the breaking properties of the glass fibres in opposition to their Young's modulus, which is more stable.

Organization and Morphology of the Stems

Figure 3 presents a section of a stem of flax (Fig. 3a) as well as a zoom on the peripheral area (Fig. 3b) where the fibres are arranged into bundles. The fibres were considered as support materials to reinforce the stem structure. For each batch, three stems were studied. Table 1 shows the results of the morphological analysis of the stems at middle height.

We can observe that the number of fibres per section and their diameter increased with the diameter of the stems. Thus, the drought had a direct impact on the number of thickened fibres per section. Either, consequently to the plant adaptation to drastic environmental conditions, there was a reduction of the division of fibre cells at the level of the meristem or a lack of cellulose deposition (hence of thickening). Such a relationship also existed in the genetic variability (Brutch et al. 2011) (R^2 = 0.74).

In parallel, the diameter of fibres increased significantly with the height. When Brutch et al. (2011) compared 13 lines, no correlation was found between the diameter of fibres and the stem height (genotypical variation). Thus, ecological regulation (this study) and genotypic relation showed opposite trends within a line (Marylin, this study). Both stem diameter and cell diameter increased with height

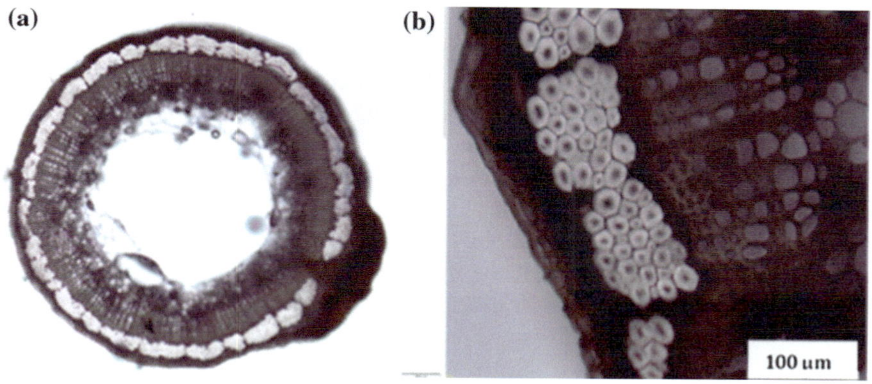

Fig. 3 Optical image of a stem section **a** (X 10) and focus on a fibre-bundle area **b** (X 40)

Table 1 Morphological analysis of the flax stems according to their height

	25 cm stem			60 cm stem			80 cm stem		
Stem number	A	B	C	D	E	F	G	H	I
Stem diameter at middle height (mm)	1.02	0.95	0.95	1.43	1.50	1.53	2.24	2.10	2.08
Number of fibres	577	464	437	643	751	716	1092	982	847
Fibre average diameter in the section (µm)	12.4	12.5	15.4	15.4	18	17.3	21.47	21.80	21.25
Fibre surface rate in the section (%)	8.9	8.3	11.4	7.8	11	9.4	11.67	11.15	9.20
Fibre surface rate in the section without the central cavity (%)	14.4	10.7	14.6	13.4	18.2	14.4	17.4	17.4	14.8
D–d (mm)	0.58	0.51	0.58	0.85	0.94	0.96	1.55	1.5	1.53

For all the stems, the studied section is located at middle height

suggesting some kind of synchronous regulation of most of the morphological parameters.

Consequently, the conjugated increase of the fibre number and of their diameter induced an increase of the diameter of the stems. Thus, the number of thickened fibres within a bundle is not a constant for a given variety: there is a strong impact due to the climate, via the regulation of the stem diameter and height ($R^2 = 0.94$). The use of image analysis showed that for a short stem of 25 cm there was an average of 2.6 fibres in the thickness of a bundle and 4.3 in the case of an 80 cm length stem. It is average information because the number of fibres in a bundle and the section of the latter are not perfectly constant. This is explained, amongst other things, by the existence of leafs fixed on the stems which could disturb the bundles organization.

In complement we can notice that the number of fibres in a section changed according to the position along the stem height, being most numerous in the middle of the stem. This was previously described for another commercial variety, Natasha (Morvan et al. 2003). Whatever the stem height, a little variation in the number of fibres is observed between the bottom and middle position whereas a significant fibre number decrease is observed for the top position. Moreover, the bibliography shows that the mechanical properties are the best in middle part of the plant (Charlet et al. 2007, 2009). The mechanical and morphological differences observed at various levels in the height of the plants might be explained by the cellulose filling of the secondary wall which occurred after the cell elongation and differentiation steps.

Correlation Between the Stem Stiffness or Morphology and the Fibre Mechanical Properties

3-point bending tests were carried out on fractions of stems taken from the central part of the plant in order to determine their flexure stiffness. The total area of the fibre bundles over a section was estimated using image analysis. Knowing the flexure stiffness of the stems and the area and thickness of the fibre crown, it was possible to estimate the Young modulus of fibres in the stems thanks to the method described in the experimental part. The hypothesis was that only the fibres contributed to the stiffness of the stems.

Figure 4 shows the calculated fibre Young's modulus for the different tested stems as a function of their length. Whatever the plant height, the calculated fibre stiffness is stable. This point confirmed the results previously obtained by elementary fibre tensile characterization (Fig. 2). The average Young modulus (deflection tests on stems) of fibres in the stems was estimated to be 63.9 ± 8.6 GPa;

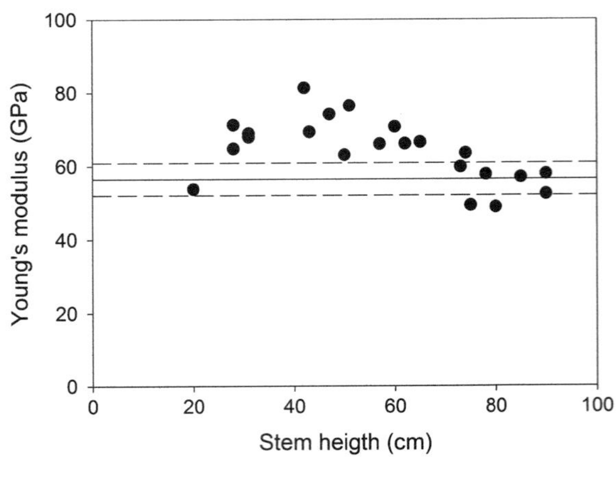

Fig. 4 Elementary fibres Young's modulus estimated from the stem stiffness and its internal organization

this value was in the same range as the stiffness measured by tensile tests (56.5 ± 4.4 GPa). The slight difference was considered to be acceptable taking account the uncertainty of the crown diameter measurements, as well as the possible modification of the fibre diameters induced by the ethanol-water immersion before cutting. Moreover, we hypothesized that only fibres contribute to the plant stiffness; the role of the woody core was neglected in our approach. On the other hand, the agro industrial processes (retting and scutching), as well as the dew retting, might have reduced the stiffness potential of the fibres.

It was shown that the fiber yield and number of fibers increased with the height of the plants, the number of fibers being maximal at medium height. In addition, the fibers' diameter decreases with the height of the stem because of the growth mechanisms. Lastly, flexural tests on stems made it possible to correlate their stiffness with the Young modulus of the elementary fibers as well as their internal organization.

In a second time, we studied two recent flax varieties (Eden and TDL25) with a distinct lodging resistance. The results of the mechanical characterization indicate that the Eden variety has a superiority concerning fiber stiffness. The analysis of the localized bending stiffness of a dried stem and of its fiber distribution in the corresponding cross section allowed us to estimate the fibers average modulus. As detailed previously for Marylin flax, results obtained by this method were very close to the tensile tests values. The same procedure was used on green stems to approach the living state of the plant and to determine the Young's modulus of green fibers. Figure 5 shows the comparison between dried or green fibers mechanical properties.

The results highlighted a variation of fiber stiffness between the green state and the dried state (around +25 %). The results enabled the use of a simplified buckling model, which confirmed the superiority of the Eden variety in lodging resistance. Thus, the analysis of the mechanical properties of flax stems and their structure could be a selection criterion.

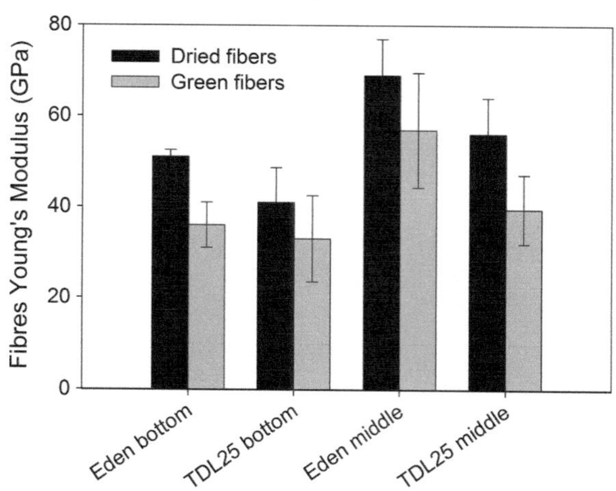

Fig. 5 Evolution of fibers' rigidity between the green and dried state

Conclusions

During this work, we highlighted important differences in the structure of the stems or flax fibres according to the years of growth; they could be linked to the climatic conditions suffered by the plants. Firstly, the study was conducted on Marylin flax. The fibre yield was strongly correlated to the fibre height and morphological differences occurred in length and diameter of the stems (both characters being highly correlated indicating a synchronous bi-dimensional regulation of the fibre dimensions). Moreover, the number of thickened fibres and their diameters were shown to vary according to the location of the plant and also according to the height of the stems. Thus, the stem diameter increased with the height of the plants as well as the thicknesses of the fibre bundles whereas the average diameter of fibres decreased from the bottom to the top of the stem. According to the environment, the plant will thus develop in a different way and adapt its morphology to be able to resist as much as possible to the external aggressive conditions and in particular, to lodging.

Interestingly, the mechanical properties of the fibres (taken from the middle) remain constant overall whatever the morphology or the structure of the stems and the diameter of the fibres. We showed that the morphology of the cell walls is directly related to the stem stiffness and that it is possible, through bending tests on portions of stems, to estimate the stiffness of elementary fibres by knowing their arrangement within the stem. These results were confirmed on two other varieties (Eden and TDL25) and the same approach was used on green stems to be close to that of a living plant. The estimate of the average stiffness of the green fibers was around 25 % lower than the dried fibers. Thus, with a simplified model, the risk of buckling (lodging) of green plants was then approached. The model takes into account the plant mass, the stem height and the flexural stem stiffness (bottom part); showing that the bending stiffness of the stem is a very important parameter for the resistance to lodging which is a buckling instability. The stem stiffness appeared to be mainly influenced by the fibers' rigidity in the present study, explaining the superiority of Eden for lodging.

In future works, it would be interesting to study the length of the elementary fibers in the plant height and the evolution (microstructure, composition) of cell walls during drying. Also, the study of the resistance to lodging in real conditions (living flax) could be very instructive.

Acknowledgments The authors would like to thank Terre de Lin and Jean Paul Trouvé for supplying flax fibers and stems, and ADEME for the financial support.

References

Lefeuvre A, Bourmaud A, Morvan C, Baley C. Tensile properties of elementary fibres of flax and glass: Analysis of reproducibility and scattering. Mater Lett 2014;130:289–91.

Lefeuvre A, Bourmaud A, Lebrun L, Morvan C, Baley C. A study of the yearly reproducibility of flax fiber tensile properties. Ind Crops Prod 2013;50:400–7.

Baley C, Bourmaud A. Average tensile properties of French elementary flax fibers. Mater Lett 2014;122:159–61.

Menoux Y, Katz E, Eyssautier A, De Parcevaux S. Résistance à la verse du lin textile: influence du milieu et critères de verse proposés. Agronomie 1982;2:173–80.

Telewski FW, Jaffe MJ. Thigmomorphogenesis: Field and laboratory studies of Abies fraseri in response to wind or mechanical perturbation. Physiol Plant 1986;66:211–8.

Biddington N. The effects of mechanically-induced stress in plants — a review. Plant Growth Regul 1986;4:103–23.

Bert F. Lin Fibre: culture et transformation. 2013.

AFNOR. Determination of tensile properties of elementary flax fibres XPT 25-501-2. 2009.

Coroller G, Lefeuvre A, Le Duigou A, Bourmaud A, Ausias G, Gaudry T, et al. Effect of flax fibres individualisation on tensile failure of flax/epoxy unidirectional composite. Compos Part A Appl Sci Manuf 2013;51:62–70.

Brutch NB, Sharov IY, Pavlov A V, Porokhovinova EA. Diversity of flax characters associated with fibre formation and environmental influence on their formation. Russ J Genet Appl Res 2011;1:361–70.

Morvan C, Andème-Onzighi C, Girault R, Himmelsbach DS, Driouich A, Akin DE. Building flax fibres: more than one brick in the walls. Plant Physiol Biochem 2003;41:935–44.

Charlet K, Jernot JP, Gomina M, Bréard J, Morvan C, Baley C. Influence of an Agatha flax fibre location in a stem on its mechanical, chemical and morphological properties. Compos Sci Technol 2009;69:1399–403.

Charlet K, Baley C, Morvan C, Jernot JP, Gomina M, Bréard J. Characteristics of Hermès flax fibres as a function of their location in the stem and properties of the derived unidirectional composites. Compos Part A Appl Sci Manuf 2007;38:1912–21.

Young's Modulus of Plant Fibers

Patricia Jouannot-Chesney, Jean-Paul Jernot, Joël Bréard and Moussa Gomina

Abstract A simple model is used to explain the decrease of the Young's modulus of plant fibers as the apparent diameter is increased. The assumption made in this work is that the fiber consists of a very thin but stiff outer shell, and a thicker but less stiff inner shell which defines the load bearing area. The model is favorably compared with the experimental data reported in the literature for flax, hemp and stinging nettle fibers.

Keywords Plant fiber · Mechanical properties · Young's modulus · Fiber diameter · Structure

Introduction

Because of their good specific mechanical properties, their natural and annual renewable character with a low environmental impact, their availability and health harmlessness, plant fibers provide an opportunity for the development of various composite materials. Thus, the literature reports encouraging results on the mechanical properties and physical performance of concrete reinforced by sisal (Agopyan et al. 2005; De Andrade Silva et al. 2009), bamboo (Ghavani 2005), hemp (Li et al. 2006) and flax (Le Hoang et al. 2011; Chafei et al. 2014). However, the behavior of the cement paste at early age (hydration, workability, setting) is strongly perturbed due to the presence of plant fibers, and maintaining the composite performance over the long term is still the subject of numerous studies

P. Jouannot-Chesney (✉) · J.-P. Jernot · M. Gomina
CRISMAT, UMR 6508 ENSICAEN/CNRS, 6 Bd Maréchal Juin,
14050 Caen Cedex 4, France
e-mail: patricia.jouannot@ensicaen.fr

J. Bréard
LOMC, UMR 6294 Université du Havre/CNRS, 53 rue de Prony,
76058 Le Havre, France

R. Fangueiro and S. Rana (eds.), *Natural Fibres: Advances in Science and Technology Towards Industrial Applications*, RILEM Bookseries 12,
DOI 10.1007/978-94-017-7515-1_5

(Chafei et al. 2015). The association of plant fibers with synthetic resins provides low biocomposites with good mechanical properties in applications where lightness is an important asset (sports, boating, transportation…). However, the development at an industrial scale of such materials still requires investigations for understanding the hydro-thermo-mechanical behavior of plant fibers (Thuault et al. 2013; Barbulée et al. 2014), and implementation of suitable elaboration and characterization methods (Destaing 2012).

The mechanisms of formation and growth of bast fibers lead to strong longitudinal variations in diameter. This, combined with the presence of the lumen brings important differences compared to synthetic fibers, and complicates the determination of the longitudinal Young's modulus and the ultimate stress: both require the knowledge of the load bearing area of the fiber. It follows a huge dispersion in the values of the mechanical properties.

Moreover a steep decrease is observed for the Young's modulus and the ultimate stress as the fiber diameter is increased. This paper addresses the question of the decrease of the Young's modulus, E, as the diameter of plant fiber, D, is increased.

Decrease of the Young's Modulus as a Function of the Apparent Diameter of the Fiber

Several papers have reported the decrease of the longitudinal Young's modulus as the apparent diameter of flax fibers increases (Baley 2002; Charlet et al. 2008) (Figs. 1 and 2). This trend is abnormal for a homogeneous isotropic material (Fig. 3) and is commonly ascribed to the complex microstructure of plant fibers which consists in different concentric walls: the primary wall (P), the first layer S1 and the thicker layer S2 (Fig. 4). All these layers can be considered as composite structures in which cellulose microfibrils are embedded in an amorphous non-cellulosic polysaccharide matrix consisting mainly in hemicelluloses and pectins.

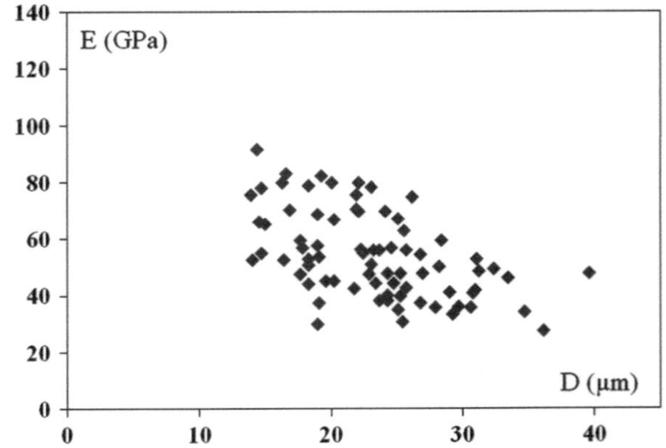

Fig. 1 Decrease of the Young's modulus as the flax fiber diameter increases (Baley 2002)

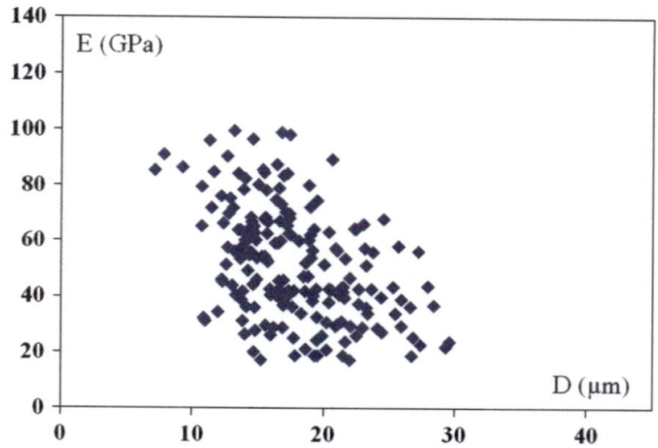

Fig. 2 Decrease of the Young's modulus as the flax fiber diameter increases (Charlet et al. 2008)

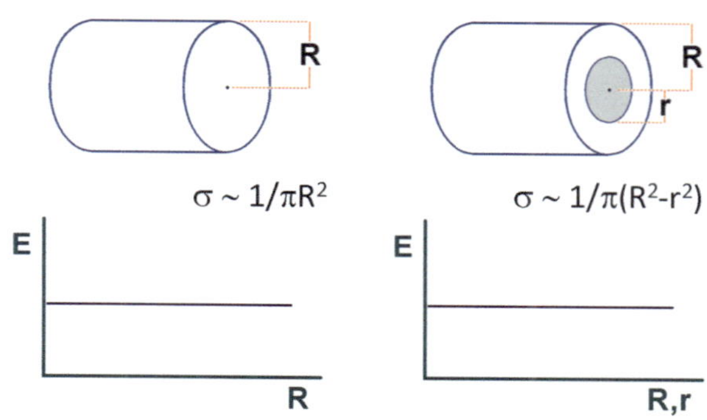

Fig. 3 Schematic representation of the relationship between the Young's modulus and the radius for a solid cylinder (*left*) and a hollow cylinder (*right*)

Fig. 4 Multilayered structure of a flax fiber

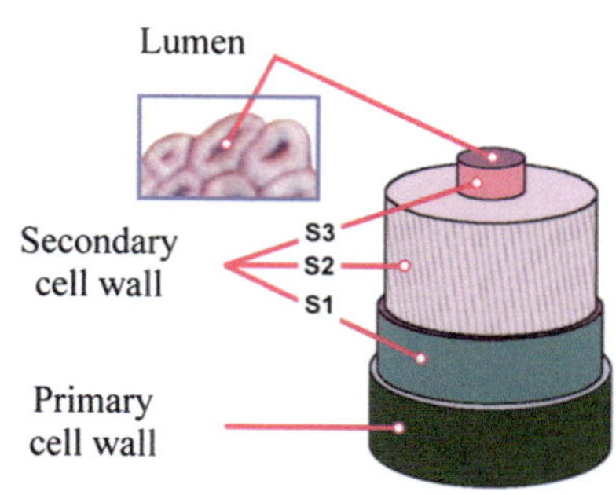

Hitherto, it was assumed that the mechanical properties of plant fibers are governed by the main layer S2. Unfortunately no clear relationship based on the microstructure of the fiber has been proposed to explain the dependence of the Young's modulus on the fiber diameter.

In fact, an experimental difficulty arising with plant fibers is the presence of a lumen. Thus, the real section of the fiber is inaccessible. For the estimation of the Young's modulus which requires the fiber's cross-sectional area, one uses the area deduced from the apparent diameter. Therefore, the value of the Young's modulus is widely underestimated and this underestimation may even vary with the fiber size (Figs. 1 and 2). If the load bearing area is taken into account, we can assume a constant value modulus.

Taking into Account the Presence of the Lumen

A statistical analysis was carried out on a batch of about 800 flax fibers of Hermes variety grown in the plain of Caen (Normandy, France) in 2004. The area of the lumen and the apparent total area of the fiber were measured by using image analysis. The ratio between the wall area and the cross-sectional area of the fiber defines the fraction area of the solid phase. This fraction is lower than 1 due to the presence of the lumen.

A linear relationship is observed on Fig. 5 between the wall area and the total cross-sectional area. This slope corresponds to an average value of about 8 % for the porosity of the fibers (Charlet et al. 2006) in accordance with the literature (Sinha 1974). Since the porosity remains almost constant whatever the fiber size, taking into account this porosity cannot explain the variation in the Young's modulus when the diameter is increased.

Other explanations have been proposed from the lack of circularity of the fibers (Aslan 2011), but they only could explain the scattering of the modulus values.

Fig. 5 Relationship between the wall area and the apparent total area of the fiber (Charlet et al. 2006)

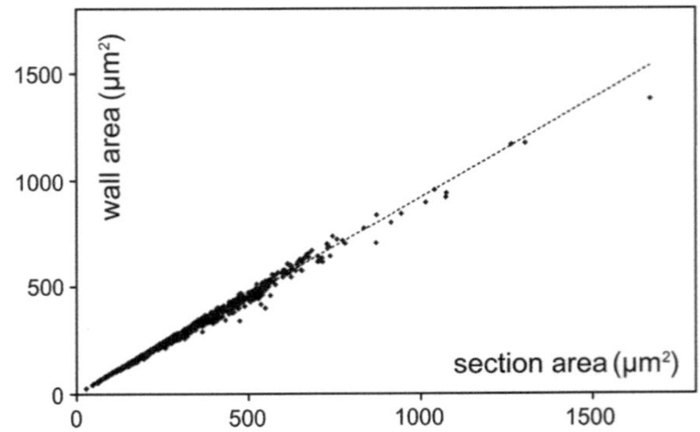

Assuming the Presence of Another Structural Heterogeneity

In order to explain the dependence of fiber modulus on diameter observed experimentally, we assumed that the structure of the fiber is such that a small proportion of its section participates in the stiffness when the diameter increases. Thus we represent the fiber by two concentric shells around the lumen. The inner shell has a radius (R − e) whereas the outer one has a thickness e (Fig. 6).

The apparent load bearing area commonly used is the wall area πR^2, whereas the effective load bearing area in the frame of this model is the area of the assumed stiffer outer shell. It can be expressed as:

$$\pi\left[R^2 - (R-e)^2\right] \approx 2\pi Re \qquad (1)$$

The Young's modulus, E, can then be expressed by:

$$E/E_0 = 4e/D \qquad (2)$$

where E_0 is the modulus of the outer shell. It is assimilated here to the longitudinal modulus of the cellulosic microfibrils for which the most common value is about 130 GPa (Salmén 2004).

The results issued from this model were compared with experimental values given above for flax fibers (Figs. 7 and 8). The best fits are obtained for e = 2.2 μm and e = 1.7 μm, respectively. The different e values obtained for the two samples of flax fibers may be linked to the history of the plants: growing conditions, retting and scutching parameters.

The model was also applied to two other plant fibers for which similar diameter dependence has been reported: hemp (Duval et al. 2011; Placet et al. 2012) and stinging nettle (Bodros and Baley 2008). Hemp fibers are among the most used in

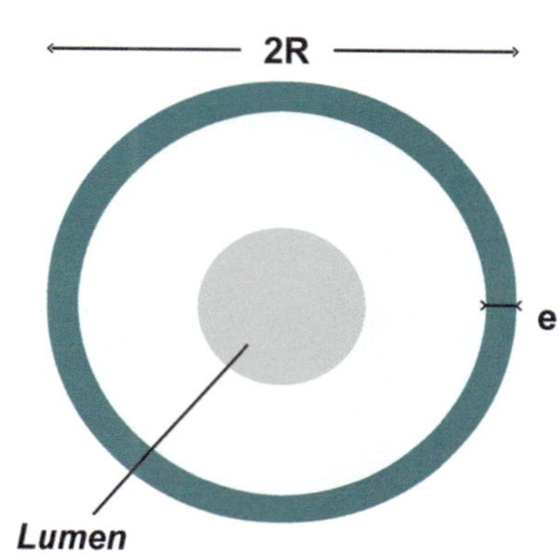

Fig. 6 Modelised fiber cross-section. In the model, e is the thickness of the outer shell load bearing area

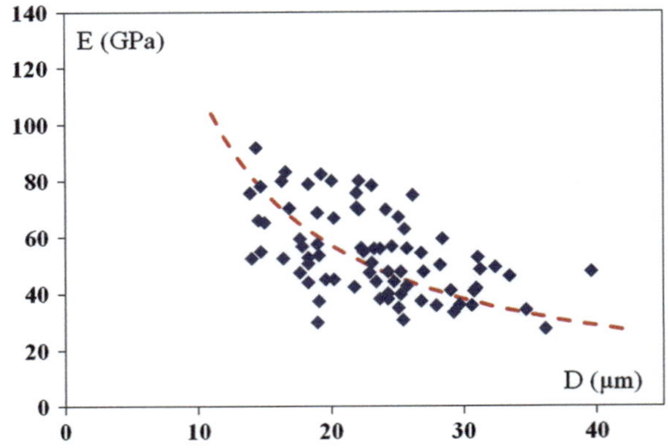

Fig. 7 Comparison of the model with the experimental values of Fig. 1 (flax)

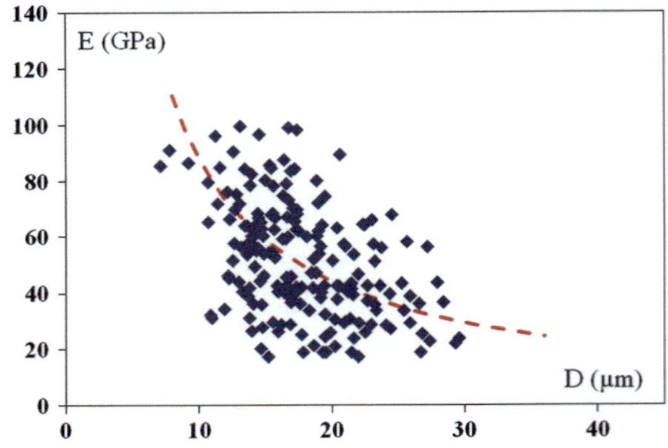

Fig. 8 Comparison of the model with the experimental values of Fig. 2 (flax)

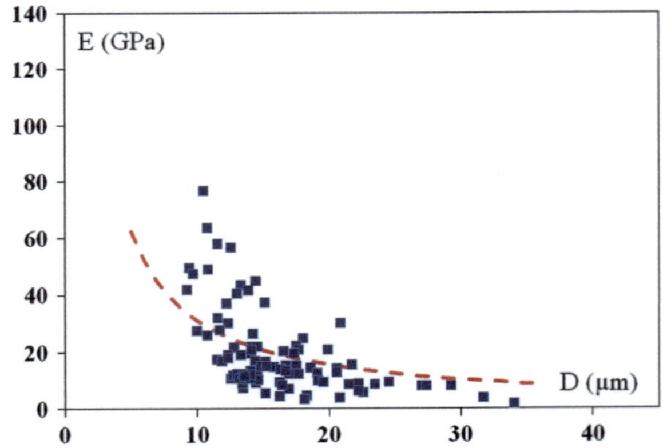

Fig. 9 Comparison of the model with the experimental values reported for hemp (Duval et al. 2011)

the manufacture of plant fiber composites whereas stinging nettle fibers are still little-used despite their good mechanical properties probably because of their low commercial availability. The best fits obtained for the reported hemp and stinging nettle fibers are e = 0.6 µm and e = 3.2 µm, respectively (Figs. 9 and 10).

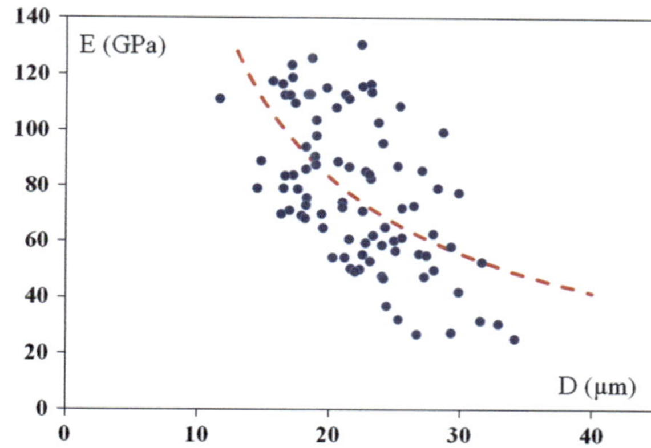

Fig. 10 Comparison of the model with the experimental values reported for stinging nettle (Bodros and Baley 2008)

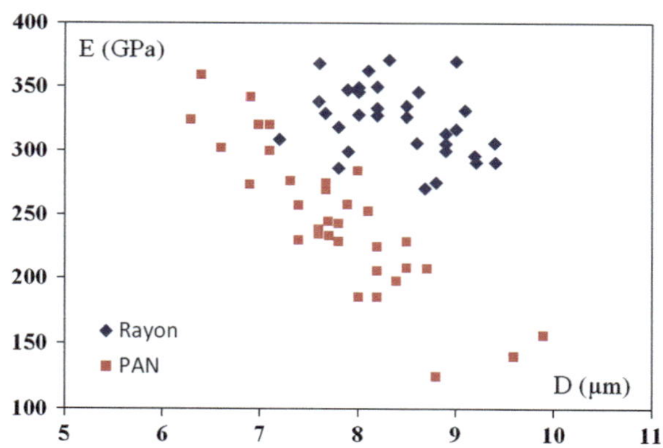

Fig. 11 Evolution of the Young's modulus versus diameter for graphite fibers made with polyacrylonitrile precursor or rayon precursor (Jones and Duncan 1971)

Low outer shell thicknesses e obtained for flax fibers are comparable to the cumulative size of the primary wall (P) and a transition zone within the first layer S1 (cf. Fig. 4). High elastic properties in the axial direction of the fiber prevail in the primary wall due to the dense entanglement of the microfibrils with high stiffness while these properties are lowered in layer S1 owing to a large microfibrillar angle.

The decrease of the Young's modulus as a function of fiber diameter had also been reported for graphite fibers manufactured from polyacrylonitrile (PAN) and rayon (Fig. 11) (Jones and Duncan 1971). This behaviour was ascribed to the presence of well aligned cristallites parallel to the fiber axis in a thin outer shell (the sheath) and a thicker interior shell of misaligned cristallites (the core). This microstructural description has also been invoked to explain the dependence of the strength of carbon fibers as a function of diameter.

Conclusion

A geometrical model was proposed to explain the observed decrease in Young's modulus of plant fibers as a function of diameter. The model adequately explains the experimental measurements reported in the literature for different plant fibers. It assumes the existence of a high Young's modulus layer located at the periphery of the fiber, which has a key role compared to the more compliant core area. This "core-shell" effect has been associated with the layered structure of the fiber and the different mechanical properties of the walls.

References

Agopyan V, Savastano Jr H, John VM, Cincotto MA. Developments on vegetable fibre–cement based materials in São Paulo, Brazil: an overview. Cement & Concrete Composites, (2005), 27, 527–536.

Aslan M, Chinga-Carrasco G, Sørensen BF, Madsen B. Strength variability of single flax fibres. J. Mater. Sci., (2011), 46, 6344–6354.

Baley C. Analysis of the flax fibres tensile behaviour and analysis of the tensile stiffness increase. Composites: Part A, (2002), 33, 939–948.

Barbulée A, Jernot, JP, Bréard, J, Gomina, M. Damage to flax fibre slivers under monotonic uniaxial tensile loading. Composites Part A: Applied Science and Manufacturing, (2014), 64, 107–114.

Bodros E, Baley C. Study of the tensile properties of stinging nettle fibres (Urtica Dioica). Materials Letters, (2008), 62, 2147–2149.

Chafei S, Khadraoui F, Boutouil M, Gomina M. Optimizing the formulation of flax fiber-reinforced cement composites. Construction and Building Materials, (2014), 54, 659–664.

Chafei S, Khadraoui F, Boutouil M, Gomina M. Effect of flax fibers treatments on the rheological and the mechanical behavior of a cement composite. Construction and Building Materials, (2015), 79, 229–235.

Charlet K, Morvan C, Bréard J, Jernot JP, Gomina M. Etude morphologique d'un composite naturel, la fibre de lin. Rév Compos Matér Av, (2006), 16, 11–24.

Charlet K, Jernot JP, Gomina M, Bizet L, Bréard J. Analyse structurale et comportement mécanique d'une fibre de lin. Rév Compos Matér Av, (2008), 18, 157–162.

Destaing F. Contribution à l'étude du comportement mécanique de matériaux composites biosourcés lin/Pa11 élaborés par thermocompression. PhD thesis at the University of Caen, Lower Normandy, France, December 2012.

Duval A, Bourmaud A, Augier L, Baley C. Influence of the sampling area of the stem on the mechanical properties of hemp fibres. Materials Letters, (2011), 65, 797–800.

De Andrade Silva F, Mobasher B, Toledo Filho RD. Cracking mechanisms in durable sisal fiber-reinforced cement composites. Cement & Concrete Composites, (2009), 31, 721–730.

Ghavai K. Bamboo as reinforcement in structural concrete elements. Cement & Concrete Composites, (2005), 27, 637–649.

Jones BF, Duncan RG. The effect of fibre diameter on the mechanical properties of graphite fibres manufactured from polyacrylonitrile and rayon. J. Mater. Sci., (1971), 6, 289–293.

Le Hoang T, Khadraoui F, Boutouil M, Gomina M. Mechanical properties of flax fibre-reinforced cement composites. International Conference on Technological Advancements in Civil Engineering (ICTACE), Chennai, India, Proceedings of ICTACE, (2011), 250–255.

Li Z, Wang X and Wang L. Properties of hemp fibre-reinforced concrete composites. Composites Part A, (2006), 37, 497–505.
Placet V, Trivaudey F, Cisse O, Guicheret-Retel V, Boubakar L. Diameter dependence of the apparent tensile modulus of hemp fibres: a morphological, structural or ultrastructural effect? Composites: Part A, (2012), 43, 275–287.
Salmén L. Micromechanical understanding of the cell-wall structure. C.R. Biologies, (2004), 327, 873–880.
Sinha MK. Rope-making with banana-plant fibre. J. Text. Inst., (1974), 65, 612–615.
Thuault A, Eve S, Blond D, Bréard J, Gomina M. Effects of the hygrothermal environment on the mechanical properties of flax fibres. Journal of Composite Materials, (2013), 48 [14], 1699–1707.

Characterization of Brazil Nut Fibers

P.Y. Inamura, F.H. Kraide, M.J.A. Armelin, M.A. Scapin, E.A. B. Moura and N.L. Mastro

Abstract The fruit from Brazil nut (*Bertholletia excelsa*) is characteristically a spherical capsule. When ripe, the capsule releases seeds through its lower portion. In this work, an attempt has been made to characterize the Brazil nut capsule or bur and shell fibers by various techniques. The organic composition was established and elementary composition was determined by Instrumental Neutron Activation Analysis (INAA) followed by gamma-ray spectrometry and Wavelength Dispersive X-ray Fluorescence Spectrometry (WDXRF). Possible applications of these fibers must be considered as novel markets for lignocellulosics have been identified in recent years, representing an exceptional opportunity for sustainable technological development.

Keywords Brazil nut · Lignocellulosics · INAA · WDXRF

Introduction

The agro-industry generates innumerable sources of biomass that are not sufficiently nor adequately utilized. The use of lignocellulosic fibers and their constituents as raw materials in the production of polymeric and composite materials represent an exceptional opportunity for sustainable technological development (Bastioli 1998; Silva et al. 2010). In order to think about potential and future exploitation and applications of lignocellulosics, properties and performance of fibers under environmental or other conditions must be known. The necessity of developing a detailed chemical composition and physical properties data bases of the many natural fibers which are potentially available in Rowell et al. (2000).

P.Y. Inamura · F.H. Kraide · M.J.A. Armelin · M.A. Scapin · E.A.B. Moura
N.L. Mastro (✉)
Instituto de Pesquisas Energéticas e Nucleares, IPEN-CNEN/SP, Av. Prof. Lineu Prestes, 2242, 05508-000 São Paulo, SP, Brazil
e-mail: nlmastro@ipen.br

R. Fangueiro and S. Rana (eds.), *Natural Fibres: Advances in Science and Technology Towards Industrial Applications*, RILEM Bookseries 12,
DOI 10.1007/978-94-017-7515-1_6

Many lignocellulosic fibers were studied and several of their properties, such as chemical and physical nature and mechanical strength have been already evaluated. The chemical composition of fibers from piassava, sisal, coir, flax, hemp, sugarcane bagasse, banana, curauá, pineapple, cotton lint, bamboo, jute, ramie and rice straw, for instance, were subjected to different important studies (D'Almeida et al. 2006; Satyanarayana et al. 2007; Tomczak et al. 2007; Moura et al. 2010; Corrêa et al. 2010). The available properties of some of these fibers reported so far can be understood in terms of their observed structure.

Besides a small utilization in artistic handicraft, very few works on Brazil nut shell and bur fibers were performed which revealed some further application. One that is possible to mention comes from our laboratory (Inamura et al. 2010).

The Brazil nut (*Bertholletia excelsa*) tree occurs only in the Amazon basin river, thus it is produced in Brazil and also in parts of Bolivia and Peru. The Brazil nut is one of the most valuable non-timber forest products of the Brazilian Amazon and one of the few which reaches the international market, and come entirely from wild collection rather than from plantations.

Brazil nuts, the commercially harvested, edible seed, are considered to be one of the most valuable products that can be harvested from undisturbed rainforest. The tree is large and stands out of other trees tops in the forest and can reach up to 50 m, with a trunk diameter that reaches from 10 to 12 m and it lives over 500 years. The fruits, known to Brazilians as *Castanha do Pará*, are characteristically a spherical capsule, with a thick, hard, dark brown surface (Fig. 1). The fruits are difficult to be

Fig. 1 Brazil nut (*Bertholletia excelsa*) shell and bur fiber

collected: each hard outer bur can weigh over 1 kg and contain between 14 and 25 seeds (nuts) inside, surrounded by yellow pulp that is rich in vitamins, fats and proteins (Vianna et al. 1972; APIZ—Zoro Indigenous People Association 2008).

Brazil nut is an angular nut with a very hard hull. Their nut almond is very white, with a dark brown tegument with high energy content and rich in proteins of high biological values. A tree may produce more than 150 kg of nuts a year and the species is propagated naturally. It is difficult to estimate the nut production because there is a significant variation in burs number from one year to another and among the trees. For example, new Brazil nut trees produce 30–50 burs per year, whereas mature trees, 200–400 years old, can reach up to 1000 burs produce in just one year (APIZ—Zoro Indigenous People Association 2008). In this work, an attempt has been made to characterize the Brazil nut bur and shell fibers by several techniques.

Materials and Methods

Material

Brazil nut shell and bur fibers from the residues disposed by the processing industries of Brazil nuts were provided by Amazon Brazil Nuts. The dry fiber was reduced to fine powder, with particle sizes equal or lower than 200 μm by using ball mill.

Fiber Characterization

The Brazil nut shell and bur fibers were evaluated following the standard procedures and all analyses were performed at least in triplicate. The extractives present in the fiber were removed via Soxhlet extraction with toluene/ethanol for 4 h/70 °C, subsequently replaced by water (100 °C/4 h).

The lignin content was determined by Klason method. This method is one among many others used to evaluate the lignin content, which is based on acid hydrolysis of polysaccharides. Subsequently, the gravimetric determination of soluble and insoluble lignin, according to TAPPI T13M-54 was carried out. For insoluble lignin content, the extractive-free fiber was weighed; after that, 72 % w/v sulfuric acid solution was added, homogenized and maintained in an incubator at 25.0 ± 0.5 °C for 24 h. The solution was transferred to a 1 L flask and 560 mL distilled water was added, followed by a 4 h reflux. The insoluble lignin was washed several times with water and dried in an oven at 105 ± 2 °C until constant weight.

The soluble lignin content was determined by the filtrate obtained from the insoluble lignin percolation. This filtrate was analyzed by UV-Visible Spectrophotometer, model UV-1601, Shimadzu and the absorbance was measured at wavelengths from 280 to 215 nm. The lignin concentrations (g L^{-1}) in diluted samples were calculated by the following expression:

$$C\left(g\,L^{-1}\right) = \frac{4.53(A_{215}) - A_{280}}{300} \qquad (1)$$

C (g L^{-1})	concentration in g/L of soluble lignin Klason in diluted samples
A_{215}	absorbance value at 215 nm
A_{280}	absorbance value at 280 nm.

The total lignin content of the sample was quantified by the sum of soluble and insoluble lignin determinations.

For cellulose content determination, 25 mL of nitric acid-glacial acetic acid were added in 1 g of extractive-free, dried Brazil nut bur and shell fiber. The sample was maintained under reflux at 120 ± 3 °C for 25 min. The system was cooled to room temperature and the cellulose residues were percolated in sintered glass, washed with 500 mL of hot water and then with 25 mL of ethanol. The sample was dried in oven at 105 ± 2 °C until constant weight and then cooled and weighed.

The moisture content of Brazil nuts bur and shell fibers was performed in triplicate. Samples of 1 g were weighed and then placed in an oven at 105 ± 2 °C until constant weight. The samples were cooled in a desiccator and then weighed and quantified.

The determination of density for porous materials of varying composition such as lignocellulosic materials, must consider the pores and microcracks of the samples. The measurement was performed in triplicate. The Brazil nuts bur and shell fibers were dried in the oven at 105 ± 2 °C until constant weight. After dried, the 5 g of samples were completely submerged in graduated cylinders partially filled with distilled water. The assays were maintained in solution for 24 h to determine the variation of the water.

Thermogravimetric Analysis (TG)

Brazil nut shell and bur fibers thermal stability was evaluated by a thermogravimetric analyzer TGA-50 (Shimadzu, Japan). Sample with 5.0 ± 1.0 mg was placed in a platinum cell, under air atmosphere with a flow rate of 50 mL min^{-1}. The experiment was conducted from ambient temperature (25 °C) to 600 °C at a heating rate of 10 °C min^{-1}.

Scanning Electron Microscopy (SEM)

The Brazil nut shell and bur fibers morphological characterizations was accomplished by means of a Philips, XL 30 SEM. Powder sample was fixed in carbon, introduced into a vacuum sputter coater until reach the critical point drying and then coated with gold. Therefore, the analysis was performed using secondary electrons. The scanning electron microscope was equipped with an energy-dispersive X-rays spectroscopy (EDS) system that gathered a spectrum of elements.

Wavelength Dispersive X-ray Fluorescence (WDXRF) Analysis

The pressed powdered samples (grain size ca 100 μm) were prepared according to Scapin et al. (2009). A Rigaku Co. X-ray fluorescence spectrometer, model RIX 3000 was used with the proper measurement conditions (Scapin et al. 2009).

Instrumental Neutron Activation Analysis (INAA)

Aliquots of approximately 200 mg were transferred to polyethylene bags, which had been previously cleaned by leaching with a diluted HNO_3 (1:5) and purified water.

Certified standard solutions (Spex Certiprep) of Br, Cr, Cs, Co, Na, Rb, Sb, Sc, Se, Th were used to prepare elemental synthetic standards. Aliquots (50–100 μL) were pipette on to small sheets of analytical filter paper (Whatman 42) for irradiation. After drying, these filter papers were placed into polyethylene bags.

Irradiations were carried out at the IEA-R1 nuclear research reactor. The thermal neutron flux utilized ranged from 1 to 4×10^{12} n cm^{-2} s^{-1}. Sample and standards (Br, Cr, Cs, Co, Na, Rb, Sb, Sc, Se, Th) were irradiated together in an aluminum container for 8 h. The 82Br and 24Na were measured after 3 days of decay time, while ^{51}Cr, ^{134}Cs, ^{60}Co, ^{86}Rb, ^{122}Sb, ^{75}Se and ^{233}Pa ($^{233}Th \rightarrow {}^{233}Pa$) were measured after, at least, 10 days of decay time.

The equipment used to measure the gamma-radiation was a Canberra Model GX2020 hyperpure Ge detector, coupled to a Model 1510 Integrated Signal Processor and MCA System 100, both from Canberra. The detector used had a resolution (FWHM) of 0.9 keV for 122 keV gamma-rays of ^{57}Co and 1.9 keV for 1332 keV gamma-ray of ^{60}Co.

Analysis of the certified reference material IAEA Soil-7 was also carried out simultaneously and, the results obtained were within the range of certified values.

Table 1 Chemical composition (% w/w) of Brazil nut bur and shell fiber

	Brazil nut bur fiber	Brazil nut shell fiber
Cellulose (%)	53.3 (1.1)	38.2 (1.0)
Lignin (%)	35.8 (0.8)	54.1 (1.1)
Extractives (%)	7.5 (0.2)	3.9 (0.2)

Data expressed as mean (standard deviation)

Results and Discussion

Fiber chemical components are distributed through outer cell wall, which is composed of primary and secondary wall layers. Chemical composition differs between different plants and also different parts of the same plant, according to the geographic location, age, climate and soil conditions, among other factors (Rowell et al. 2000).

The Brazil nut shell and bur fibers chemical composition is presented in Table 1. The cellulose content found to Brazil nut bur fiber, 53.3 %, is in the range for jute fiber, 45–63 % (Rowell et al. 2000) and somewhat higher was found for sugarcane bagasse, 54.3–55.2 % (Satyanarayana et al. 2007). For Brazil nut shell fiber, the cellulose content of 40.0 % was similar to the coir fiber (Bledzki et al. 1996).

The lignin content of Brazil nut bur and shell fiber, 35.8 % and 59.5 %, respectively, found in the present work was considerably higher than that from bamboo, 20–25 % (Kozlowski et al. 2004) and 36.14 % for coir fiber (Espert et al. 2004), but inferior to that found in piassava, ~48 % (D'Almeida et al. 2006).

The moisture content and density of Brazil fibers are presented in Table 2. Similar results for moisture were found for cotton, jute or flax, 10.0 % (Bledzki et al. 1996) and curauá fiber, 9.10 % (Spinacé et al. 2009). The density values were in the same range found for banana fiber, 1350 kg m^{-3} (Idicula et al. 2005) and not so far from that of curauá fiber, 1100 ± 91 kg m^{-3} (Spinacé et al. 2009). A major limitation for all cellulosic fibers is the hydrophilic nature if used as reinforcement in plastic. This hydrophilicity influences the all mechanical properties as well as other physical properties of the fiber itself (Bledzki et al. 1996). For that reason moisture determination is an important issue as moisture presence can prevent composite formation (Spinacé et al. 2009).

In general terms, the chemical composition and density values obtained here are consistent with data from other fibers reported in the literature (Rowell et al. 2000; Satyanarayana et al. 2007; Tomczak et al. 2007).

Table 2 Moisture content and density of Brazil fibers

	Moisture content (%)	Density (kg m^{-3})
Brazil nut shell fiber	14.73 (0.09)	1336 (1)
Brazil nut bur fiber	10.29 (0.03)	1299 (1)

Data expressed as mean (standard deviation)

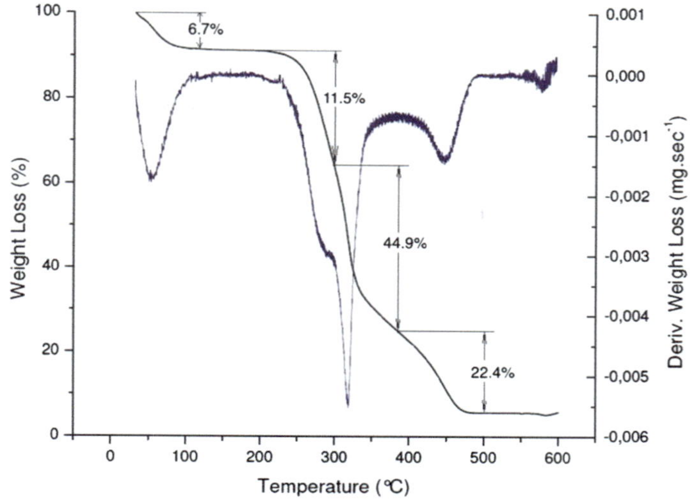

Fig. 2 Thermogravimetric analysis of Brazil nut bur fiber

The thermogravimetric analyses (TG) of Brazil nut bur and shell fiber are presented in Figs. 2 and 3, respectively, as an instrumental output of mass versus increasing temperature. The numerical derivative TG trace is also shown. Distinct transformation regions, all indicating mass loss, can be observed. The first one indicates a weight loss at the beginning of the analysis (about 55–75 °C) and then a thermal stability until 230 °C. The weight loss of 6.7 % at the lower temperatures

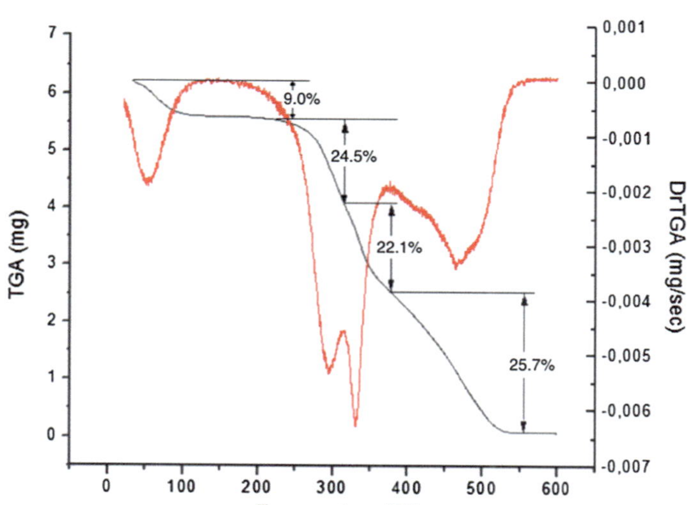

Fig. 3 Thermogravimetric analysis of Brazil nut shell fiber

can be related to water evaporation (D'Almeida et al. 2006; Tomczak et al. 2007; Corrêa et al. 2010). Weight losses (11.5 %) were showed at 260 and 310 °C (44.9 %), and at 426 °C, another weight loss of 22.4 % can also be seen. According to the literature (D'Almeida et al. 2006; Tomczak et al. 2007; Ramiah 1970) the degradation temperature showed at about 260 °C can be attributed to decomposition of hemicellulose. The lignin onset temperatures at 426 °C (Fig. 2) and 451 °C (Fig. 3) are similar as those reported before for lignin degradation (Ramiah 1970). Silva et al. (2010) showed similar degradation temperature stages for sisal fiber/polyurethane resin-based composites and other authors obtained similar results for other lignocellulosic fibers (Satyanarayana et al. 2007; Tomczak et al. 2007).

Figures 4 and 5 shows some cross and longitudinal section microphotographs of the Brazil nut shell fiber. Can be observed that shell fiber present different morphology, depending of the fiber layers. And Figs. 6 and 7 presents cross and longitudinal section microphotographs of the Brazil nut bur fiber. In Figs. 6b and 7 could be observed that the bur fiber presents irregular fibers, in different directions, like polymers cross-linking. And this quality can be given by the difficulty to fracture the fiber. It is possible to see an array of protruded features. These protrusions, shown in detail in Fig. 4d, would be, according to some authors, Si rich particles (Calado et al. 2000).

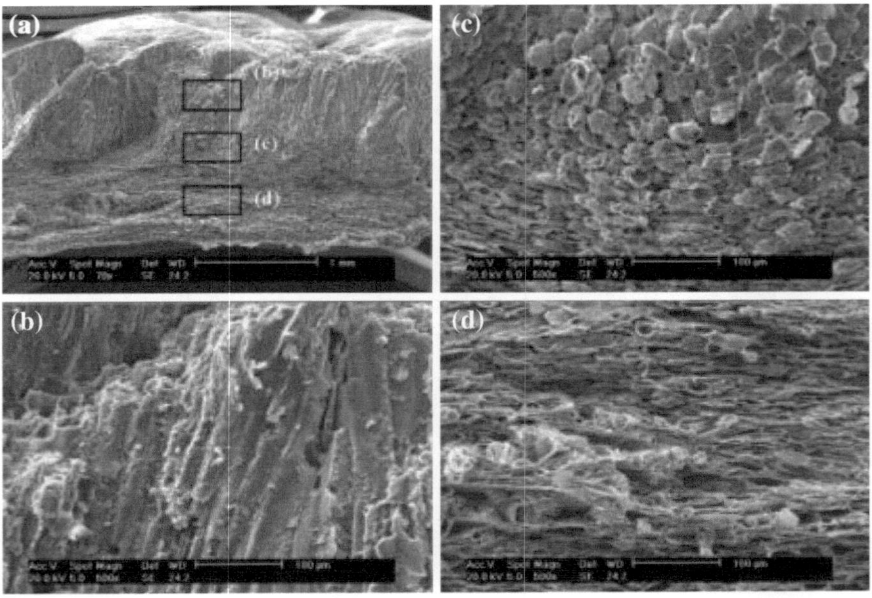

Fig. 4 SEM cross-section images of Brazil nut shell fiber: **a** 70×; **b** superior, **c** central and **d** inferior parts of fiber, 500×

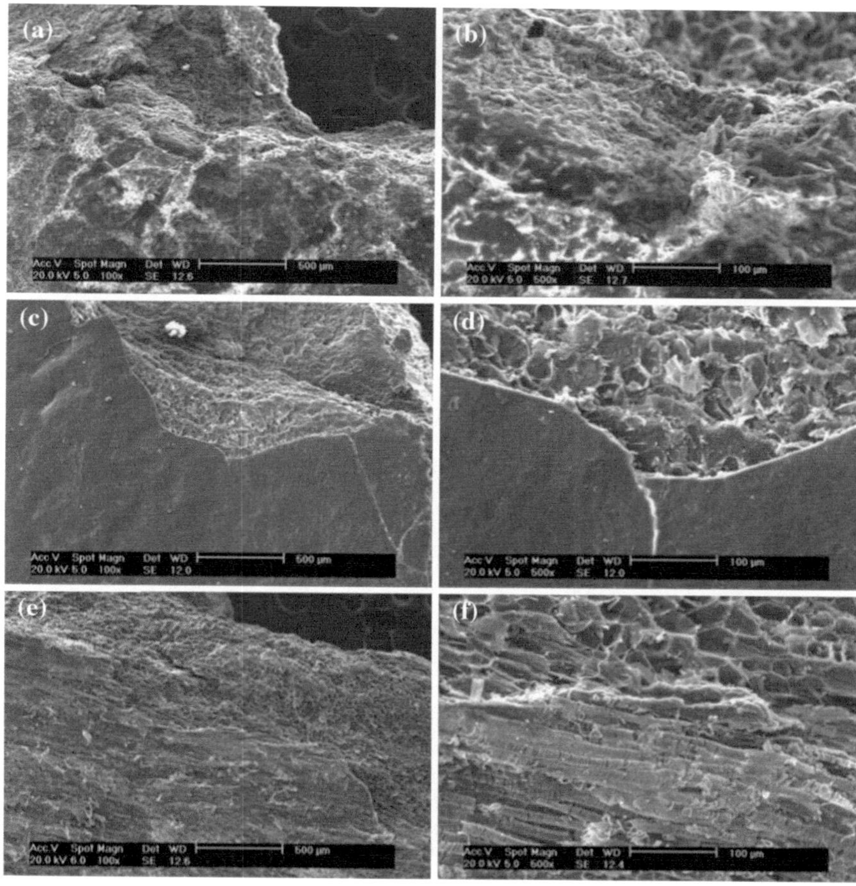

Fig. 5 SEM longitudinal section images of Brazil nut shell fiber: **a**, **b** external micrograph of fiber, 100× and 500×; **c**, **d** internal micrograph of fiber, 100× and 500×; **e**, **f** central micrograph of fiber, 100× and 500×, respectively

Although EDS has reduced accuracy in inhomogeneous and rough samples, the elements identified (Figs. 8 and 9) are consistent with the data found by other methods presented as follows in Tables 3 and 4.

The results obtained by INAA and WDXRF for the elemental composition in both kinds of fiber from Brazilian nut are shown in Tables 3 and 4. Each value is expressed with its uncertainty in parenthesis. In the case of INAA, the uncertainty was evaluated considering the error from statistical counting.

As Skinner and Jahren (2003) pointed out, the process of biomineralization creates heterogeneous accumulations in vegetal tissues with inhomogeneous

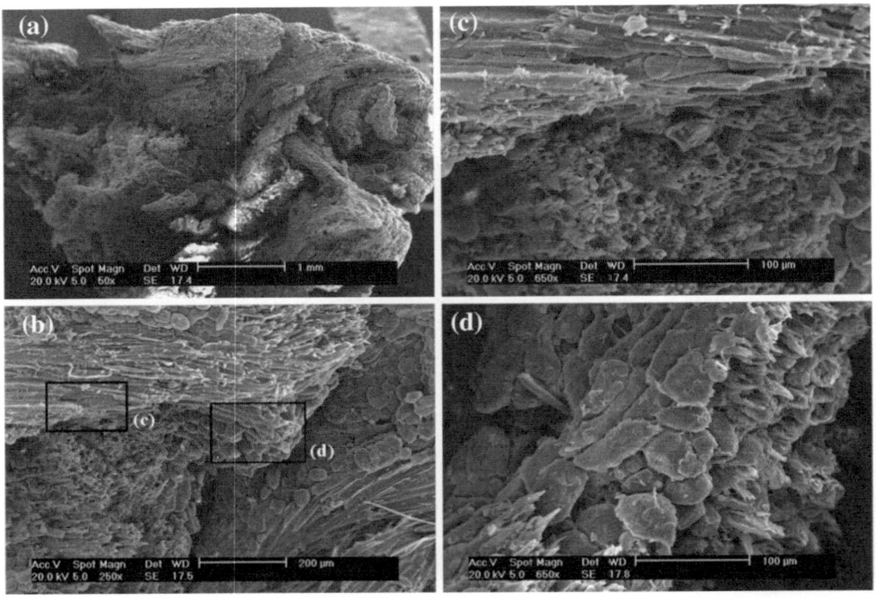

Fig. 6 SEM cross-section images of Brazil nut bur fiber: **a** 50×; **b** 250×; **c** 650×; and **d** 650×

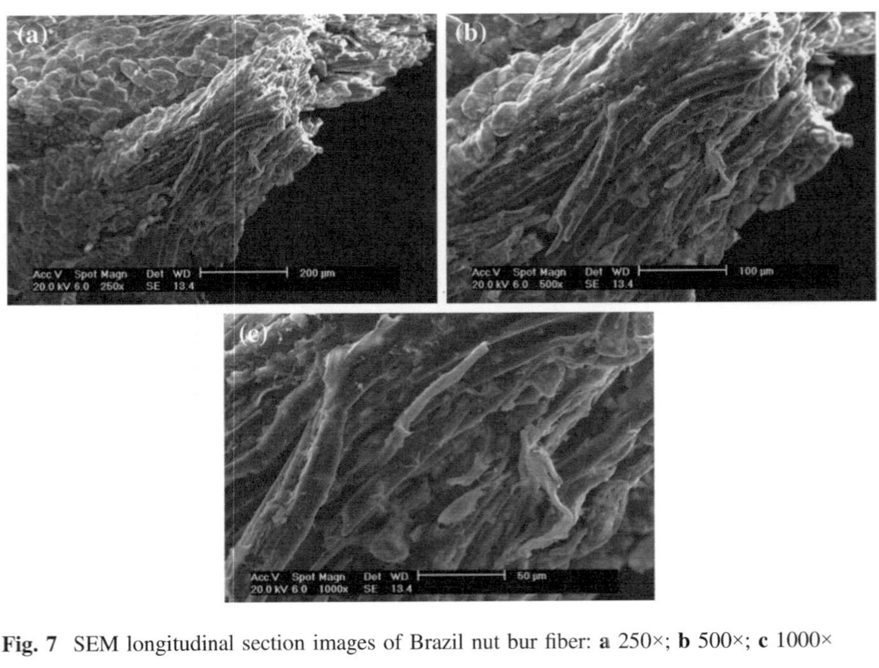

Fig. 7 SEM longitudinal section images of Brazil nut bur fiber: **a** 250×; **b** 500×; **c** 1000×

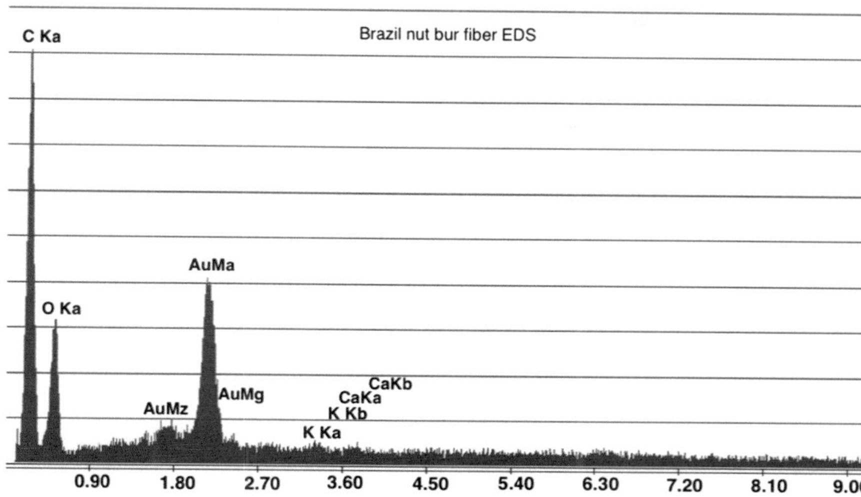

Fig. 8 EDS spectrum of Brazil nut bur fiber

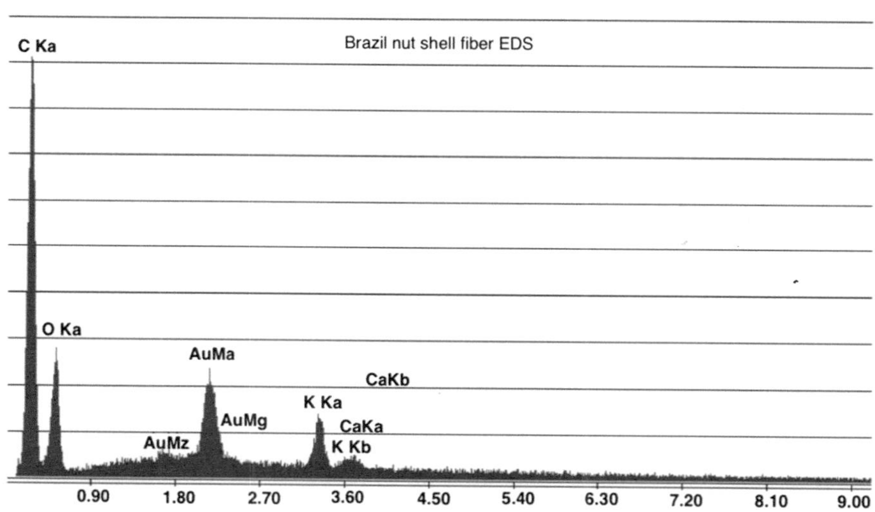

Fig. 9 EDS spectrum of Brazil nut shell fiber

distributions that reflect the environment in which they are formed. Vegetal material can contain carbonate, phosphate, oxalate, silica, iron, or sulfur-containing minerals with particular chemistries. The knowledge of the mineral composition of the fiber

Table 3 Inorganic composition of Brazil nut bur fiber

Element/compound (unit)	Concentration (uncertainty)
Al_2O_3 (μg/g)[a]	1651 (100)
As_2O_3 (μg/g)[a]	16 (10)
Br (μg/g)[b]	87 (4)
CaO (μg/g)[a]	1080 (100)
Cl (μg/g)[a]	697 (30)
Cr (μg/g)[b]	39 (1)
Cs (μg/g)[b]	1.74 (0.04)
Co (μg/kg)[b]	1609 (58)
CuO (μg/g)[a]	25 (10)
Fe_2O_3 (μg/g)[a]	14462 (300)
K_2O (μg/g)[a]	8229 (100)
MgO (μg/g)[a]	555 (30)
MnO (μg/g)[a]	67 (10)
Na (μg/g)[b]	49 (5)
NiO (μg/g)[a]	31 (10)
P_2O_5 (μg/g)[a]	447 (30)
Rb (μg/g)[b]	16.3 (0.6)
Sb (μg/kg)[b]	36 (4)
Sc (μg/kg)[b]	9.7 (0.6)
Se (μg/kg)[b]	198 (44)
SiO_2 (μg/g)[a]	625 (30)
SO_3 (μg/g)[a]	1996 (100)
SnO_2 (μg/g)[a]	<10
SrO (μg/g)[a]	<10
Th (μg/kg)[b]	19 (4)
ZnO (μg/g)[a]	21 (10)
ZrO_2 (μg/g)[a]	<10

[a]Determined by WDXRF
[b]Determined by INAA

can be important for the correct analysis of some properties of this material, as inorganic elements are related to metabolism of living organisms.

Comparing present data from bur and shell fibers, it is possible to perceive that the first is richer in Fe_2O_3, and this fact correlate with a stronger red color of that fiber. On the other hand, shell fiber presented a higher content of CaO and K_2O and SiO_2.

Table 4 Inorganic composition of Brazil nut shell fiber

Element/compound (unit)	Concentration (uncertainty)
Al_2O_3 (μg/g)[a]	6533 (100)
As_2O_3 (μg/g)[a]	37 (10)
Br (μg/g)[b]	30 (1)
CaO (μg/g)[a]	6684 (100)
Cl (μg/g)[a]	290 (30)
Cr (μg/g)[b]	2.83 (0.09)
Cs (μg/g)[b]	1.1 (0.1)
Co (μg/kg)[b]	1272 (22)
CuO (μg/g)[a]	111 (20)
Fe_2O_3 (μg/g)[a]	6380 (100)
K_2O (μg/g)[a]	34381 (500)
MgO (μg/g)[a]	3214 (100)
MnO (μg/g)[a]	160 (20)
Na (μg/g)[b]	62.3 (0.5)
NiO (μg/g)[a]	83 (10)
P_2O_5 (μg/g)[a]	1735 (100)
Rb (μg/g)[b]	41 (1)
Sb (μg/kg)[b]	20 (2)
Sc (μg/kg)[b]	12.4 (0.4)
Se (μg/kg)[b]	456 (46)
SiO_2 (μg/g)[a]	2019 (100)
SO_3 (μg/g)[a]	9356 (100)
SnO_2 (μg/g)[a]	<10
SrO (μg/g)[a]	34 (10)
Th (μg/kg)[b]	19 (3)
ZnO (μg/g)[a]	87 (10)
ZrO_2 (μg/g)[a]	<10

[a]Determined by WDXRF
[b]Determined by INAA

Conclusion

This work presents an investigation on the thermal behavior, chemical and elementary composition of Brazil nut fibers. Their characteristics are comparable to those of fibers already studied by other groups. The chemical analysis showed that Brazil nut bur fiber is a lignin-rich fiber. This material, nowadays a leftover, could result to be useful as other lignin-rich fibers, considering that many lignocellulosic materials are suited for composite material development. As poor information is available on this kind of fiber until now, the present paper is a contribution to elucidate characteristics of Brazil nut shell and bur fibers.

Acknowledgements The authors want to express their thanks to Eng. Elizabeth S.R. Somessari and Eng. Carlos Gaia da Silveira for guidance in the irradiation operation. Financial support from the University of Sao Paulo, CAPES, CNPq is also acknowledged.

References

BASTIOLI C. "Biodegradable Materials - Present Situation and Future Perspectives", *Macromol. Symp.*, v. 135, pp. 193-204, 1998.

SILVA, R.V.; UEKI, M,M,; SPINELLI, D.; BOSE FILHO, W.W..; TARPANI, J.R. "Thermal, Mechanical, and Hygroscopic Behavior of Sisal Fiber/Polyurethane Resin-based Composites", *J. Reinf. Plast Compos.*, v. 29, pp. 1399–1417, 2010.

ROWELL, R.M.; HAN, J.S.; ROWELL, J.S. "Characterization and factors effecting fiber properties", In: Frollini, E., Leão, A.L., Mattoso, L.H.C. (eds), *Natural Polymers and Agrofibers Based Composites*, Embrapa Instrumentação Agropecuária: São Carlos, pp.115-132, 2000.

D'ALMEIDA, J.R.M.; AQUINO, R.C.M.P.; MONTEIRO, S.N. "Tensile mechanical properties, morphological aspects and chemical characterization of piassava (Attalea funifera) fibers", *Compos. Part A.*, v. 37, pp. 1473–1479, 2006.

SATYANARAYANA, K.G.; GUIMARÃES, J.L.; WYPYCH, F. "Studies on lignocellulosic fibers of Brazil. Part I: Source, production, morphology, properties and applications", *Compos. Part A*, v. 38, pp. 1694–1709, 2007.

TOMCZAK, F.; SATYANARAYANA, K.G.; SYDENSTRICKE,R T.H.D. "Studies on lignocellulosic fibers of Brazil: Part III – Morphology and properties of Brazilian curauá fibers", *Compos. Part A*, v. 38, pp. 2227–2236, 2007.

MOURA, E.A.B.; NOGUEIRA, B.R.; ORTIZ, A.V. "Changes in physicochemical, morphological and thermal properties of electron-beam irradiated ethylene-vinyl alcohol copolymer (EVOH) as a function of radiation dose", In: *Proceedings of AccApp*, Vienna, IAEA, 2010.

CORRÊA, A.C.; TEIXEIRA, E.M.; PESSAN, L.A.; MATTOSO, L.H.C. "Cellulose nanofibers from curaua fibers", *Cellul.*, v. 17, pp. 1183–1192, 2010.

INAMURA, P.Y.; KASAWA, C.T.S.; COLOMBO, M.A.; MOURA, E.A.B.; MASTRO, N.L. DEL. "Changes in Gelatin/Brazil Nut Shell Fiber Composite Treated by Electron Beam Radiation", In: *Proceedings of APSRC*, India, September, 2010.

VIANNA, P.R. *Estudo da castanha-do-Brasil*, In: Ministério da Agricultura, Comissão de Financiamento da Produção, Brasília, 1972.

APIZ - Zoro Indigenous People Association. *Good practices of harvest, storage and trade of Brazil nut*. In: Defanti, (eds.), Brasil, pp. 10, 2008. *(In Portuguese)*.

SCAPIN, M.A.; SALVADOR, V.L.R.; COTRIM, M.E.B.; PIRES, M.A.F.; SATO, I.M. "Uncertainty Measurement Evaluation of WDXRF and EDXRF Techniques for the Si and U_{total} Determination in U_3Si_2 used as Nuclear Fuel Material", In: *Proceedings of INAC*, Rio de Janeiro, Brazil, 2009.

BLEDZKI, A.K.; REIHMANE, S.; GASSAN, J. "Properties and modification methods for vegetable fibers for natural fiber composites", *J. Appl. Polym. Sci.*, v. 59, pp. 1329–36, 1996.

KOZLOWSKI, R.; RAWLUK, M.; BARRIGA, J. "State of the art-production, processing and applications of fibrous plants", In: *Proceedings of SINTEX-2004*, Natal, Brasil, 2004; Sivam, R.L., Araújo, R.C, (eds), [CDROM], Paper No.TIP-1-00, 2004.

ESPERT, A.; VILAPLANA, F.; KARLSSON, S. "Comparison of water absorption in natural cellulosic fibres from wood and one-year crops in polypropylene composites and its influence on their mechanical properties", *Compos. Part A*, v. 35, pp. 1267–1276, 2004.

SPINACÉ, M.A.S.; LAMBERT, C.S.; FERMOSELLI, K.K.G.; DE PAOLI, M.A. "Characterization of lignocellulosic curaua fibres", *Carbohydr. Polym.*, v. 77, pp. 47–53, 2009.

IDICULA, M.; MALHOTRA, S.K.; JOSEPH, K.; THOMAS, S. "Dynamic mechanical analysis of randomly oriented intimately mixed short banana/sisal hybrid fiber reinforced polyester composites", *Compos. Sci. Technol.,* v. 65, pp. 1077–1087, 2005.
RAMIAH, M.V. "Thermogravimetric and differential thermal analysis of cellulose, hemicellulose, and lignin", *J. Appl. Polym. Sci.*, v. 14, pp. 1323–37, 1970.
CALADO, V.; BARRETO, D.W.; D'ALMEIDA, J.R.M. "The effect of a chemical treatment on the structure and morphology of coir fibers", *J. Mater. Sci. Lett.*, v. 19, pp. 2151–2153, 2000.
SKINNER, H.C.W.; JAHREN, H.C.W. "Biomineralization", In: *Treatise on Geochemistry*; ISBN (set): 0-08-043751-6, Elsevier, v. 8, pp. 117–184, 2003.

Part II
New, Functional and Nanodimensional Natural Fibres

Brazilian Buriti Palm Fiber (*Mauritia flexuosa* Mart.)

Ivete Maria Cattani and Júlia Baruque-Ramos

Abstract The buriti (*Mauritia flexuosa* Mart.) is a palm tree present in the Brazilian territory, mainly in the Amazon Forest and Cerrado biomes, and throughout South America. It is known as the "Tree of Life" because all of its parts can be fully used, being the main source of income for many people in rural communities. In different Brazilian states, the fibers removed from the young leaves (buds) of the Buriti palm tree are employed in handicraft manufacturing. These fibers are popularly called Buriti "linen or silk". This paper presents the characteristics of the Buriti palm tree; its collection, processing, physicochemical properties, traditional and new employments. The employment of buriti fiber in traditional and new applications in design and industry is promising. Moreover, responsible and sustainable management of this natural resource can collaborate in the construction of social benefits. The encouragement of recovery and strengthening of local cultural identity and the organization of communities around a social economy will lead to the generation of income for extractive communities.

Keywords Buriti fiber · Collection · Processing · Physicochemical properties · Applications

Introduction

Brazil is one of the countries included among the so-called megadiverse countries, a group of 12 nations which house 70 % of the total biodiversity of the planet (IBGE 2012). Because of its territorial extension, approximately 8.5 million km^2, the country has large differences in its climate and ecology, forming distinct biogeographical

I.M. Cattani (✉) · J. Baruque-Ramos
University of São Paulo, School of Arts, Sciences and Humanities, São Paulo, SP, Brazil
e-mail: ivetecattani@usp.br

J. Baruque-Ramos
e-mail: jbaruque@usp.br

R. Fangueiro and S. Rana (eds.), *Natural Fibres: Advances in Science and Technology Towards Industrial Applications*, RILEM Bookseries 12,
DOI 10.1007/978-94-017-7515-1_7

zones or biomes (Amazon Forest, Cerrado, Caatinga, Atlantic Forest, Pantanal and Pampa). Also noteworthy is the socio-biodiversity found in Brazil represented by 200 indigenous folks and various communities, such as "quilombolas" (descendants of runaway slaves), "caiçaras" (fishermen), and "seringueiros" (latex collectors)—which bring together an invaluable collection of traditional knowledge of conservation of biodiversity (MMA 2014). This natural multiplicity and knowledge found in the country highlights the importance of the development of research aiming at making better use of Brazilian biodiversity and socio-biodiversity. In recent years, growing attention to environmental issues has promoted the increase in research on composites using natural fibers as reinforcing fillers (El Banna et al. 2011). However, these new demands require radical reconsideration of the current model of consumption and point out to the need of verifying the environmental, economic and competitive benefits for the market so to intervene with a systemic view, individualizing every 'eco-efficiency' opportunity (Pires 2008).

The Brazilian industry has, in general, invested in the redirection of its production mode and considers the transition to an environmentally sustainable and socially inclusive economy compatible with the goals of economic growth and competitive improvement (CNI 2012). In this scenario of new opportunities for a sustainable employment of Brazilian biodiversity, the creative impulse of the entrepreneur is of great importance once it is up to him/her to imagine a product or service that meets the needs with less resources and work; it is creativity that will allow the researcher to find elegant solutions to increasingly complex problems (Kazazian 2005). It is within this framework that this study presents the characteristics of the buriti palm tree (*Mauritia flexuosa* Mart.), its collection, processing and traditional employment of fibers, as well as its physicochemical properties and new employments.

Buriti Palm Tree

Considered the most beautiful plants in the plant kingdom, the palm trees are represented by about 1500 species, grouped in about 200 genera (Henderson et al. 1995).

The Buriti Palm tree (*Mauritia flexuosa* Mart.) is one of the most important palm trees and, perhaps, the one which has been taken greatest advantage of since prehistoric times, being its ripe fruit cheerfully greeted in scattered tribes in the Amazon nowadays (Corrêa 1984; Bondar 1964) (Fig. 2a). Its botanical classification is (Henderson et al. 1995): Kingdom: Plantae; Division: Magnoliophyta; Class: Liliopsida; Subclass: Arecales; Family: Arecaceae (Palmae); Subfamily: Calamoideae; Tribu: Lepidocaryeae; Genus: *Mauritia*; Species: *Mauritia flexuosa* Martius.

According to Lorenzi (2004), the Buriti is the most common and abundant palm tree in the Brazilian territory and is usually found in homogeneous groups of trees called "buritizais". In Brazil, it is known as "miriti", "buriti", "buriti-do-brejo", "moriti", "moriti-do-brejo", and "muritizeiro" (Medina 1959). It is a gender of robust palms, large, dioecious, solitary, with huge fan-shaped leaves and segments of equal width found in South America (Colombia, Ecuador, Peru, Venezuela, Brazil and Guyana) and Trinidad, growing in immense form of homogeneous populations in seasonally flooded lowlands (Lorenzi et al. 2010). It has erect, cylindrical, hairless and helpless stipe; it can reach up to 50 m in height and 50 cm in diameter, with 20–30 leaves, measuring 5 m long and up to 3 m wide (Corrêa 1984). This species is a source of food, beverages, cosmetics and raw material for crafts and handiwork. The consumption of its fruits can be in the form of juice or sweet. The petiole is used to make toys. The buriti is ornamental and can be grown for landscaping (Lorenzi et al. 2010). It is widely used in Brazilian handicraft, especially in the North, Northeast and Midwest regions (most abundant in regions of the biomes of the Amazon Forest and Cerrado).

Collection, Processing and Traditional Employment

The cultural and economic contribution of craftsmen is expressed through handmade artifacts produced from all the riches Brazilian nature offers. The buriti is one of those important sources of natural resources. This palm is known as the "Tree of Life" because all of its parts can be used for different purposes and has become the main source of income for many residents in rural communities.

Buriti leaves can display up to 5 m in length allowing the production of mats, baskets, brooms, nets, walls, house roofs, boats, sidings, slats and cushions for beds and tables, strings and bags (Fig. 1a and b). The petioles of the leaves are employed to make cages, traps, toys, furniture as well as ferries and paddles; they are also tied up to make a kind of raft for fishing and, due to their spongy feature, they can be used to make corks for bottles and pads for bed and table (Fig. 1c and d) (Rios and Pastore 2011).

So as to make delicate handicraft objects and products, the inhabitants of the Brazilian countryside remove the fiber of the new buriti leaves, locally known as "eye". The collection at higher palms is performed by men, while women and children remove the buds of young buriti palms (Fig. 2a–c). The withdrawal of new buds is performed following the tradition of each region and according to the experience and observation of the artisans. In order to preserve the palm, it is necessary to keep some new buds and perform the next collection only after the birth of the new buds. This period can range from three to six months. In addition to the care management of palm trees, the whole procedure of fiber processing is performed with attention to all its stages until it becomes a product.

The Buriti "linen" is a thin film of this new shredded bud. Once collected, the Buriti "linen" is extracted from the opening of the leaflets of new leaves from this bud (Fig. 2d, f and g).

(a) (b) (c) (d)

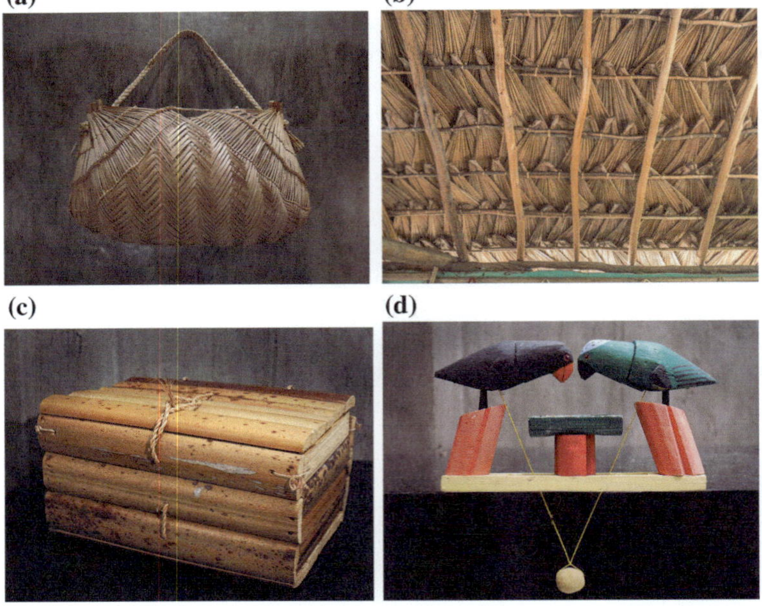

Fig. 1 **a** Basket produced from buriti leaves by Xavante ethnicity women from Etenhiritipá hamlet, Pimentel Barbosa reserve, Mato Grosso. It is used for collecting fruits, carrying children, storing clothes and keeping water; **b** Roof of buriti leaves in the village of Marcelino, Barreirinhas, Maranhão; **c** Box produced from the buriti leaf petiole by handcrafters of "quilombola" village located in Nova Aurora, Goiás; **d** Toy made from buriti leaf stalk by artisans of the city of Abaetetuba, Pará (World Capital of Miriti Toy). *Source* The photos (**a**, **c**, **d**) are courtesy by Manolo Pacheco and (**b**) of Mauro Goulart, 2014

With the aid of a knife, the craftsman performs a light incision on the surface of the leaflets, allowing the removal of a thin film that is pulled one by one and a handful is placed on the floor (Fig. 3a–c). The remaining part of the bud, i.e., the residue (called "borra") (Fig. 3d) is also used for making accessories and home objects (Keller 2011).

For the application of buriti "linen" in handcrafts, it is washed and dyed with dyes extracted from leaves, seeds, flowers and bark of native plants of the region. Ashes are used to enhance the color and to retain the dye in the fiber (Fig. 4a–c). After dried, the fiber is prepared according to the technique to be applied.

Different craft techniques are practiced in Barreirinhas city and each community has its own characteristics of production so that certain craft techniques are assigned to the reputation of some localities (Figueiredo 2012).

The crochet technique (Fig. 5a–d), characteristic of communities of Cebola and Palmeira dos Eduardos, which consists of a braided knitting similar to a lacy knitting (Guimarães 2014), is performed with special needle, endowed with a hook in one of its ends.

Fig. 2 Marcelino village, Barreirinhas city, Maranhão. **a** Buriti palm tree; **b–c** Buriti bud withdrawal; **d–g** Separation of leaflets from the bud. *Source* The photos are courtesy by Mauro Goulart, 2014

Fig. 3 **a–b** Withdrawal of buriti "linen"; **c** Buriti "linen"; **d** Buriti "borra" (leaves left over after the withdrawal of "linen"). *Source* The photos are courtesy by Mauro Goulart, 2014

The technique "malha de cascudo", done through an elaborate weaving, presents a pattern that resembles the scales of a freshwater fish called "cascudo". Because it has a double embroidering (like two overlapping tissue) it is also known as "duas capas" ("two covers") (Medeiros 2015) (Fig. 6a).

Fig. 4 **a–b** Buriti "linen" dyeing with the bark of Gonçalo-alves (*Astronium fraxinifolium* Schott—tropical arboreal species); **c** Application of ashes over the dyed fiber; **d** Handicraft of artisan women. *Source* The photos are courtesy by Mauro Goulart, 2014

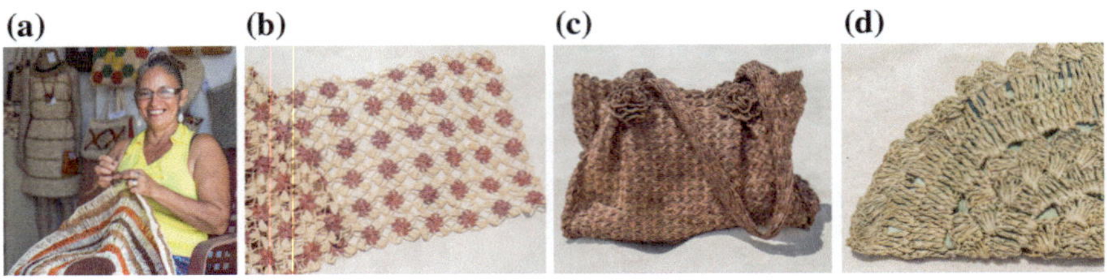

Fig. 5 Crochet technique: **a** Artisan work; **b** Tablecloth; **c** Bag; **d** Detail of a wallet. *Source* The photos are courtesy by Mauro Goulart, 2014

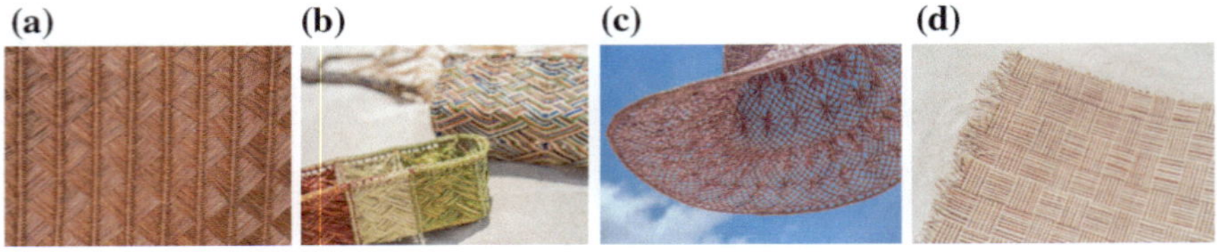

Fig. 6 **a** "Malha de Cascudo"—detail of a bag; **b** "Malha de Cofo"—kitchen baskets; **c** Hat (macramé); **d** Mat (weaving). *Source* The photos are courtesy by Mauro Goulart, 2014

The "malha de cofo" has the basis the nodes of macramé to present the embroidery of a traditional basket called "cofo", which is also made from other vegetable fibers and is used in the region to transport products in general (Medeiros 2015) (Fig. 6b). These two techniques are typical of the Barreirinhas region and the Marcelino village.

The macramé (Fig. 6c) is a technique that does not use any machinery or tools, just embroidery and nodes. Weaving (Fig. 6d) is a technique held in small looms (Guimarães 2014). Some products are made with mixed techniques (Fig. 7a–d).

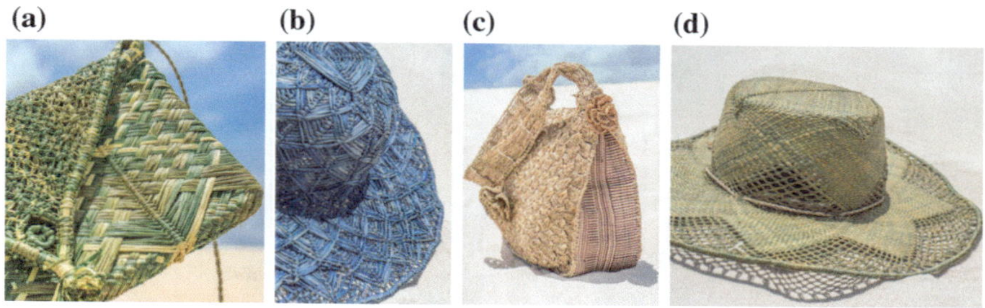

Fig. 7 **a** Wallet (macramé and "malha de cofo"); **b** Hat ("malha de cascudo" and macramé; **c** Bag (crochet and weaving); **d** Hat (weaving and macramé). *Source* The photos are courtesy by Mauro Goulart, 2014

Physicochemical Properties and New Employments

The main tensile properties of buriti fiber were reported by Monteiro et al. (2011) as shown in Table 1.

According to previous studies performed by the present authors, the values of tensile strength of buriti fiber ("linen") is approximately twice higher than buriti "borra" (the residue after the extraction from the buds) (data not shown). This is related to the fact that the count number of "linen" fibers (employed in the production of the finest handicraft articles) are minor (meaning thinner and more resistant fibers) than the "borra" fibers.

In addition, according to preliminary studies of the present authors, the length of the in nature fiber withdrawal from buriti bud can exceed 2 m, the elongation ranges from 6 to 9 %, regain is near 9 % and the cellular diameter is near 8.5 μm. These two last values are compatible with other lignocellulosic fibers such as cotton, jute and flax. Longitudinal (Fig. 8a) and cross microscopy (Fig. 8b) performed by the authors present, respectively, porosity and no microfibrils, and multicellular structure, polygonal cell shape with round or oval lumen.

Recently, Polleto et al. (2014) reported the relationship between cellulose crystallinity, the influence of extractive content on lignocellulosic fiber degradation, the correlation between chemical composition and the physical properties of ten types of natural fibers, including buriti fiber. These results were obtained through FTIR spectroscopy, X-ray diffraction and thermogravimetry techniques. Their results for chemical composition and thermal degradation are presented in Tables 2 and 3.

Table 1 Density and mechanical properties of buriti (*Mauritia flexuosa*) fiber (Monteiro et al. 2011)

Density ρ (g/cm^3)	Tensile strength σ (MPa)	Young's modulus (GPa)	Maximum σ/ρ (MPa cm^3/g)
0.63–1.12	129–254	–	403

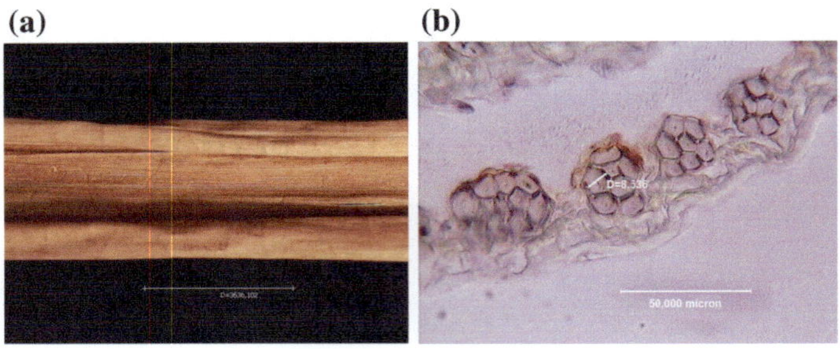

Fig. 8 Buriti fiber. **a** Longitudinal microscopy (51×; 3.6 mm scale in the image); **b** Cross microscopy (1280×; 50 μm scale in the image)

Table 2 Chemical composition of buriti (Polleto et al. 2014)

Holocellulose[a] (wt%)	Lignin (wt%)	Pectin (wt%)	Waxes (wt%)	Extractives (wt%)
65–71	21–27	–	–	5.4–6.0

[a]The sum of the contents of cellulose and hemicellulose

Table 3 Thermal degradation temperature and residue at 800 °C for the buriti fiber (Polleto et al. 2014)

Ti (°C) 3 wt% loss	*T* shoulder (°C)	DTG peak (°C)	Residue at 800 °C (%)
92	291	334	17.5

In recent years, a significant number of research enrolling buriti fiber for composites development and other applications were developed. Among them, it is worthy of mention: (i) the use of biotextiles to recuperate degraded areas by erosion (Furtado et al. 2005); (ii) Cardanol–formaldehyde thermoset composites reinforced with buriti fibers (Santos et al. 2010; Monteiro et al. 2009, 2012); (iii) buriti petiole/polyester composites (Portela et al. 2010); and (iv) PLA (polilactic acid)/ buriti fiber composites (Brambilla 2013). However, until this moment, the present authors have no knowledge of effective large scale industrial processes or applications employing this fiber.

Despite the fact that buriti fiber is a renewable resource (Monteiro et al. 2012), it is important to take into account its limitations regarding collection for industrial employment. A study performed by Saraiva (2009) described these limitations and potentialities and, although all the system variables were not evaluated, he points out the potential value in the extraction of buriti in the region of Lençóis Maranhenses and throughout the Cerrado biome. According to this author, investment in buriti fiber processing by the municipality or in a production center, the strengthening of communities and the opening and stimulation of consumer markets are among the main actions to be encouraged.

Conclusion

The employment of buriti fiber in traditional and new applications for design and industry is promising. The buriti "linen" presents results compatible with other fibers that are successfully used in textiles, being propitious for new employments in this sector. Responsible and sustainable management of this natural resource can enable the permanence of these extractive families in their communities, as well as stimulate the maintenance and strengthening of their cultural identities.

Acknowledgments We gratefully acknowledge **CAPES** (Coordination for the Improvement of Higher Education Personnel of Brazilian Education Ministry), **USP** (University of São Paulo), **SEBRAE-MA** (Brazilian Micro and Small Business Support Service), **ArteCoop** (Cooperative of Artisans from Lençóis Maranhenses), **Lençóis Maranhenses National Park** and the friends **Mario Goulart** and **Manolo Pacheco**, who generously provided the photos taken in communities.

References

BONDAR, G. **Palmeiras do Brasil**. Secretaria da Agricultura do Estado de São Paulo – Instituto de Botânica, São Paulo, 1964.39p.

BRAMBILLA, V. C. **Avaliação das propriedades mecânicas, térmicas, morfológicas e degradação de compósitos de poli (ácido lático)/buriti**. 163 f. MSc. Dissertation. University of Caxias do Sul, Postgraduation Program in Processes and Technology Engineering. Caxias do Sul (RS), Brazil. 2013. 163 p.

CNI - Confederação Nacional da Indústria. **A Indústria Brasileira no caminho da sustentabilidade – Encontro da Indústria para a sustentabilidade**, 2012. 49p. Available in http://www.fiemt.com.br/arquivos/916_cni_parte_1_rio20_web.pdf. Accessed Set 29, 2014.

CORRÊA, M.P. **Dicionário das plantas úteis do Brasil e das exóticas cultivadas**. Rio de Janeiro, Imprensa Nacional, 1926-1978. 6 v. ilust. Reedited in 1984 by Ministério da Agricultura – Instituto Brasileiro de Desenvolvimento Florestal. 4329 p.

EL BANNA, W.R. et al. Fibras de bananeira como material de engenharia. In: **XXXIX Congresso Brasileiro de Educação em Engenharia**, Blumenau (SC), Anais. Grupo de Pós-Graduação em Engenharia Mecânica da Universidade do Pará, 2011.

FIGUEIREDO, W. (Coord.) **Fibras e trama de Barreirinhas**. Rio de Janeiro: IPHAN, CNFCP, 2012. 44 p.

FURTADO, M. S. et al. The use of biotextiles to recuperate degraded areas by erosion. **Revista Sociedade & Natureza**, Special Issue, 2005.

GUIMARÃES, B.M.G. **Estudo das características físico-químicas de fibras têxteis vegetais de espécies de Malvaceae**. 169p. MSc. dissertation. Postgraduation in Textile and Fashion. School of Arts, Sciences and Humanities, University of São Paulo. São Paulo, 2014.

HENDERSON, A.; GALEANO, G. BERNAL, R. (1995).**Field guide to the Palms of the Americas**. Princeton University Press, New Jersey, 414p.

IBGE - INSTITUTO BRASILEIRO DE GEOGRAFIA E ESTATÍSTICA. **Indicadores de Desenvolvimento Sustentável – Brasil 2012. Estudos & Pesquisas**. Available in http://www.ibge.gov.br/home/geociencias/recursosnaturais/levantamento/default.shtm?c=2. Accessed Sep 28, 2014.

KAZAZIAN, T. (Org.). **Haverá a idade das coisas leves: Design para o desenvolvimento sustentável**. SENAC, São Paulo, 2005, 194p.

KELLER, P. F. Trabalho Artesanal Em Fibra De Buriti No Maranhão. **Cadernos de Pesquisa**, v. 18, n. 3, 2011.

LORENZI, H.; NOBLICK, L.R.; KAHN, F.; FERREIRA, E. **Flora Brasileira - Aracaceae (Palmeiras)**. Editora Plantarum, Nova Odessa, 2010, 384 p.

LORENZI, H.; SOUZA, H.M.; MEDEIROS-COSTA, J.T.; CERQUEIRA, L.S.C.DE; FERREIRA, E. **Palmeiras Brasileiras e Exóticas Cultivadas**. Nova Odessa, Editora Plantarum, 2004. 242 p.

MEDEIROS, M. Tecendo com fibras. Available in http://www.Artesanatocomdesign.blogspot.com.br Accessed Mar 06, 2015.

MEDINA, J. C. **Plantas Fibrosas da Flora Mundial**. Printed by Instituto Agronômico de Campinas, 1959. 913 p.

MMA - Ministério do Meio Ambiente. **Biodiversidade Brasileira**. Available in http://www.mma.gov.br/biodiversidade/biodiversidade-brasileira. Accessed Oct 02, 2014.

MONTEIRO, S. N. et al. Natural lignocellulosic fibers as engineering materials—an overview. **Metallurgical and Materials Transactions A**, v. 42, n. 10, p. 2963-2974, 2011.

MONTEIRO, S. N. et al. Natural-fiber polymer-matrix composites: cheaper, tougher, and environmentally friendly. **Jom**, v. 61, n. 1, p. 17-22, 2009.

MONTEIRO, S. N. et al. Thermogravimetric stability of polymer composites reinforced with less common lignocellulosic fibers–an Overview. **Journal of Materials Research and Technology**, v. 1, n. 2, p. 117-126, 2012.

PIRES, D.B. (Org.). **Design de Moda - Olhares Diversos**. SENAC, São Paulo, 2008. 205p.

POLETTO, M. et al. Native Cellulose: Structure, Characterization and Thermal Properties. **Materials**, v. 7, n. 9, p. 6105-6119, 2014.

PORTELA, T. G. R. et al. Tensile behavior of lignocellulosic fiber reinforced polymer composites: Part II buriti petiole/polyester. **Matéria (Rio de Janeiro)**, v. 15, n. 2, p. 195-201, 2010.

RIOS,M.N.S; PASTORE JR.F. (Org). Plantas da Amazônia: 450 espécies de uso em geral. Universidade de Brasília (UnB), EBook, p. 459-480.Brasília, 2011.161p.

SANTOS, R. S. et al. Cardanol–formaldehyde thermoset composites reinforced with buriti fibers: preparation and characterization. **Composites Part A: Applied Science and Manufacturing**, v. 41, n. 9, p. 1123-1129, 2010.

SARAIVA, N. A. **Manejo sustentável e potencial econômico da extração do buriti nos Lençóis Maranhenses**.129 f. MSc. Dissertation. University of Brasilia, Center of Sustainable Development, Political and Environmental Management. Brasilia (DF), Brazil. 2009. 129 p.

Degradation of Dyes Using Plantain Fibers Modified with Nanoparticles

Yuliana Cadavid, Edith M. Cadena, Juan M. Velez and Juan F. Santa

Abstract Natural fibers are a good option to degrade dyes. In this work, nanoparticles of MnO_2 were obtained onto the surface of natural fibers (Plantain Pseudostem) via in situ synthesis in order to perform a new treatment for pollution and dyes. The fibers were treated with Hydrochloric Acid and Sodium Hydroxide to modify their surface (cationization), and then they were immersed in a $KMnO_4$ solution (2 mM). The treatment was done in an ultrasonic bath and the dye degradation was monitored by UV-Vis with a xenon lamp by detecting the maximum dye peak and the Color Removal Percentage (CR%) after 15 min of soaking. The results showed that the removal of dyes was 90 % after rinsing. The morphology of the fibers was studied in an electron microscope (FE-SEM) with a coupled microanalysis detector (EDS). The test confirmed that MnO_2 nanoparticles were synthetized onto the surface of the natural fibers.

Keywords Indigo blue · $KMnO_4$ · Natural fibers · Chemical functionalization

Abbreviations

FF	Functionalized Fibers
PP	Plantain Pseudostem
EDS	Energy Dispersive X-ray Spectroscopy
FE-SEM	Field Emission-Scanning Electron Microscopy
FTIR	Fourier Transformed InfraRed
UV-Vis	Ultraviolet-Visible Spectroscopy

Y. Cadavid (✉) · E.M. Cadena · J.M. Velez · J.F. Santa
Grupo de Biofibras y derivados, Universidad Nacional de Colombia, Medellín, Colombia
e-mail: ycadavidm@unal.edu.co

J.F. Santa
Grupo de Materiales Avanzados y Energía– MATyER, Instituto Tecnológico Metropolitano, Medellín, Colombia

R. Fangueiro and S. Rana (eds.), *Natural Fibres: Advances in Science and Technology Towards Industrial Applications*, RILEM Bookseries 12,
DOI 10.1007/978-94-017-7515-1_8

Introduction

Plantain Crops in Colombia

Clean technologies promote the use of agro-industrial waste in new materials with potential applications in several fields (e.g. food and pharmaceutical sector, industrial packaging). The use of lignocellulose residues is now an excellent alternative to decrease agricultural wastes and their environmental impact, since it is a good way to add value to leftovers. Colombia is an important producer of plantains and it has around 381,000 ha cropped producing around 3,178,000 tons (Agronet, 2013).

Plantain is cataloged as a perennial substitutable crop, since after obtaining the fruit (about one year) the plant must be replaced (Betancurt et al. 2011). Plantain pseudostem (PP) is discarded and left on the ground causing problems to the soil. Recently, the plantain's crops are being incorporated in different industries to solve several issues or just to create new composites (Węcławski et al. 2014; Aguilar 2007; Ray et al. 2001; Sosa et al. 2011; Valadez-Gonzalez et al. 1999). PP components include cellulose (56 %), hemicellulose (43 %), lignin (19 %), pectin, waxes and water soluble substances (Mora et al. 2014), being cellulose among the most abundant biomaterial on earth, having a low cost, environmentally friendly and having biodegradable properties.

Plantain crops are very important for some countries such as Philippines, Ecuador, Costa Rica, Colombia, and Virgin Islands, among others (Chacón et al. 2013). From those crops in Colombia, four million metric tons of residues are generated every year and 95 % of the residues (leaves, buds, stems) are used as compost and potential applications of those residues are still unknown. Consequently, the use of plantain stems to extract lignocellulose fibers as raw material for the production of composites is an interesting area for researchers and producers.

A good option would be to create new materials (biocomposites) based on vegetable fibers since their internal structure can be used as a vehicle to deposit nanoparticles and functionalize the fibers to remove dyes from wastewater (Chacón et al. 2013). Water contamination is an important issue and many chemical and biological processes are used to remove dyes in the textile industries (Chacón et al. 2013; Ong et al. 2011; Singh and Arora 2011; Rosemal et al. 2009; Bruchet et al. 2004; Moyers and Wu 1985; Chen and Yeh 2005).

Current Applications of Plantain Fibers

Plantain fibers have been widely studied by other authors (Sabyasachi et al. 2009; Akpabio and Akpakpan 2012; MUN and IM 2014; Li et al. 2007; Alvarez et al. 2015), and they are currently used in several applications such as composites (MUN and IM 2014; Li et al. 2007), paper (Akpabio and Akpakpan 2012), fiberboards (Alvarez et al. 2015), textiles (Gallego et al. 2015) and others (Sabyasachi et al. 2009). Some of the applications of natural fibers found in the literature are summarized in Fig. 1. Additionally, new applications are emerging for natural fibers. In those applications, fibers are used as a vehicle to carry nanoparticles since the porosity and surface area of the fibers can be used to promote chemical reactions on their surface. After performing chemical reactions the functionalized fiber are used for dye degradation (Chacón et al. 2013).

New applications of fibers can be useful to increase their lifecycle (Fig. 2). In those cases, instead of producing agricultural wastes, plantain crops can be used to decrease the water contamination from textile industries by dyes. At the end of the process, a non-toxic residue (fibers), can be used to degrade more dyes or it can be discarded with no impact to the environment.

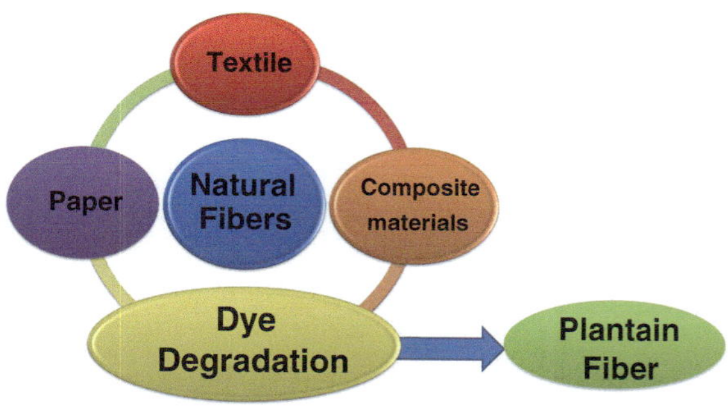

Fig. 1 Some of the current applications of natural fibers

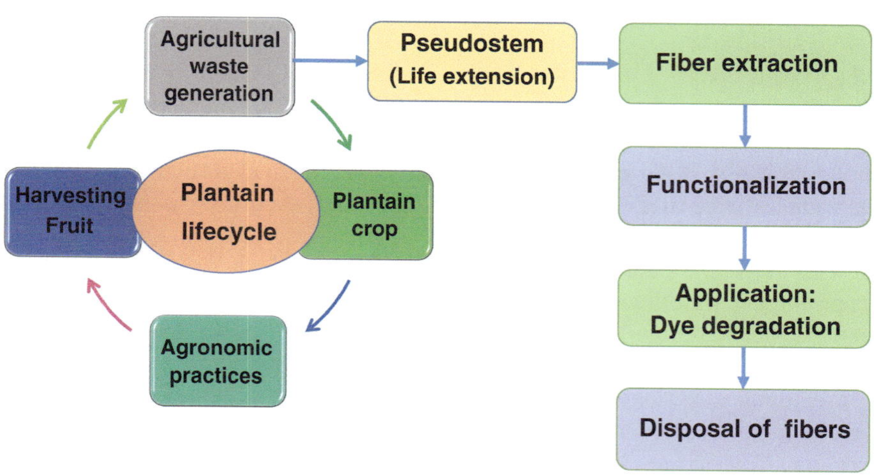

Fig. 2 Process of extension of lifecycle of plantain pseudostem

Contamination by Chemical Dyes

Chemical dyes used in the textile industry generate wastewaters that are normally disposed to the streams causing variations in the fluids by the presence of suspended solids. Chemicals also decrease photosynthetic activity and produce toxicity with negative consequences for human health and the environment.

Remediation alternatives for this problem have been studied by other authors (Singh and Arora 2011; Rosemal et al. 2009). The most recognized treatments are those using TiO_2 particles treated with UV light, electrochemical treatment, nanofiltration, cation exchange, activated carbon, flocculation, among others. Those treatments are commercially expensive but they have been technically proved (Bharathi and Ramesh 2013; Pérez et al. 1999).

Chacón et al. (2013) have studied natural fibers (Fique fibers) for efficient dye degradation. They found that Fique fibers can be used for the in situ synthesis of nanostructured MnO_2 to produce a genuine bionanocomposite and after 5 min of treatment, a maximum efficiency of 98.8 % was obtained.

However, since the morphology (surface area, pores, microfibrillar angle, diameter, etc.) of every natural fiber is different, if another natural fiber is going to be used for dye degradation new tests must be performed. Accordingly, in this work, functionalized fibers obtained from plantain (Pseudostem) were tested to degrade dyes (indigo blue). MnO_2 nanoparticles were obtained onto the surface of natural fibers via in situ synthesis. This process could become an alternative to degrade textile dyes, since it could be cost-effective and environmentally friendly.

Methodology

Extraction of Fibers from Plantain Pseudostem

Fibers were extracted from Pseudostem of *Musa × Paradisiaca* from Barbosa, (elevation above sea level 1359 m, Latitude: 6° 25′ 58.5″N, Longitude: 1° 15′ 30.0″ W) a small town near to Medellín, Colombia. Several pseudostems collected are shown in Fig. 3. After collecting pseudostems, longitudinal sections were cut and the samples were unsheathed. During this operation, each layer is separated (see Fig. 4) and the core of the pseudostem was discarded. This part (approximately 26 % of the total weight of the plant) corresponds to the floral stem, responsible for supporting the fruit (bunch). That part was discarded since it is non-fibrous and the material breaks during the extraction process (see Table 1 for details about efficiency during extraction).

Longitudinal sections were put into a fique extraction machine. The machine has a diesel motor coupled to rotating blades with a flat edge to prevent the fibers from breaking. Considering fique's fibers are stronger than plantain′s fibers, the power required to extract fibers form the pseudostem is lower and they can be extracted in

Fig. 3 Collected pseudostems

a fique extraction machine. Sheaths were passed through the machine to remove the vegetal material and to leave uncovered the fibers. During the extraction process, fibers kept the same length of the sheath.

Afterwards, fibers were soaked into water for one (1) day to remove the cellulose residues between the fibers. In the field, this process is called fermentation by commercial producers of fibers. After one day of fermentation, the fibers were cleaned and dried at the sun for another day and finally they were manually combed to remove some impurities remaining after the process. The extraction sequence can be illustrated as shown in Fig. 4.

Cationization Process

Cleaned fibers were immersed into an ultrasonic bath at 42 kHz and 100 W during 1 h to remove some impurities and then they were dried at ambient temperature. After cleaning, fibers were treated with a solution of 5 % HCl w/v. during 3 h at ambient temperature to impregnate completely the fibers and then they were washed in deionized water.

Thereafter fibers were immersed in 6 % NaOH w/v. for 3 h at 60 °C in a heated bath and then they were washed in deionized water and finally dried at 80 °C in an oven. The color of the fibers changed due to the removal of the lignin as shown in Fig. 5.

Fig. 4 Mechanical extraction process of the plantain fibers. **a** Pseudostem after harvesting the fruit. **b** Collected pseudostems. **c** Unsheathing. **d** Extraction of fiber. **e** Wet fiber extracted. **f** Fermentation process. **g** Drying. **h** Combing of the fibers

Table 1 Parameters during extraction process

Parameters	Pseudostems		
	N°1	N°2	N°3
Length (m)	1.21	1.4	1.51
Weight (kg)	14	13	24
Residues (%)	28.6	29.2	20
Total weight sheaths (kg)	10	9.2	19.2
% Average moisture (sheaths)	89.5		
Total weight (sheaths) (kg) d.w	1.05	0.97	2.02
Extracted fiber weight (kg)	0.16	0.11	0.10
% Average fiber moisture	7.60		
Extracted fiber weight (kg) d.w	0.15	0.10	0.09
Yield (%)	14.3	10.3	4.4

d.w dry weight

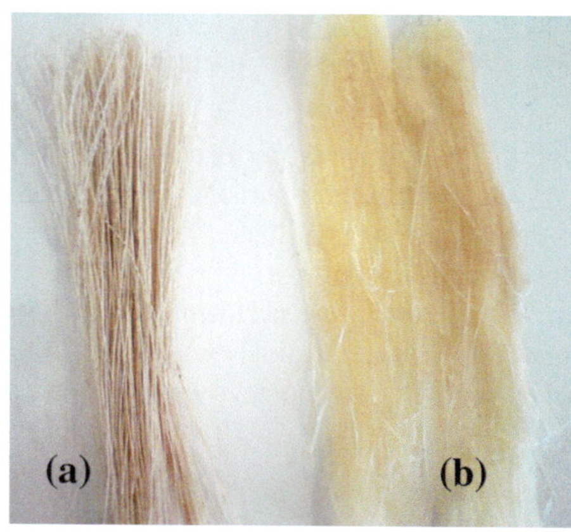

Fig. 5 Fibers after surface modification. **a**. Ultrasonic bath. **b** Cationization

Functionalization of Fibers

Cationized fibers were immersed into in a $KMnO_4$ solution (2 mM) assisted by the ultrasonic bath (42 kHz, 100 W) during 45 min. Fibers were washed after the process with deionized water and they were left inside an oven at 60 °C during 24 h. The fibers are shown in Fig. 6. A scheme to represent this process is shown in Fig. 7.

Fig. 6 Functionalized fibers

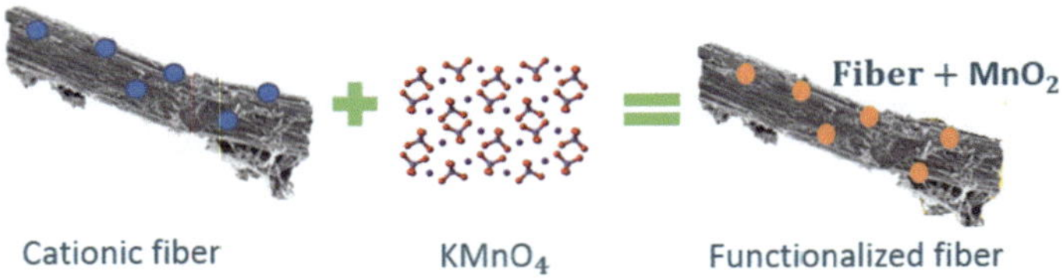

Fig. 7 Scheme of functionalization

Functionalized Fibers (FF) were characterized by a Scanning electron microscope (JSM-7100F FE-SEM) and the samples were previously coated with gold. Energy Dispersive Spectrometry (EDS) was performed to evaluate the functionalization of the fibers.

Dye Degradation

Dye degradation was performed using 1.0 g of the FF immersed in 50 mL of the dye solution (indigo blue) at 40 ppm during 30 min under constant stirring. The pH of the solution was adjusted at a value of 2.5 by using CH_3COOH and it was kept constant during the test.

UV-Vis tests were performed in a Thermo Scientific Evolution 60 with a xenon lamp. Absorbance analysis was used to evaluate kinetic degradation of the dye and color removal percentage—% CR was calculated by the below equation,

$$CR\,(\%) = 1 - \frac{Dye\,abs_t}{Dye\,abs_0} \times 100$$

where Dye abs_t is the absorbance value at the time t and Dye abs_0 is the initial absorbance value. The fiber's color was measured by a Sphere Spectrophotometer X-rite SP60 with CIELab coordinates, D65/10° (illuminant/observer conditions) and opacity constant SPEX (Specular Excluded).

Results and Discussion

Fibers Characterization

Pseudostems have vascular bundles to provide support to the plant and transport the nutrients. The inner structure of the fibers is characterized by large pores leading to microfibrills. Each fiber is formed by several hollow microfibrills as shown in Fig. 8. Several microfibrills completely aligned can be observed. This feature of plantain fibers is very important since it increases the surface area.

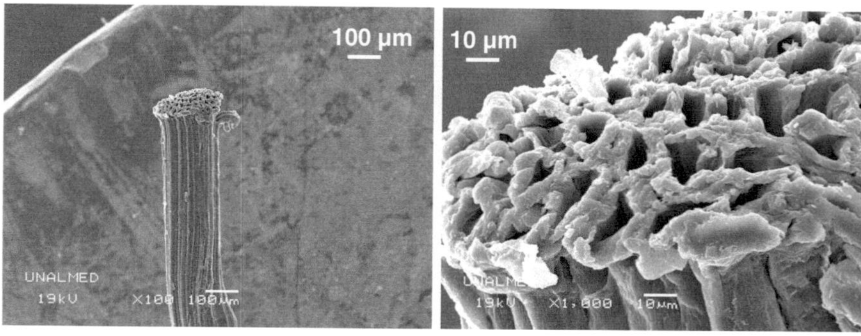

Fig. 8 Microstructure of the plantain fiber

Tubular structure of fibers provides potential uses in physical and chemical processes used in the treatment of gases and liquids such as gas filters, humidity control, fluid exchange or adhesion of particles, among others, where the surface area is a crucial parameter.

The surface of the fibers (FE-SEM) is shown in Fig. 9. From the results it can be concluded that the fibers were effectively functionalized since after functionalization, nanoparticles were found onto the fibers. Figure 9c shows a detail of the nanoparticles. The particles have a circular shape and the EDS spectrum (Fig. 9d) shows high contents of Mn and K, meaning that the functionalization was successful.

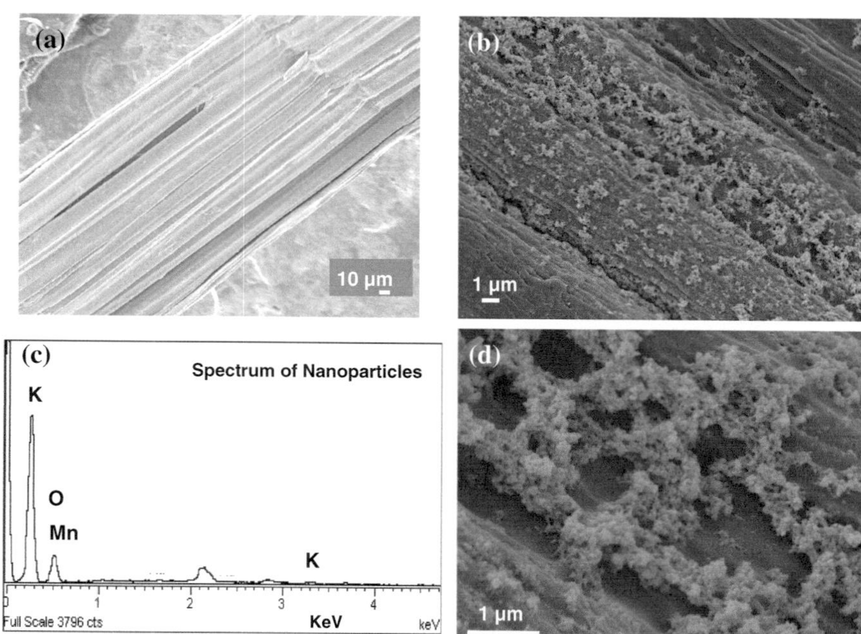

Fig. 9 FE-SEM. **a** Raw fibers with no nanoparticles. **b** Nanoparticles onto the fiber (2 mM). **c** EDS of functionalized fiber evidencing the presence of Mn and K. **d** Details of nanoparticles

Dye Degradation

Functionalized fibers (FF) changed their structure due to lignin removal by acid-alkaline procedure (HCl-NaOH) affecting their tensile strength and appearance. After functionalization, the fibers were immersed into the dye solution with a pH = 2.5. This adjustment was performed to facilitate ions interaction.

Results of dyes removal monitored by UV–Vis at λ = 566 nm is shown in Fig. 10. From this figure was concluded that, after 15 min, the absorbance reached a minimum lower than 0.01 of absorbance. After 15 min of soaking, absence of the dye's peak is confirmed by Fig. 10.

The results of kinetic degradation and color removal (CR) tested every 5 min is shown in Fig. 11. From the pictures, the change of color during the treatment with FF can be observed. The biocomposite is effective to perform the discoloration process because after 15 min of continuous stirring the solution turned visibly clear. From the results it also can be concluded that the removal had an efficiency of 90 % and after 30 min a color removal of the 98 % was reached.

After functionalization treatment, fibers changed its color and it turned into a dark brown color, confirming that the functionalization took place. The data shown in Table 2 indicates the results of the color modification monitored by the CIELab

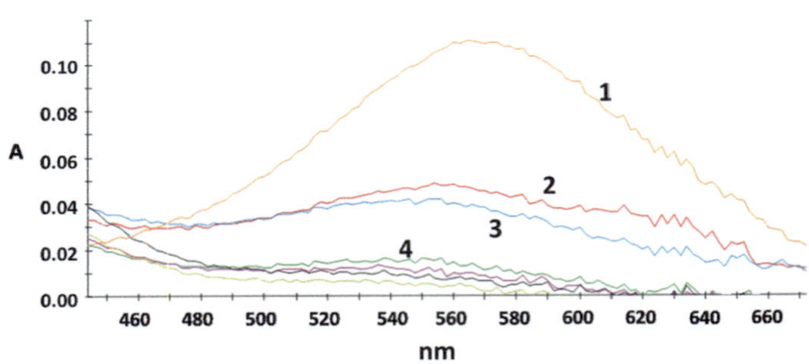

Fig. 10 Dye degradation. Absorbances at *1.* 0 min; *2.* 5 min; *3.* 10 min; *4.* 15 min

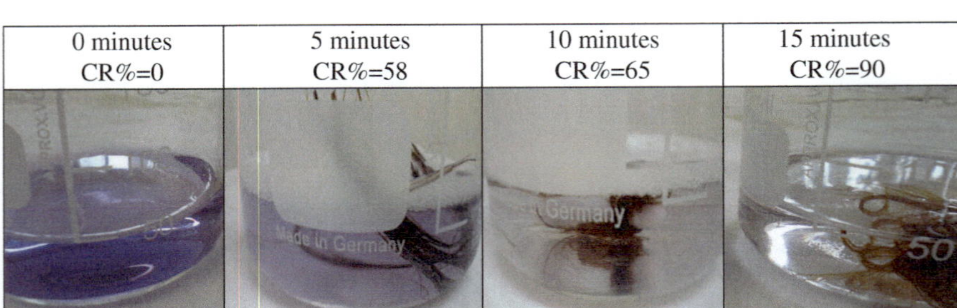

Fig. 11 Evolution of dye degradation

Table 2 Color parameters of the fiber

Parameter	Raw fiber	FF before removal	FF after removal
L*	81	80	61
a*	2	2	6
b*	14	14	26
Result			

L*: lightness
a*: red/green (+/-) opponent colors
b*: yellow/blue (+/-) opponent colors

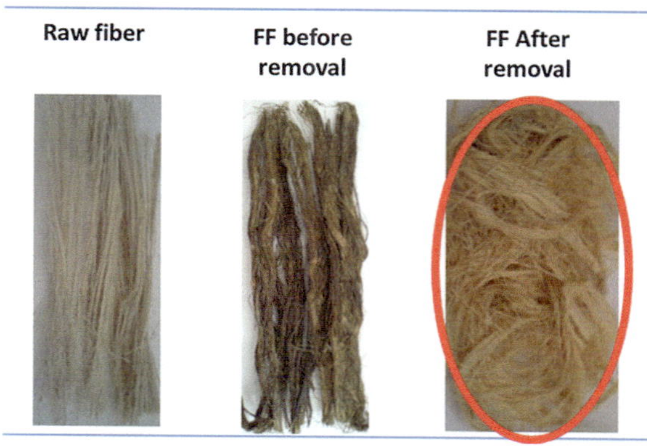

Fig. 12 Fiber's color

coordinates. Accordingly, after removing the fibers from the solution, the fiber's color was darker than the color of the raw fiber (Fig. 12 Fiber's color).

This behavior could be caused by the remaining oxide particles onto the surface of the fibers. This result implies that the fibers used in this article can be used several times for the dye's removal procedure as some authors previously reported (Chacón et al. 2013). From the results, it also can be concluded that the functionalization process is not a complex treatment and it is environmentally-friendly considering that a non-toxic fiber is obtained with the ability to degrade the color molecule. The products obtained after this treatment are molecules that can be degraded easily (Chacón et al. 2013; Moyers and Wu 1985).

Discussion

After removing the fibers from the solution, the fiber's color was darker than the color of the raw fiber. This behavior could be caused by the remaining oxide particles onto the surface of the fibers. This result implies that the fibers used in this research can be used several times for the dye's removal procedure as some authors previously reported (Chacón et al. 2013).

The mechanism of dye degradation is being investigated with different techniques to find the model of action of the functionalized fibers. It is very important to compare the results with chemical methods using oxidizing agents, photochemical removal, electrochemical destruction; or physical methods as adsorption, activated carbon, silica gel, etc.

Several treatments have been used to assess the degradation of the dyes. Biologic treatments have noteworthy results but they are not optimum. However, a combination between chemical and biological treatment could be an option to obtain a significant response.

Tubular structure of the fibers reveled by SEM suggested that several geometrical arranges of the fibers could be studied in order to manufacture filters with improved efficiency trying to obtain a high specific area and a good surface to volume ratio.

Conclusions

Plantain Fiber is a good alternative to degrade dyes because of its physical properties such as structure, porosity and chemical composition. Also the degradation time was relatively short and the materials have a low cost and they are easily obtained as it is an agricultural residue. A good strategy would be to make a material suitable to removal dyes such as filters, membranes, among others, and the functionalized plantain fibers could be a proper choice, given that they degraded indigo blue just in around 15 min.

Functionalization of plantain fibers proved to be effective to remove dyes. It is important to continue researching to obtain new functional properties (e.g. antimicrobial and fungicidal fibers) using coupling and grafting of nanoparticles onto fibers from agricultural waste through eco-friendly process. Another interesting topic is the reuse of functionalized fiber to perform several cycles of dyes degradation. Additionally, degradation of different dyes could be studied to evaluate the performance of functionalized fibers under different dyes molecules.

References

Agronet, "Producción nacional por producto.," 2013. [Online]. Available: http://www.agronet.gov.co/agronetweb1/Estadísticas.aspx.

Á. M. Betancurt, "Residuos del plátano, renta para productores," *UNperiódico*, 2011.

B. T. Węcławski, M. Fan, and D. Hui, "Compressive behaviour of natural fibre composite," *Compos. Part B Eng.*, vol. 67, pp. 183–191, Dec. 2014.

S. Aguilar, "Extracción de fibras no leñosas: cabuya (Furcraea Andin trel.) y banano (Musa Paradisiaca L.) para estandarizar un proceso tecnológico destinado," *Rev. Iberoam. polímeros*, vol. 8, no. 2, pp. 89–98, 2007.
D. Ray, B. K. Sarkar, A. K. Rana, and N. R. Bose, "The mechanical properties of vinylester resin matrix composites reinforced with alkali-treated jute fibres," *Compos. Part A Appl. Sci. Manuf.*, vol. 32, no. 1, pp. 119–127, Jan. 2001.
A. Sosa, J. Rivas, G. Mogollon, I. Gutierrez, and A. Aguilera, "Evaluación papelera del pseudotallo de platano (*Musa Paradisiaca*) en formulaciones con Hevea brasiliensis, Eucalyptus urophylla y Pinus caribaea var. hondurensis," *La Rev. For. Venez.*, no. 1, p. 9, 2011.
A. Valadez-Gonzalez, J. M. Cervantes-Uc, R. Olayo, and P. J. Herrera-Franco, "Effect of fiber surface treatment on the fiber–matrix bond strength of natural fiber reinforced composites," *Compos. Part B Eng.*, vol. 30, no. 3, pp. 309–320, Apr. 1999.
Y. Cadavid, J. M. Velez, J. Felipe, Santa, and E. M. Cadena, "Transformación del pseudotallo de plátano (Musa Paradisiaca) para la obtención de fibras: Análisis Morfológico," ENID, Universidad Nacional de Colombia, pp. 1–5, 2014.
M. L. Chacón, C. Blanco, J. P. Hinestroza, and M. Y. Combariza, "Biocomposite of nanostructured MnO2 and Fique fibers for efficient dye degradation," *Green Chem.*, no. 15, pp. 2920–2928, 2013.
S. T. Ong, P. S. Keng, W. N. Lee, S. T. Ha, and Y. T. Hung, "Dye Waste Treatment," *Water*, vol. 3, no. 4, pp. 157–176, Feb. 2011.
K. Singh and S. Arora, "Removal of Synthetic Textile Dyes From Wastewaters: A Critical Review on Present Treatment Technologies," *Crit. Rev. Environ. Sci. Technol.*, vol. 41, no. 9, pp. 807–878, Apr. 2011.
M. Rosemal, M. Haris, and K. Sathasivam, "The Removal of Methyl Red from Aqueous Solutions Using Banana," *Am. J. Appl. Sci.*, vol. 6, no. 9, pp. 1690–1700, 2009.
A. Bruchet, J. P. Duguet, and I. H. M. Suffet, "Role of oxidants and disinfectants on the removal, masking and generation of tastes and odours," *Re/Views Environ. Sci. Bio/Techonology*, vol. 3, pp. 33–41, 2004.
B. Moyers and J. S. Wu, "Removal of organic precursors by permanganate oxidation and alum coagulation," *Water Res.*, vol. 19, no. 3, 1985.
J. J. Chen and H. H. Yeh, "The mechanisms of potassium permanganate on algae removal.," *Water Res.*, vol. 39, no. 18, pp. 4420–8, Nov. 2005.
M. D. Sabyasachi and M. Venkatesh, "Plantains and their postharvest uses: An overview," *Stewart Postharvest Rev.*, vol. 5, no. 5, pp. 1–11, 2009.
U. D. Akpabio and A. E. Akpakpan, "Pulp and Paper from Agricultural Wastes: Plantain Pseudostem Wastes and Screw Pine Leaves," *Int. J. Mod. Chem.*, vol. 2, no. 3, pp. 100–107, 2012.
A. Mun and H. Im, "The use of banana and abaca fibres as reinforcements in composites," *Biofiber Reinf. Compos. Mater.*, no. 51, p. 236, 2014.
X. Li, L. G. Tabil, and S. Panigrahi, "Chemical Treatments of Natural Fiber for Use in Natural Fiber-Reinforced Composites: A Review," *J. Polym. Environ.*, vol. 15, no. 1, pp. 25–33, Jan. 2007.
C. Alvarez, O. Rojas, B. Rojano, and P. Gañán, "Development of self-bonded fiberboards from fiber of leaf plantain: effect of water and organic extractives removal," *BioResources*, vol. 10, no. 1, pp. 672–683, 2015.
C. Gallego, F. Zuleta, C. Alvarez, and A. Restrepo, "2nd International Conference of Natural Fibers," in *Softening of colombian plantain fiber fabrics*, 2015, pp. 367–368.
K. S. Bharathi and S. T. Ramesh, "Removal of dyes using agricultural waste as low-cost adsorbents: a review," *Appl. Water Sci.*, vol. 3, no. 4, pp. 773–790, Jul. 2013.
A. G. Pérez, C. Sanz, J. J. Ríos, R. Olías, and J. M. Olías, "Effects of ozone treatment on postharvest strawberry quality," *J. Agric. Food Chem.*, vol. 47, no. 4, pp. 1652–1656, Apr. 1999.

Removal of Crude Oil Using a New Natural Fibre—*Calotropis procera*

J.H.O. Nascimento, M.P.G. Coelho, A.P. Silva, K.K.O.S. Silva, A.R.L. dos Santos, C.F. Campos, J.P.S. Morais and R.L. Sivam

Abstract Human evolution and the increase in population are responsible for the emergence of different energy sources and the development of new products and technologies. One of these sources, the result of the industrial revolution and of great importance to the development of humanity, is petroleum, a substance composed primarily of hydrocarbons, which give rise to several other products such as fuels, lubricants, polymers, solvents, cooking gas, asphalt, fertilizers, and paints, among others. However, the incorrect storage and transport of this product can cause leaks and spills that generate huge damages to the environment and the economy. In environmental accidents involving oil spills in an aqueous medium, the difficulty of removing the oil has been one of the major challenges for the environment. Some of the materials used for this function are the synthetic non-biodegradable textile fibres, resulting in another environmental residue. The present research has been developed by a multidisciplinary team; *Calotropis procera*, a new fibre has been studied to be used in the sorption of oil spills, and an

J.H.O. Nascimento (✉) · K.K.O.S. Silva · C.F. Campos · R.L. Sivam
Chemical Processes, Nanoscience and Functional Textiles Group,
Department of Textile Engineering, Center of Technology,
Federal University of Rio Grande do Norte, Natal-RN, Brazil
e-mail: heribertoo@ufrnet.br

M.P.G. Coelho · A.P. Silva · R.L. Sivam
Postgraduate in Mechanical Engineering, Center of Technology,
Federal University of Rio Grande do Norte, Natal-RN, Brazil

A.R.L. dos Santos
Federal Institute of the Sertão of Pernambuco-Petrolina Campus,
Pernambuco, Brazil

J.P.S. Morais
Embrapa Tropical Agroindustry, Fortaleza-CE, Brazil

J.P.S. Morais
Department of Plant and Soil Science, Texas Tech University, Lubbock, TX, USA

R. Fangueiro and S. Rana (eds.), *Natural Fibres: Advances in Science and Technology Towards Industrial Applications*, RILEM Bookseries 12,
DOI 10.1007/978-94-017-7515-1_9

intensive research has been carried out to explore the possible potential of *Calotropis procera* fibres as bioadsorbents of petroleum in water. The choice of these fibers is due to their surface properties such as oleophylic and hydrophobic surface, their buoyancy and yet, being biodegradable natural polymers derived from the Brazilian Northeast region. An experimental planning was done and response surface methodology (RSM) was used by *Design Expert* software. The results were statistically efficient, obtaining a $R^2 = 0.9995$. The fibers showed high adsorption efficiency, removing more than 90 % petroleum in water in both static and dynamic state.

Keywords Bioadsorbents · *Calotropis procera* · Crude oil · Oleophylic · Textile fibres

Introduction

Oil is a raw material known since ancient times, as bitumen. It originates from organic waste of plants and animals that suffered decay during many centuries and that, when comes in contact with environmental temperatures and pressures, are transformed into oil. Petroleum is an oily substance, flammable, less dense than water and with color ranging from light brown to black. Today, oil is a commodity which is highly essential for industrial development, although much research has been developed so that in the future, it can be replaced by environmentally favourable energy sources.

Oil spills can bring great economic losses, providing environmental problems and potentially leading to economic crises. When this occurs in the ocean and lakes, causes contamination of coastlines, rivers, cliffs and mangrove swamps, affecting the local wildlife, environment, fishing and tourism site (Xing et al. 2015).

Oil spills in the environment can occur due to various causes such as hurricanes, earthquakes, vandalism, illegal eviction, mechanical equipment leaks, and accidents, among others. However, the larger and more frequent natural disasters are related to the stages of drilling, transport and storage of the product where numerous leaks are caused by mechanical and human failures. In this context, various methods are being studied and developed to assist in the cleaning of contaminated areas (Wilson et al. 2015).

The method of in situ burning is one such technique in which the combustion of oil in water reduces its concentration (Bruce et al. 1990). Another method is bioremediation which uses organisms such as fungi, bacteria and yeast that degrade the oil complex substances into smaller substances for their food (Bragg et al. 1994). Dispersing agents are also used for the removal of spilled oil (Pietroski et al. 2015).

Petroleum adsorption using different adsorbents is a low-cost and efficient method. One of the most used methods is the adsorption using nanostructured materials as: Surfactant-Loaded halloysite clay nanotube (Nyankson et al. 2015), graphene based sponge (Cho et al. 2015) and multi-wall carbon nanotubes immobilized on the surface of polyurethane foam (Keshavarz et al. 2015).

In addition to the synthetic materials, organic minerals such as perlite, exfoliated graphite, vermiculites, zeolites, silica aerogel, diatomite, are also being used as adsorbents; however, these have little buoyancy, which is important for its use in adsorption of spilled oil.

Developing adsorbent materials of low cost such as natural and synthetic textile fibres have also been the objective of research and development. Some of them are already being commercially used for this purpose, such as polypropylene fibres, polyurethane, polystyrene and polyester, which are synthetic materials with great affinity with the oil and are hydrophobic; but these materials are not biodegradable and generate waste after use (Wei et al. 2003; Wrześniewska-Tosik et al. 2012; Lin et al. 2013).

Due to some disadvantages of existing products, there exists a strong demand for materials which are biodegradable, abundant, low-cost and easy to use. One such material which has these advantages is the cellulose fibres such as Kapok (Nascimento et al. 2015; Wang et al. 2012a, b; Rengasamy et al. 2012), Banana fibres (Teli and Valia 2013), Bagasse sugarcane fibres (Bayat et al. 2005), EPFB fibres (Empty Palm Fruit Bunch) (Idris et al. 2014), Cellulose fibres from Corn Straw (Wang et al. 2015), Curaua fibres (Elias et al. 2015), Cotton fibres (Deschamps et al. 2003; Wang et al. 2015a, b) and so on.

Among these natural fibers, *Calotropis procera* (*silk flower*) is also a potential candidate to be explored as fiber adsorbent. However there is no study that has been carried out for this purpose. Only its root and leaf powder has been disclosed as adsorbent of Zn (II), Cu (II), copper and methylene blue dye (Vinod et al. 2011; Mubeen et al. 2010; Oyelude and Owusu 2011). The use of *Calotropis procera* (*silk flower*) has been an attractive and technically viable option for the Northeast region of Brazil in the diet of confined lambs. The substitution of the commonly used foraging broom corn hay (Sorghum bicolor L) by this led to good results on the growth and quality of the lamb meat (Madruga et al. 2008).

In the present work, the potentiality of *Calotropis procera* (*silk flower*) fibres as bioadsorbents in petroleum/water system was analyzed. The fibre was characterized by different physico-chemical and microstructural analysis. The adsorption kinetics that dominate this thermodynamic system was evaluated and an experimental planning was used to assess the variables using Design Expert® software.

Experimental Part

Materials

Calotropis procera Fibres

The *Calotropis procera* fibres (silk flower) were provided by Federal Institute of the Sertão of Pernambuco, Petrolina Campus, located in northeastern Brazil.

The fibres were manually separated from their hulls and seeds (Fig. 1) and the size and diameter of the fibres were observed through an Optical Microscope and Scanning Electron Microscope (SEM) TM 3000 equipment.

The fibres were analyzed by Fourier transform infrared spectroscopy (FTIR). FTIR was conducted on a thin and translucent sample to obtain spectra in the infrared region.

Water Drop Absorption Test

The measured contact angle between water, petroleum and fibre was evaluated by Drop SHPE analysis (DSASEM) The drop size was of 10 μl, which takes 2 or 3 s for each image. The model used was the Young-Laplace. The model that was used to determine the contact angle is provided in Eq. 1 (Young-Laplace equation).

$$\cos_{\theta} = \frac{\gamma_{sv} - \gamma_{sl}}{\gamma_{lv}} \quad (1)$$

where γ_{sv}, γ_{sl} and γ_{lv} represent the interfacial tensions of liquid (L), solid (s) and vapor (V) and θ is the contact angle. When the value of the angle is less than 90° it indicates that the surface is hydrophilic, when the angle is greater than 90° it indicates that the surface of the material is hydrophobic, if it is equal to 0° it indicates that the surface is fully wettable, and when is equal to or greater than 150°, it indicates that the surface is superhydrophobic.

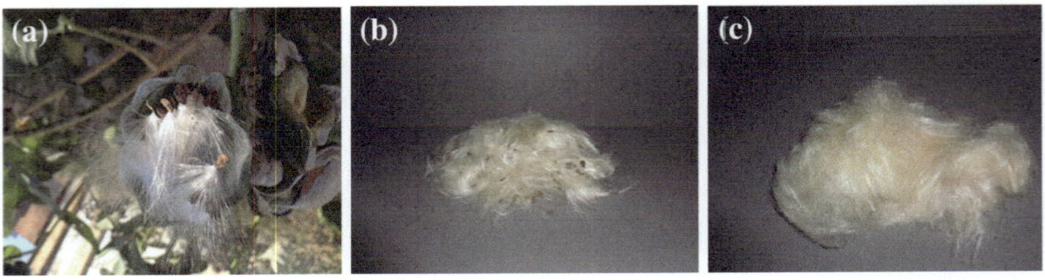

Fig. 1 The *Calotropis procera* (*silk flower*): **a** Plant and fruits. **b** Fibres and seeds. **c** Clean fibres

Petroleum

Petroleum used in this study was provided by Resources of the Central Brazil, offshore platform of Ponta do Mel, located in Areia Branca, Alagamar formation, Rio Grande do Norte, Brazil. The density of the oil was measured in a Pycnometer, and its viscosity was measured in a viscometer (Brookfield RS Rheometer).

Experimental Planning by Statistical Analysis

The experimental planning through design expert software has been used for the optimization of processes. This planning decreases costs and time for testing in research. In addition, this shows clearly the influence of each parameter set out previously as well as the statistical significance (Behnood et al. 2014).

The fibres were placed in a glass beaker containing 100 ml of distilled water, with the volume of petroleum of 1.16; 3.49 and 5.82 ml and fibre mass of 0.5; 0.3 and 0.1 g. The adsorption time studied ranged between 5, 10 and 15 min and temperature varied between 25 and 35 and 45 °C.

In planning, the variables of this work were 4, fiber mass, volume, time and temperature (Table 1). The possible interactions between the minimum and maximum values, as well as the central points of the variables were: +1 represent the maximum values; −1 represent the minimum values, and 0 represent the center points.

Kinetic Study

The fibres were weighed before and after the contact between fibres and petroleum. The amount of adsorbed petroleum was calculated using Eq. 2 (Wang et al. 2012a, b):

$$Q = \frac{m_f - m_0}{m_0} \tag{2}$$

Table 1 Experimental planning variables

Variables	−1	0	+1
Fibre mass (g)	0.1	0.3	0.5
Petroleum volume (ml)	1.16	3.49	5.82
Time (s)	5	10	15
Temperature (°C)	25	35	45

where
Q Quantity of petroleum adsorbed in the fibre;
m_f Fibre mass + petroleum mass adsorbed;
m_0 Fibre mass.

The removal rate (%) was determined using Eq. 3.

$$Removal\,rate\,(\%) = \frac{(C_i - C_f) * 100}{C_i} \tag{3}$$

where
C_i Initial concentration of petroleum (mL);
C_f final concentration of petroleum (mL).

The kinetic study was performed using the pseudo-second order model with 1 g of oil and 0,1 g of fibers in 1, 2, 4, 6 and 8 s. The ability of adsorption was determined using Eq. 4.

$$q = \frac{(C_i - C_f) * V}{m} \tag{4}$$

where
q adsorption capacity.
C_i Initial concentration of petroleum (mL);
C_f final concentration of petroleum (mL).
V volume of water + oil (mL).
m Fibre mass (g).

Results and Discussion

Microstructural Characterization of Calotropis procera *Fibre*

The results obtained in the analysis of optical microscopy and scanning electron microscopy (SEM) can verify the microstructure of *Calotropis procera* fibers. The images show that *Calotropis procera* fibres feature a hollow structure with internal diameter of 12.5 ± 2 μm and outside diameter of 16.6 ± 3 μm (Figs. 2 and 3). This hollow microstructure can help in fibre buoyancy.

When applying the drop of petroleum on *Calotropis procera* fibre, the oil spread rapidly and it was not possible to get the measurement of contact angle, due to the limitations of the equipment used. This phenomenon indicates that this type of

Fig. 2 *Calotropis procera* fibre image by optical micoscopy

Fig. 3 *Calotropis procera* fibre image by SEM

petroleum is quickly adsorbed by the fiber demonstrating the oleophilic behavior of fibre. Unlike petroleum, the drop of water exhibited a 123° contact angle (Fig. 4). The result of the contact angles between fiber and water proves that the fiber shows hydrophobic behavior.

In the FTIR curve of Calotro*pis procera* fibre (Fig. 5) the peak at 3347.4 cm^{-1} suggests the presence of hydroxyl groups (OH), and near 2920.2 cm^{-1}, the peaks are for the aliphatic carbon and hydrogen bonds (CH_2 and CH_3); the existence of these groups are associated with the presence of natural fats and waxes on the surface of the fiber due to the presence of n-alkanes and small amounts of fatty acids, alcohols, aldehydes, ketones and esters of n-alkyl. The peaks near 1742 cm^{-1} represent the presence of carbonyl, which characterizes the presence of ester and the link between carbon and oxygen (C–O) representing the acetyl groups responsible

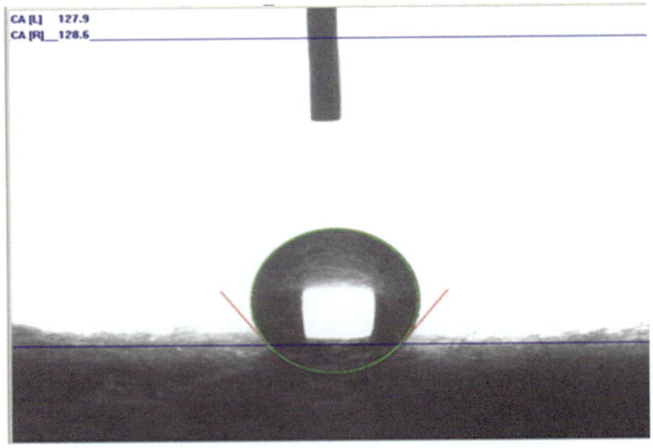

Fig. 4 Contact angle measurement between water and fibre

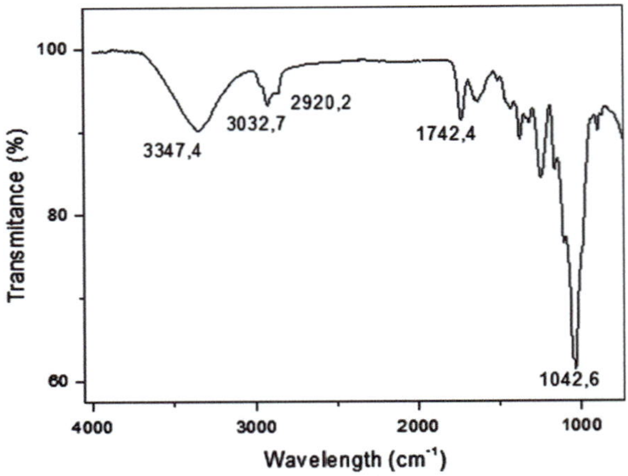

Fig. 5 FTIR spectrum of *Calotropis procera* fibre

for the presence of wax. In addition, the peak 1042.6 cm^{-1} is associated with the presence of cellulose, hemicellulose and lignin (Rengasamy et al. 2012).

Characterization of Petroleum

Petroleum was characterized as mild, its density was 0.86 $g \cdot cm^{-3}$ at 15.55 °C (60 ° F). The API degree was calculated and obtained a value of 33° which characterizes a light crude oil (Sarapardeh et al. 2015).

The viscosity was measured at temperatures of 20, 40, 60 °C and obtained results are presented in Table 2.

Table 2 Experimental planning variables

Temperature (°C)	Viscosity (Pas)	Viscosity (cp)
20	0.01575	15.75
40	0.0085	3.85
60	0.00375	3.75

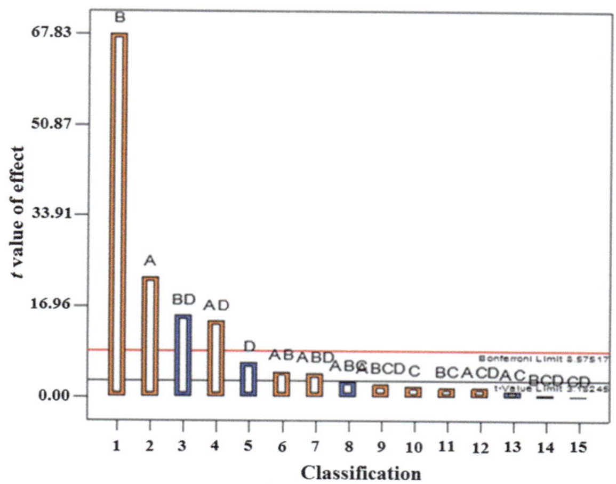

Fig. 6 Pareto chart showing the main effects and interactions of the parameters to the petroleum adsorbed as response

Statistical Analysis by RSM (Response Surface Methodology)

The process controls of adsorption were: (A) fibre mass, (B) oil volume, (C) time and (D) temperature. On Pareto chart image (Fig. 6), the orange color represents the positive effects and the blue color represents the negative effects.

It is observed that the parameters have really statistically significant influence. According to the Pareto chart the variables B and A, D, and interaction BD, AD, AB and ABD showed statistical significance of 95 %. The relevant variables in the process were oil volume (B) and fibre mass (A) both with positive influence in the process. The time (C) showed no significant influence on the adsorption process, which means that the oil is quickly adsorbed by fibres in the shortest time.

The analyses presented a standard deviation: 0.080; R^2:0.9995; R^2 adjusted: 0.9968; R^2 Predictable: 0.95; CV: 3.11; Accuracy: 51.211. According to the results obtained, the coefficients of determination R^2, R^2 adjusted, R^2 predictable were also close to 1, which indicates that the values obtained experimentally agree with the values of the statistical model proposed. The appropriate precision was also quite high exceeding 4 indicating that the square model is suitable. The second-order surface model can be used to predict the experimental values, with a probability error of less than 0.02.

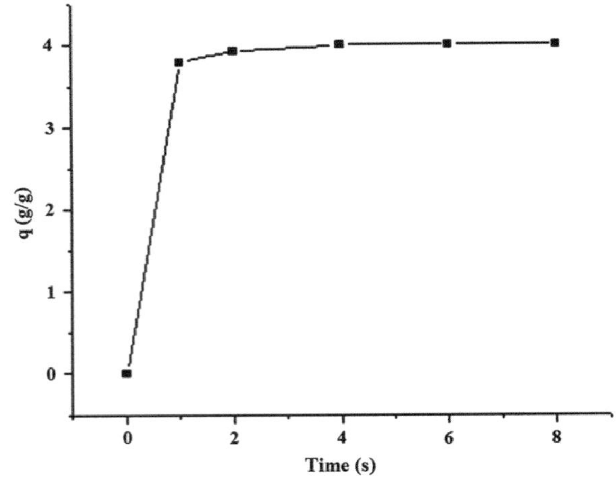

Fig. 7 Petroleum amount adsorbed on *Calotropis procera* fibre

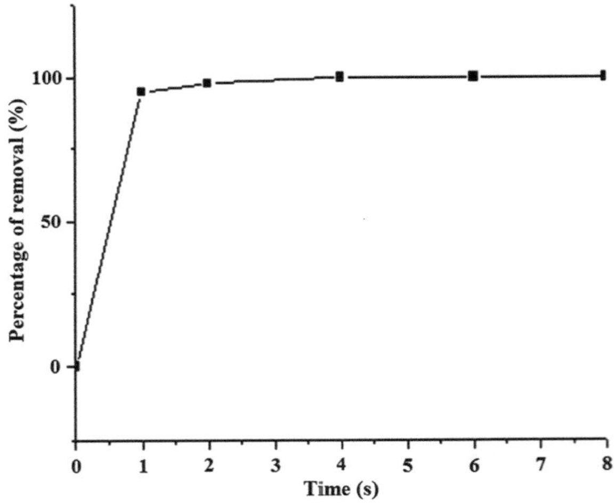

Fig. 8 Petroleum percentage adsorbed on *Calotropis procera* fibre

Figures 7, 8 and 9 showed that the adsorption of petroleum by *Calotropis procera* fibers occurs quickly. The kinetic balance and maximum removal (98.6 %) of petroleum were obtained in 4 s.

In Fig. 10, according to the result (R^2 = 0.999) the adsorption process can be described by the equation of pseudo second order, as shown in the equation $t/q^t = 0.24895t + 0.00768$, with a constant kinetics of $K_2 = 8.14\ gg^{-1}\ s^{-1}$.

Fig. 9 Image of petroleum adsorbed by *Calotropis procera* fibres

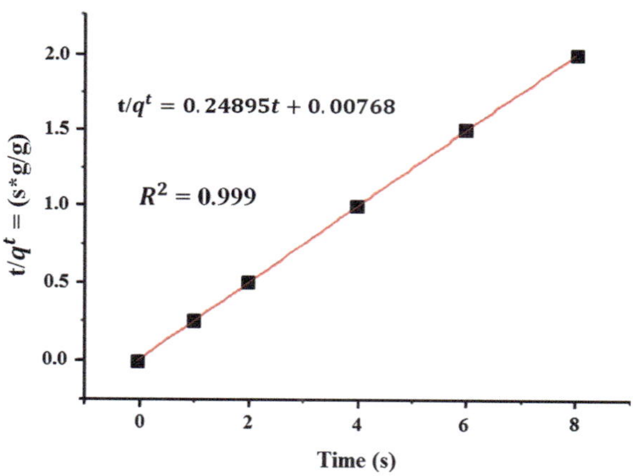

Fig. 10 Petroleum percentage adsorbed by the *Calotropis procera* fibre

Conclusion

The present work demonstrated that *Calotropis procera* (silk flower) fibres are suitable for the adsorption of petroleum in water because they are hydrophobic, oleophilic, biodegradable, possess good buoyancy and hollow physical structure that provides an adsorption on the surfaces and interiors of fibres. Additionally, they are low-cost and abundant in tropical and semi-arid regions of Brazil. The experimental planning used in the static system to model the data of fibre mass, volume, time and temperature in order to analyze the concentration of petroleum removed in relation to oil spill initially obtained a $R^2 = 0.9995$, which showed that the model does not require adjustments. In kinetic study it was found that the adsorption process can be described by the equation of pseudo second order, with the maximum removal (98.6 %) of petroleum obtained in 4 s.

References

Q. Xing, R. Meng, M. Lou, L. Bing, X. Liu. 2015. *Remote Sensing of Ships and Offshore Oil Platforms and Mapping the Marine Oil Spill Risk Source in the Bohai Sea.* Aquatic Procedia. Vol. 3. Pag. 127 – 132.

M. J. Wilson, S. Frickel, D. Nguyen, T. Bui, S. Echsner, B. R. Simon, J. L. Howard, K. Miller, J. K. Wickliffe. 2015. *A Targeted Health Risk Assessment Following the Deepwater Horizon Oil Spill: Polycyclic Aromatic Hydrocarbon Exposure in Vietnamese-American Shrimp Consumers.* Environmental Health Perspectives. Vol. 123. Nº2. Pag. 152-159.

B. A. Benner Jr., N. P. Bryner, S. A. Wise, G. W. Mulholland, R. C. Lao, M. F. Finga. 1990. *Polycyclic aromatic hydrocarbon emissions from the combustion of crude oil on water.* Environ. Sci. Technol. *Vol.* 24. Nº 9. Pag. 1418–1427.

J. R. Bragg, R. C. Prince, E. J. Harner, R. M. Atlas. 1994. *Effectiveness of bioremediation for the Exxon Valdez oil spill.* Nature. Vol. 368. Pag. 413 – 418.

J. P. Pietroski, J. R. White, R. D. De Laune. 2015. *Effects of dispersant used for oil spill remediation on nitrogen cycling in Louisiana coastal salt marsh soil.* Chemosphere. Vol.119. Pag. 562–567.

E. Nyankson, O. Olasehinde, V. T. John, R. B. Gupta. 2015. *Surfactant-Loaded Halloysite Clay Nanotube Dispersants for Crude Oil Spill Remediation.* Ind. Eng. Chem. Res. Vol. *54. Nº 38. Pag. 9328–9341.*

E-C. Cho, Y-S. Hsiao, K-C. Lee, J-H. Huang. 2015. *Few-layer graphene based sponge as a highly efficient, recyclable and selective sorbent for organic solvents and oils.* RSC Adv. Vol. 5. Pag. 53741-53748.

A. Keshavarz, H. Zilouei, A. Abdolmaleki, A. Asadinezhad. 2015. *Enhancing oil removal from water by immobilizing multi-wall carbon nanotubes on the surface of polyurethane foam.* Journal of Environmental Management. Vol. 157. Nº1. Pag. 279–286.

Q.F. Wei, R. R. Mather, A.F. Otheringham, R.D. Yang. 2003. *Evaluation of nonwoven polypropylene oil sorbents in marine oil-spill recovery.* Marine Pollution Bulletin. Vol. 46. Nº6. Pag. 780–783.

K. Wrześniewska-Tosik, O. Marchut-Mikołajczyk, T. Mik, Dorota Wieczorek, Michalina Pałczyńska. 2012. *Mats for Removing Technical Oil Contamination.* Fibres & Textiles in Eastern Europe. Vol. 20.Pag. 101-106.

J. Lin, F. Tian, Y. Shang, F. Wang, B. Ding, J. Yu, Z. Guo. 2013. *Co-axial electrospun polystyrene/polyurethane fibres for oil collection from water surface.* Nanoscale. Vol. 5. Pag.2745-2755.

J. Wang, Y. Zheng, A. Wang. 2012. *Effect of Kapok fiber treated with various solvents on oil absorbency.* Industrial crops and products.Vol.40. Pag.178-184.

R.S. Rengasamy, D. Dipayan, C. P. Karan. 2012. *Study of sorption behavior of filled and structured fiber assemblies made from Polypropylene, Kapok and Milkweed fibers.* Journal of Hazardous Materials. Vol. 186. Pag.526 – 532.

M.D. Teli, S.P. Valia. 2013. *Acetylation of banana fibre to improve oil absorbency.* Carbohydrate Polymers. Vol. 92. Pag. 328-333.

A. Bayat, S.F. Aghamiri, A. Moheb, G. R. Vakili-Nezhaad. 2005. *Oil Spill Cleanup from Sea Water by Sorbent Materials.* Chemical Engineering & Technology. Vol. 28. Pag. 1525-1528.

J. Idris, G. D. Eyu, A. M. Mansor, Z. Ahmad, and C. S. Chukwuekezie. 2014. *A Preliminary Study of Biodegradable Waste as Sorbent Material for Oil-Spill Cleanup.* The Scientific World Journal. Vol. Article ID 638687.

Z. Wang, J.P. Barford, C. W. Hui, G. McKay. 2015a. *Kinetic and equilibrium studies of hydrophilic and hydrophobic rice husk cellulosic fibers used as oil spill sorbents.* Chemical Engineering Journal. Vol. 281. Pag. 961–969.

E. Elias, R. Costa, F. Marques, G. Oliveira, Q. Guo, S. Thomas, F. G. Souza Jr. 2015. *Oil-spill cleanup: The influence of acetylated curaua fibers on the oil-removal capability of magnetic composites.* Journal of Applied Polymer Science. Vol. 132. Pag. 41732–41739

G. Deschamps, H. Caruel, M.E. Borredon, C. Bonnin, C. Vignoles. 2003. *Oil Removal from Water by Selective Sorption on Hydrophobic Cotton Fibers. 1. Study of Sorption Properties and Comparison with Other Cotton Fiber-Based Sorbents*. Environ. Sci. Technol. Vol. 37. Pag. 1013-1015.

J. Wang, G. Geng, A. Wang, X. Liu, J. Du, Z. Zou, S. Zhang, F. Han. *Double biomimetic fabrication of robustly superhydrophobic cotton fiber and its application in oil spill cleanup*. 2015b. Industrial Crops and Products. Vol. 77. Pag. 36–43.

J.H.O.Nascimento, A.P.Silva, M. P. G. Coelho, R. Ladchumananandasivam, K.K.O.S. Silva, C. F. Campos, F. R. Oliveira, J.P. M. Saraiva. 2015. *Optimizing the oil adsorption in kapok fibers by response surface methodology*. Book of Abstracts: 2nd International Conference on Natural Fibers – From Nature to Market. Pag. 371. Azores, Portugal. 27-29 April, 2015.

V. Vinod, D. Kailash, C. Suresh, L. Madan. 2011. *Adsorption Studies of Zn (II) ions from Wastewater using Calotropis procera as an Adsorbent* Research Journal of Recent Sciences. Vol. 1. Pag.160-165.

H. Mubeen, I. Naeem, A. Taskeen. 2010. *Phytoremediation of Cu (II) by Calotropis Procera Roots*. New York Science Journal. Vol. 3. Pag. 1-5.

E. O. Oyelude and U. R. Owusu. 2011. *Adsorption of Methylene Blue from aqueous solution using acid modified Calotropis procera leaf power*. Journal of Applied Sciences in Environmental Sanitation. Vol.6. Pag. 477-484.

M.S. Madruga, R.G. Costa, A.M. Silva, A.V.M.S. Marques, R.N. Cavalcanti, N. Narain, C.L.C. Albuquerque, G.E. Lira Filho. 2008. *Effect of silk flower hay (Calotropis procera Sw) feeding on the physical and chemical quality of Longissimus dorsi muscle of Santa Inez lambs*. Meat Science. Vol. 78. Pag. 469–474.

M. Behnood, B. Nasernejad, M. Nicarjar. 2014. *Application of experimental design in optimization crude oil adsorption from saline waste water using raw bagasse*. Journal of Central South University.V.21. Pag.684-693.

J. Wang, Y. Zheng, A. Wang. 2012. Investigation of acetylated kapok fibers on the sorption of oil in water. Journal of environmental sciences. Vol. 25. Pag. 246-253.

A.H. Sarapardeh, B. Aminshahidy, A. Pajouhandeh, S. H. Yousefi, S. A. H.Kaldozakh. 2015. *A soft computing approach for the determination of crude oil viscosity: Light and intermediate crude oil systems*. Journal of the Taiwan Institute of Chemical Engineers 000. Pag. 1–10.

Amazonian Tururi Palm Fiber Material (*Manicaria saccifera* Gaertn.)

Amanda Sousa Monteiro and Júlia Baruque-Ramos

Abstract Ubuçu (*Manicaria saccifera Gaertn.*) is a palm tree found in the Brazilian territory and throughout South America, mainly in the Amazon Forest biome. Tururi, a sack that protects the fruits and hangs from the branches of Ubuçu, is a vegetal fiber material with great use potential. Ecodesign studies are focusing on new sustainable products and there is a growing demand for the use of biodegradable and sustainable materials. The Ubuçu palm tree and tururi's characteristics; its collection, processing and traditional employments; its physicochemical properties and new applicability are presented in this chapter. The employment of tururi in traditional and new applications in design and industry is promising. The work with natural fibers involves social implications of major importance to society, such as the strengthening of rural cultures and the possibility of qualified products made from fibers by these communities.

Keywords Tururi fiber · Collection · Processing · Physicochemical properties · Applications

Introduction

There is a growing demand for the use of biodegradable materials such as natural fibers. A vegetal material fiber with great potential that fits these criteria is the Tururi, a sack that hangs from the Ubuçu palm tree (*Manicaria saccifera* Gaertn.) from the Amazon region. "The combination of the vast size of Amazonia, high levels of primary productivity, spatial variability in edaphic and climatic conditions

---

A.S. Monteiro (✉) · J. Baruque-Ramos
School of Arts, Sciences and Humanities, University of São Paulo, São Paulo-SP, Brazil
e-mail: amandasousamonteiro@usp.br

J. Baruque-Ramos
e-mail: jbaruque@usp.br

R. Fangueiro and S. Rana (eds.), *Natural Fibres: Advances in Science and Technology Towards Industrial Applications*, RILEM Bookseries 12,
DOI 10.1007/978-94-017-7515-1_10

and the relative stability of the ecosystem over a long time period has turned Amazonia into one of the most species-rich areas on Earth" (Malhado et al. 2013).

Nowadays, designers can select several materials and should consider their various technologies of transformation and processing. In the pursuance of environmental sustainability, their decisions to choose materials with a lower environmental impact can preserve resources for future generations, considering how very important the degree of renovation is. Ecodesign is a project model oriented by technological criteria and, through the redesign of products, aims to face the problems concerning environmental issues (Manzini and Vezzoli 2008).

The present chapter explores the Ubuçu palm tree and tururi's characteristics; its collection, processing and traditional employments; its physicochemical properties and new applicability. This study was conducted in Muaná-Pará (Brazil), located in the Brazilian Amazon biome and "Flor do Marajó" cooperative authorized all the observations and photographs taken.

The Ubuçu Palm Tree

The Amazon is home for the greatest diversity of palm trees in Brazil, where 35 out of 42 genera and 150 out of 193–208 species recognized in the country are found. Among the huge biodiversity of the Amazon forest, palm trees have great social and economic importance, as they serve as food source as well as raw material for the construction of houses and utilitarian objects by the traditional population of the region (Henderson et al. 1995).

The Ubuçu palm tree (*Manicaria saccifera* Gaertn) is found in the Brazilian Amazon Forest and it is very important in this context. Hanging from the palm tree, there are fruit bunches protected by an enclosure formed by a fibrous, flexible and durable fabric called Tururi. The fiber is popularly known by characteristics such as durability, impermeability and resistance (Corrêa 1984; Bondar 1964). A short stature of the Ubuçu palm tree and the sacks before their extraction are shown in Fig. 1a.

(a) (b)

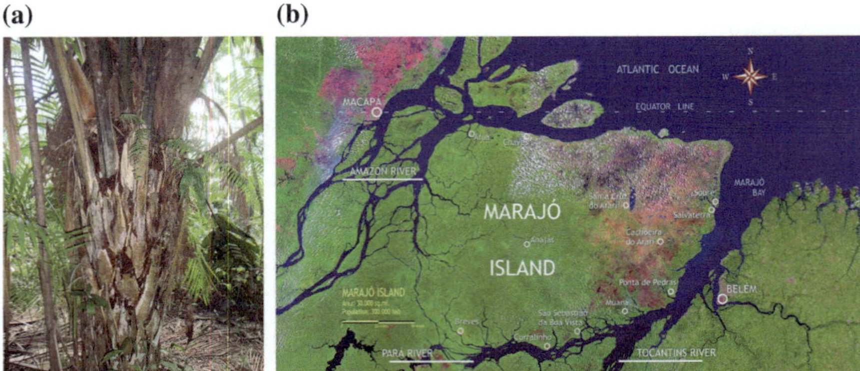

Fig. 1 **a** The Ubuçu palm and sacks on its upper part. **b** Pará state (Brazil). *Sources* **a** Photograph taken by authors; **b** Machado (2013)

The Ubuçu palm is native from Central America, Venezuela, Colombia, Guyana and Brazil. In the latter, it is native from the Amazon region and it is found mainly in the states of Amazonas, Pará and Amapá. In these states this palm tree is abundant in floodplains forests, a temporary flood of the forest caused by the constant flow of the rivers (Corrêa 1984; Bondar 1964).

Figure 1b shows a map indicating the localization of Muaná city (Marajó Island, state of Pará, Brazil), where the present study was conducted. The island of Marajo is located at the mouth of the Amazon and the Tocantins Rivers. It is distinguished by artificial hills, named "tesos", built on its pre-Columbian past by local Indians. With an area of approximately 40,100 km^2, it is the largest fluvial island in the world (Machado 2013).

The scientific name of the Ubuçu palm is *Manicaria saccifera* Gaertn. and its scientific classification is (Henderson et al. 1995): Kingdom—Plantae; Division—Magnoliophyta; Class—Liliopsid; Order—Arecales; Family—Arecaceae (Palmae); Gender—Manicaria; Species—*Manicaria saccifera* Gaertn.

The Ubuçu palm stipe has 3–4 m of height, while the entire palm tree reaches 3–6 m. It grows in vertical direction with a wavy surface and 30 cm in diameter. The leaves are 4–8 m long by 1.5 m wide. The fruit are covered with a spline outer coating. A ripe fruit weighs about 38 g. Each tree produces four fruit bunches per year, which is about 6–7 kg of fruit. Harvesting is done seasonally during the period from December to February (Balick 1979).

Traditional Amazonian population and craftsmen constantly employ the Ubuçu palm in different ways. Almost every part of it is used: mostly the sheets and the sack called "tururi" (Medina 1959). The palm-tree is a cheap alternative and is easily handled by the population (Maia 2009).

Collection, Processing and Traditional Employment

The extraction and processing of the fiber is carried out by Amazon local people using traditional methods. On visits to Muaná city (Brazil), radius of 5000 m from GPS S −1° 20′ 40.3506″ W−49° 17′ 45.3948″, the process of gathering the plant was observed. Using small boats, city residents go to parts of the forest where the Ubuçu palm grows, called lowland forests. After locating it, the tururi bunches are cut with machetes and sickles. If the Ubuçu palm is too high, they use ladders or climb the trees in order to reach the bags of tururi. Local people carry out this process, since the palm is dispersed in the forest, which is relatively dense. The extraction of the sacks from the Ubuçu palm tree and the tururi fiber are shown in Fig. 2a, b respectively.

After being collected in the floodplains, the sacks are placed in bundles and transported to the city by boat. Subsequently the fiber material passes through a classification process, followed by immersion in water. The classification of the tururi fiber material is made according to its size, width, color and quality. The fiber

Fig. 2 **a** Cutting of the sacks from the Ubuçu palm; **b** tururi sack with fruits inside presenting approximately 1 m in length and 10 cm in diameter. *Sources* **a**, **b** Photograph taken by authors

material does not have an official standard size or width (Leal 2012). After a period of about 3 h, the fiber material is washed in water to remove its impurities.

Afterwards, the processes of washing, drying and softening are carried out, in which the fiber material is then ironed with steam and stretched according to the its end use requirements. The natural fiber material is dark brown in color and, after bleached, can be dyed in various colors and hues. Its natural color can be modified by solar exposure or/and bleached by hydrogen peroxide (30 % v/v). Dyeing is carried out employing direct dyes, followed by drying at ambient conditions. There may be differences in the processing of the fiber depending on its end use. In Fig. 3a, b the process of bleaching and drying of the fiber are shown respectively.

This fiber is used in the fabrication of bags and utilitarian objects which are sold at large regional markets. However, the improvements in the processing and dyeing techniques are not disseminated among regional communities (Maia 2009).

Fig. 3 **a** Bleaching process of the fiber employing hydrogen peroxide (30 % v/v); **b** drying of the fiber at ambient conditions after dyeing. *Sources* **a**, **b** Photograph taken by author

According to Jardim et al. (2010), small pieces are used in the production of paper biomass, generally recycled from other productions. The process begins at the boiling step, carried out for softening of the fiber. Boiling lasts from two to three hours, but may vary according to the processor's crushing capacity in the subsequent phase. In order to improve the softening process, sodium hydroxide (NaOH) is applied (approximately 30 ml per liter of water). After boiling, vinegar is added (4 % acetic acid CH_3–COOH) in order to neutralize the alkalinity. After softened, the fiber is then shredded so it can be taken to the blender in small pieces and to produce the biomass.

The cooperative "Flor do Marajó" (Cooperaflomar) is composed by artisans from the municipality of Muaná on Marajó Island (PA, Brazil). The group was created in 1994 and established itself as an association since 2000. It designs and produces hats, handbags, jewelry, household items and more. Figure 4a, b show wallets and purses made by the "Flor do Marajó" cooperative.

Local craftsmen apply diverse techniques for making products using the tururi fiber. After observing products made by the artisans from the group "Flor do Marajó", it was possible to point out the most common working techniques used with the fiber, such as weaving and thermal press (Fig. 5a, b).

After washing, the fiber is stretched by ironing and sometimes it is cut into smaller pieces in order to avail most part of the material. Through modeling, bags are made, and the use of the fiber in conjunction with other materials such as woven fabrics (as lining or apparent part of the product) or animal leather (generally employing the sewing technique) is also common. A difficulty was observed in the use of some materials like leather, as it is not easy to acquire this kind of material in Pará (Brazilian state). Furthermore, the fiber is used in two ways: in its entirety and with different parts of the material united through the seam. Another technique is the thermal press to create objects such as hats.

The technique which uses tururi glued with a plain fabric is commonly applied by the cooperative. The fiber is stretched by ironing to increase its size (Fig. 6a) and glued together with woven fabrics. Figure 6b shows a bag made with the aforementioned technique in addition to sewing.

(a) (b)

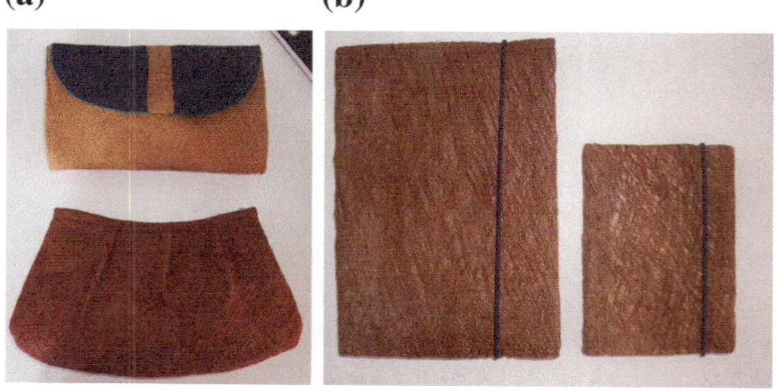

Fig. 4 "Flor do Marajó" cooperative products. **a** Wallets; **b** Purses. *Sources* **a**, **b** Photograph taken by authors

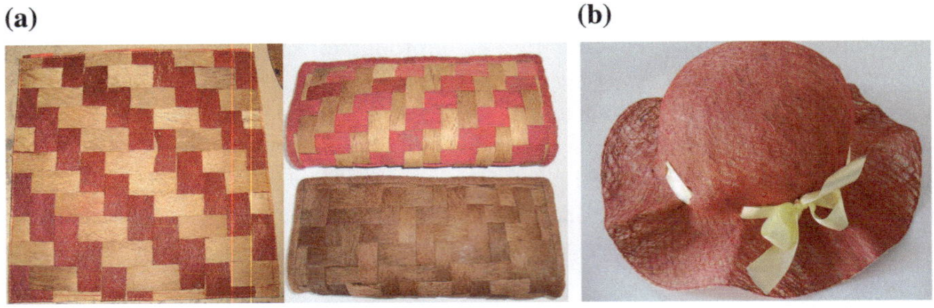

Fig. 5 Handicraft working techniques: **a** weaving; **b** thermal press. *Sources* **a**, **b** Photograph taken by authors

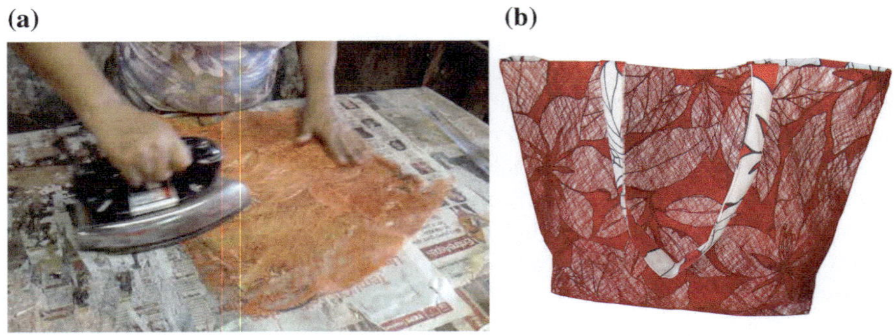

Fig. 6 **a** Technique to increase the size of tururi; **b** bag made with the described technique with tururi and woven fabric. *Sources* **a**, **b** Photograph taken by authors

These are the most common techniques performed by the cooperative, "Flor do Marajó". They are in accordance with the review on the employment of South American palm fibers, which indicates the tururi for applications such as crafts, hats, baskets and twisted (Isaza et al. 2013).

Physicochemical Properties and New Employments

The main physicochemical results related to tururi are presented as follows.

The average weight value (grams per square meter) of tururi reported by Porras et al. (2015) ranges from 366 to 583 g/cm^2. Duarte (2011) reported 246.37 g/m^2 and Oliveira and D'Almeida (2014), 204.7 g/m^2.

The tururi density reported by Porras et al. (2015) ranges from 0.84 to 1.07 g/cm^3; by Duarte (2011), 1.18 g/cm^3 and by Oliveira and D'Almeida (2014), 0.97 g/cm^3.

According to Reedy and Yang (2005), tensile tests express the behavior of the fibers when a deformation force is applied along the fibre axis. The values of main tensile properties of the tururi fiber are listed in Table 1.

By stretching tururi fibre (as shown in Fig. 6a), its properties are modified, as reported by Duarte (2011). According to this author, the weight of the tururi without stretching was 246.37; 151.30 g/m^2 when stretched 50 % more than its original size; and 104.71 g/m^2 when stretched 100 %. The maximum load without stretching was 1617.3 N; 558.3 N at 50 %; 309.5 N at 100 %. The tensile strength without stretching was 29.95 MPa; 12.27 at 50 %; and 9.38 MPa at 100 %.

Recently Porras et al. (2015) reported the characterization of *Manicaria saccifera* fiber by chemical composition analysis, infrared spectroscopy (FTIR) analysis, morphological studies (SEM), thermo-gravimetric analysis (TGA) and physical/mechanical studies. Their results for chemical composition are presented in Table 2.

In addition, according to Porras et al. (2015) the moisture content (regain) value is near 10 %. This value is comparable with the ones from lignocellulosic fibers such as jute and flax. Longitudinal (Fig. 7a) and cross microscopy (Fig. 7b) performed by the authors presented respectively, porosity and no microfibrils; and multicellular net structure with round cell lumens.

Considerable attention has been given to the extraction, processing and use of natural fibers as reinforcement of polymeric composite materials. The main advantages of natural fibers in comparison to synthetic fibers are their low cost, natural abundance, biodegradability and good specific mechanical properties (Fuentes et al. 2014). Natural materials are considerably more economical and their availability is also an important advantage. Composite materials reinforced with natural fibers have advantages, such as: constructional simplicity and versatility. Besides their high mechanical properties in terms of tensile strength and lightness,

Table 1 Tensile properties of tururi (*Manicaria saccifera*) fiber (Porras et al. 2015; Oliveira and D' Almeida 2014; Duarte 2011)

Reference	Load at break (N)	Tensile stress at maximum load (MPa)	Elongation at break (%)	Young's modulus (GPa)
Porras et al. (2015)	391	–	–	1.8–2.4
Oliveira and D'Almeida (2014)	432	–	9.35	–
Duarte (2011)	558.3	12.27	–	–

Table 2 Chemical composition of tururi (Porras et al. 2015)

Cellulose (%)	Hemicellulose (%)	Lignin (%)	Ash (%)	Extractives (%)
74.1	12	31.1	2.5	0.5

(a) (b)

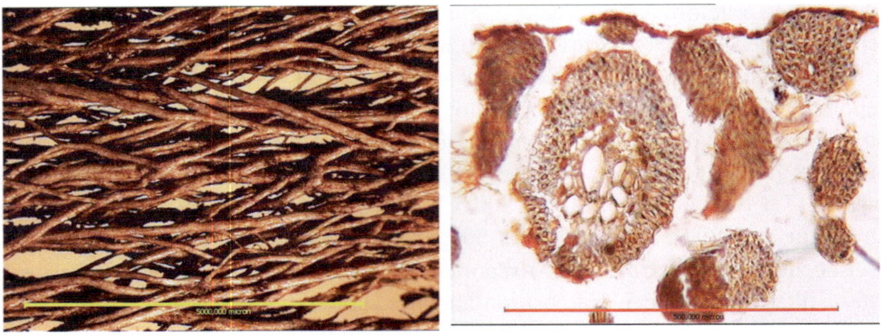

Fig. 7 Tururi structure. **a** Longitudinal microscopy (32×; 5.0 mm scale in the image); **b** Cross microscopy (320×; 500 μm scale in the image)

the natural fibers have advantages in terms of production costs, emissions of pollutants and consumption of energy for their production and disposal. The costs for the production and disposal at the end of their life are generally significantly lower than for an artificial composite material (Borri et al. 2013).

In recent years, a significant number of researches concerning tururi in the development of composites and other applications was performed.

Porras et al. (2015) reported the characterization of the cellulose fabric from the *Manicaria saccifera* palm as a possible reinforcement of composite materials. According to them, from the SEM analysis, globular protrusions that spread uniformly over the fiber were identified, which could help the mechanical interlocking with the resin. Also, the *Manicaria* fabric showed good thermal stability, low density, low moisture content and good tensile properties. Furthermore, their properties are comparable to most natural cellulose fabrics and some synthetic fabrics, such as fiber glass fabrics. In conclusion, these authors stated that the *Manicaria saccifera* fabric was observed to be a suitable candidate as natural reinforcement material for the development of biocomposites.

Oliveira and D'Almeida (2013) reported the mechanical behavior (in flexion and compression) of composites made with thermoset polymeric matrix and reinforced with tururi fiber. According to them, the results are comparable with the data from other polymer matrix composites with lignocellulosic fibers, which demonstrate the feasibility of using *Manicaria saccifera* fibers as reinforcement. The composite's wear behavior was also analyzed and its use as a parquet floor was identified as a possible application by those authors.

Saraiva (2013) developed a wooden laminated matrix composite reinforced with polyester and natural fibers, being tururi one of these fiber materials. The wood laminates were assessed at 90° and 0° to fiber orientation, respectively. The laminates made of wood/tururi without stretching and 100 % stretched presented tensile strength of 61.3 and 57.3 MPa, respectively.

Porras and Maranon (2012) reported the development and mechanical characterization of a core sandwich panel fabricated from both renewable resources, and biodegradable materials: Manicaria fiber (*Manicaria saccifera*) as reinforcement and Polilactic Acid (PLA) as matrix material. The potential of the Manicaria/PLA core sandwich panel for structural applications was evaluated by subjecting the core to compressive loads according to ASTM C365 standard. It was found that the average compressive strength and elastic modulus of the Manicaria/PLA sandwich core were 2.85 (±0.17) MPa and 32.68 (±3.68) MPa, respectively. In contrast, the mechanical properties of un-reinforced sandwich core were 1.28 (±0.45) MPa and 12.49 (±4.7) MPa, respectively. The results indicated that the mechanical properties of Manicaria fiber reinforced cores increased more than twice in comparison to those of unreinforced ones.

Duarte (2011) presented a manufacturing and characterization of a composite material with polyester resin reinforced with tururi fiber.

An alternative employment of tururi was presented by Melo et al. (2014), who investigated the removal of Cd^{2+}, Cu^{2+}, Ni^{2+}, and Pb^{2+} ions from aqueous solutions using tururi fiber as an adsorbent under both batchwise and fixed-bed conditions. It was discovered that the modification of tururi fibers with sodium hydroxide increased the adsorption efficiencies of all metal ions studied.

Despite its diverse uses and a number of academic researches, this fiber is still little known in Brazil and worldwide. Its use is limited to regions where the palm tree grows. Until this moment, the present authors have no knowledge on the effective large scale industrial processes or applications employing this particular fiber material. However, despite the fact that the tururi fiber is a renewable resource, it is important to take into account its limitations about its collection in an eventual large scale and/or its industrial employment.

Conclusion

According to the observed results, tururi fiber presents good application potential in different types of products, such as textiles, fashion, crafts and composites, which could be used in furniture articles or building materials.

Is very important to observe and preserve the work of traditional population with the fiber. Craft cooperatives such as "Flor do Marajó" have great potential and it is believed that the knowledge of technical information about the tururi fiber will generate positive impacts on its structure and functioning, as well as for designers and industry in general.

Acknowledgments We gratefully acknowledge **CAPES** (Coordination for the Improvement of Higher Education Personnel of Brazilian Education Ministry), **USP** (University of São Paulo) and **"Flor do Marajó" (Cooperaflomar)** cooperative artisans.

References

BALICK, M.J. Amazonian oil palms of promise: a survey. **Economic Botany**, v.33, n.1, p.11-28, 1979.

BONDAR, G. **Palmeiras do Brasil**. Secretaria da Agricultura do Estado de São Paulo – Instituto de Botânica, São Paulo, 1964.39p.

BORRI et al., Reinforcement of wood with natural fibers. **Composites Part B-Engineering**, v.53, p. 1-8, 2013.

CORRÊA, M.P. **Dicionário das plantas úteis do Brasil e das exóticas cultivadas**. Rio de Janeiro, Imprensa Nacional, 1926-1978. 6 v. ilust. Reedited in 1984 by Ministério da Agricultura – Instituto Brasileiro de Desenvolvimento Florestal. 4329 p.

DUARTE, C. **Production and Characterization of polyester resin composite material reinforced by tururi fiber extracted from Ubuçu palm**. Dissertation (Master in Mechanical Engineering). Institute of Technology, Federal University of Pará, Belém, 2011. 76p.

HENDERSON, A.; GALEANO, G. BERNAL, R. **Field guide to the Palms of the Americas**. Princeton University Press, New Jersey, 1995. 414p.

ISAZA, C.; BERNAL, R.; HOWARD, P. Use, production and conservation of palm fiber in South America: A review. **Journal of Human Ecology**, v. 42, n. 1, p. 69-93, 2013.

JARDIM, N.R; SANTOS, M.R; FERREIRA, M.F; RODRIGUES, N.F; RABELO, H.J.C. Desenvolvimento de tecnologias voltadas à fibra de tururi aplicada ao projeto de eco embalagens. In: **9º P&D Design Congresso Brasileiro de Pesquisa e Desenvolvimento em Design**, 2010, São Paulo. 9º P&D Design Congresso Brasileiro de Pesquisa e Desenvolvimento em Design. São Paulo: Universidade Anhembi Morumbi, 2010. v. 1.

LEAL, R.M.D.C. The Diversity of Tururi Fiber: applicability and usability. In: **8th Colloquium of Fashion,** Rio de Janeiro, 2012. Proceedings of the 8th Conference of Fashion. Rio de Janeiro 2012.

MACHADO, J. **Ilha do Marajó – Na Foz do Amazonas. A Viagem dos Argonautas.** (May 12, 2013). Available in http://aviagemdosargonautas.net/2013/05/12/ilha-do-marajo-na-foz-do-amazonas-por-joao-machado/. Accessed June 11, 2015.

MAIA, F. A. **Fibras da Amazônia na produção de moda: uma proposta de indicação geográfica**. Aparecida, SP: Ideias & Letras, 2009. 104p.

MALHADO et al., The ecological biogeography of Amazonia. **Frontiers of Biogeography**, v.5, n.2, 2013.

MANZINI, E.; VEZZOLI, C. **O Desenvolvimento de Produtos Sustentáveis: os requisitos ambientais dos produtos industriais** ("Lo sviluppo di prodotti sostenibili: i requisiti ambientali dei proditti industriali"). Traduction by Astrid de Carvalho. São Paulo: Editora da Universidade de São Paulo, 2008. 367p.

MEDINA, J. C. **Plantas Fibrosas da Flora Mundial**. Printed by Instituto Agronômico de Campinas, 1959. 913 p.

MELO, D. Q. et al. Removal of Cd^{2+}, Cu^{2+}, Ni^{2+}, and Pb^{2+} ions from aqueous solutions using tururi fibers as an adsorbent. **Journal of Applied Polymer Science**, v. 131, n. 20, 2014.

OLIVEIRA, A. K. F.; D'ALMEIDA, J. R. M. Description of the mechanical behavior of different thermoset composites reinforced with *Manicaria saccifera* fibers. **Journal of Composite Materials**, p. 0021998313484622, 2013.

OLIVEIRA, A. K. F.; D'ALMEIDA, J. R. M. Characterization of Ubuçu (*Manicaria saccifera*) Natural Fiber Mat. **Polymers from Renewable Resources**, v. 5, n. 1, p. 13, 2014.

PORRAS, A.; MARANON, A. Eco friendly core sandwich panel reinforced with Manicaria fiber and PLA matrix. In: **Proceedings of ECCM15 - 15th European Conference On Composite Materials**, Venice, Italy, 24-28 June 2012.

PORRAS, A.; MARANON, A.; ASHCROFT, I. A. Characterization of a novel natural cellulose fabric from *Manicaria saccifera* palm as possible reinforcement of composite materials. **Composites Part B: Engineering**, v. 74, p. 66-73, 2015.

REDDY, N.; YANG, Y. Properties and potential applications of natural cellulose fibers from cornhusks. **Green Chemistry**, v. 7, n. 4, p. 190-195, 2005.

SARAIVA, M. P. **Laminado de madeira reforçado com fibras naturais**. Dissertation (Master in Mechanical Engineering). Institute of Technology, Federal University of Pará, Belém, 2013. 71p.

FUENTES et al., Equilibrium contact angle measurements of natural fibers by an acoustic vibration technique, **Colloids and Surfaces A: Physicochemical and Engineering Aspects,** v. 455, p.164–173, 2014

Nanoindentation Measurements of Jute/Poly Lactic Acid Composites

Vijay Baheti and Jiri Militky

Abstract The aim of present work was to use jute fibrous waste from textile industries for the preparation of cellulose nanoparticles/nanofibrils suitable as reinforcement of biodegradable composites for applications in food packaging, agriculture mulch films, automotive plastics, etc. In the present study, waste jute fibers were wet pulverized to the scale of nanofibrils of 50 nm diameter using high energy planetary ball milling for 3 h. Subsequently, 3 wt% of jute fibers of millimeter, micrometer and nanometer scales were incorporated into poly lactic acid (PLA) matrix to prepare composite films by solvent casting. The reinforcement potentials of fibers at different sizes from millimeter to nanometer were investigated from the improvements in elastic modulus of composite films based on nanoindentation measurements. The increase in elastic modulus was observed to be 65.21, 25.40 and 13.04 % for composite films of nanofibrils, microfibrils and milifibers, respectively compared to neat PLA film. These improvements were attributed to the increased interaction between fibers and matrix as well as to the increased crystallinity of PLA in composites.

Keywords Textile waste · Recycling · Nanocellulose · Nanofibrils · Green composites · Nano-indentation

Overview of Present Situation

The increased demands of textiles brought the challenges to dispose significant amount of wastes generated during the processing (Wang 2006; Horrocks 1996). In the context of environment protection and current disposal of textile wastes, it becomes essential to recover useful products from them for economic reasons. Traditionally, textile wastes are converted to individual fiber stage through cutting,

V. Baheti (✉) · J. Militky
Department of Material Engineering, Faculty of Textile Engineering, Technical University of Liberec, Liberec, Czech Republic
e-mail: vijaykumar.baheti@gmail.com

R. Fangueiro and S. Rana (eds.), *Natural Fibres: Advances in Science and Technology Towards Industrial Applications*, RILEM Bookseries 12,
DOI 10.1007/978-94-017-7515-1_11

shredding, carding, and other mechanical processes. The fibers are then rearranged into products for applications in garment linings, household items, furniture upholstery, automotive carpeting, automobile sound absorption materials, carpet underlays, building materials for insulation and roofing felt, and low-end blankets (Wang 2006; Horrocks 1996). However, due to recent increase in competition and reduced profit margins in these industries, it has became important to search for new recycling techniques of waste textiles in order to utilize them for high end applications. One such interesting way is to separate the nanofibrils or nanocrystals from the textile wastes and subsequently incorporate them as fillers into high performance composite materials (Yuen et al. 2009; Khalil et al. 2012; Klemm et al. 2006).

Cellulose fibers are popularly used in the textile industry due to their high aspect ratio, acceptable density, good tensile strength and modulus (Yuen et al. 2009). These properties make them attractive class of textile materials traditionally used in manufacture of yarn by spinning process. But, due to certain limitations of the spinning process, shorter fibers (i.e. less than 10 mm) generated during mechanical processing are not suitable to reuse in yarn manufacture and consequently result into the waste (Yuen et al. 2009). In order to exploit the intrinsic mechanical properties of short cellulose fibers in textile industries, the idea of separating nanofibrils or nanocrystals of cellulose could provide interesting applications in other fields. The previous studies have reported the remarkable properties of cellulose materials at nanoscale dimensions (Khalil et al. 2012; Klemm et al. 2006). The extreme improvement in mechanical properties, in the range of 130–160 GPa, of cellulose nanofibrils is attributed to their increased rigidity obtained from parallel arrangement of molecular chains present without folding (Klemm et al. 2006). As a result cellulose nanofibrils are increasingly used in applications of reinforced biodegradable nanocomposites, foams, aerogels, optically transparent functional materials, and oxygen-barrier layers (Dufresne et al. 1999).

The utilization of different types of cellulosic wastes has been studied in the past in order to obtain cellulose nanofibrils at reasonably lower cost. The variety of agricultural wastes like coconut husk fibers (Rosa et al. 2010), cassava bagasse (Pasquini et al. 2010), banana rachis (Zuluaga et al. 2009), mulberry bark (Li et al. 2009), soybean pods (Wang and Sain 2007), wheat straw and soy hulls (Alemdar and Sain 2008) and cornstalks (Reddy and Yang 2005), woods are investigated for extraction of cellulose nanofibrils. However, there is no information available in literature on utilization of cellulosic wastes in textile industries in spite of large amount of short fibers are generated during the mechanical processing of yarn manufacture.

Although the cellulose nanostructures have a great potential for reinforcement into biopolymers, the major challenge in order to use them is the extraction. The variety of techniques like acid hydrolysis (Liu and Yuan 2010), enzymatic hydrolysis (Satyamurthy et al. 2011), ultrasonication (Li et al. 2012), high pressure homogenization (Leitner et al. 2007), etc. have been employed. However, most of these techniques used in the extraction are time consuming, expensive in nature and low in yields (Klemm et al. 2006). The commonly used strong acid hydrolysis method has a number of important drawbacks such as potential degradation of cellulose, corrosivity and environmental incompatibility (Thomas and Pothan 2008). In order to promote the commercialization of cellulose nanofibrils, the development of more flexible and industrially viable processing technique is needed.

Present State of Problems

Over the last two decades, reinforcement potentials of lignicellulose fibers have been investigated in numerous studies of biocomposites made from poly lactic acid (PLA) (Lunt 1998; Jonnobi et al. 2010; Petersen et al. 2001; Petersson and Oksman 2006; Sanchez-Garcia et al. 2008). However, the properties are found not consistent to meet demands of value added applications. This pattern of non consistent improvements in properties of lignocellulosic fiber composites are explained due to the variations in properties of lignocellulosic fibers derived from different resources (Klemm et al. 2006; Mussig and Stevens 2010; Dufresne et al. 1999; Rosa et al. 2010). As the individual lignocellulosic fibers are made from the packing of several micro/nano cellulose fibrils together, the number of defects present in the structure varies from source to source. One of the basic ideas to further improve fiber and composite properties is to eliminate the macroscopic flaws by disintegrating the fibers, and separating the almost defect-free, highly crystalline nanofibrils. This can be achieved by exploiting the hierarchical structure of the natural fibers (Yuen et al. 2009; Khalil et al. 2012; Klemm et al. 2006).

In theory, the performance of reinforced materials relies on the efficiency with which mechanical stress is transferred from an external energy source to the reinforcing phase through the matrix (Lunt 1998). The efficiency of transfer is a function of the amount and quality of the interfacial area between the reinforcing agent and the matrix. For composite materials, the interphase region is usually thin but plays an important role in their overall mechanical properties. The interphase is the region between the surface of the reinforcement and the polymer matrix where the chemistry is different from that of the bulk matrix (Lunt 1998). The interphase has a heterogeneous nature, with different morphological features, chemical compositions and mechanical properties from those of the reinforcing fiber or the matrix polymer.

Nowadays as the optical-electronic industry developed, polymer materials are gradually increasing their importance in the area of thin film structures. For such kind of thin film structures, conventional test techniques, e.g., tensile, compression and bending tests, are inapplicable due to the difficulties of meeting sample requirements for such tests (Li and Bhushan 2002). Therefore, there is a need to propose novel techniques to study the mechanical properties of thin films, or small volumes of materials. Nanoindentation technique has been developed for this purpose. However, relatively few nanoindentation studies of polymeric film have been presented. Recently, Beake and Leggett (2002) studied the nanomechanical behavior on uniaxially and biaxially drawn poly (ethylene terephthalate) films using nanoindentation and nanoscratch tests. Ngan and Tang (2002) studied the viscoelastic effects during in depth-sensing indentation. The objective of present study is to evaluate the reinforcement potentials of garneted jute fibers of millimeter scale (GJF), dry milled jute fibers of micrometer scale (DMJF) and wet milled jute fibers of nanometer scale (WMJF) in PLA matrix using nanoindentation testing.

Experimental Methods

Materials

Short waste jute fibers were obtained from India. The fibers were measured to have a density of 1.58 g/cm^3, modulus of 20 GPa, tensile strength of 440 MPa and elongation of 2 %. The chemical composition of fibers was reported as cellulose (60 %), hemicelluloses (20 %), lignin (10 %) and others (10 %). Poly lactic acid (PLA) was purchased from NatureWorks LLC, USA through local supplier Resinex, Czech Republic. The PLA had a density of 1.25 g/cm^3 and the average molecular weight (Mw) of 200,000. The chloroform which was used as solvent, purchased from Thermofisher, Czech Republic.

Pulverization of Jute Fibers to Nanofibers

Removal of non-cellulosic contents. Chemical pre-treatment of waste jute fibers was carried out before wet pulverization, sequentially with 4 wt% sodium hydroxide (NaOH) at 80 °C for 1 h and with 7 g/l sodium hypochlorite (NaOCl) at room temperature for 2 h under pH 10–11. Subsequently, the fibers were antichlor treated with 0.1 % sodium sulphite at 50 °C for 20 min.

Preparation of nanoscale jute fibers. High-energy planetary ball mill of Fritsch pulverisette 7 was used for the separation of individual jute fibrils. The pulverization was carried out using sintered corundum container of 80 ml capacity and zirconium balls of 10 mm diameter. During wet milling, 10 mm diameter balls were replaced by 3 mm diameter balls. The ball mill was loaded with ball to material ratio of 10:1. The rotational speed of the disc was kept at 850 rpm with reverse rotation of containers. The jute microfibrils (DMJF) were obtained after one hour of dry milling, whereas jute nanofibrils (WMJF) were obtained after three hour of wet milling in distilled water [12]. The jute nanofibrils were separated from water by centrifugation at 4000 rpm and simultaneously transferred in solvent isopropanol to avoid hornification during drying. For comparison purpose, jute millifibers (GJF) were also prepared by garneting of jute fibrous wastes. At the end of wet milling, jute particles were separated from water by centrifugation at 4000 rpm and simultaneously transferred in solvent isopropanol to avoid hornification during drying.

Characterization of nanoscale jute fibers. The particle size distribution after each hour of milling was studied on Malvern zetasizer nano series, based on dynamic light scattering (DLS) principle of brownian motion of particles. The measured quantity of DMJF and WMJF was dispersed in deionized water using ultrasonication for 5 min with the help of Bandelin ultrasonic probe. The refractive index of 1.52 was used to calculate the particle size.

The morphologies of GJF, DMJF and WMJF were observed on scanning electron microscope of TS5130-Tescan at 30 kV accelerated voltage and on field emission scanning electron microscope of Zeiss at 5 kV accelerated voltage. The amount of 0.01 g of DMJF and WMJF was dispersed in 100 ml acetone and then a drop of the dispersed solution was placed on aluminum foil, which was gold coated after drying. The specific surface area of DMJF and WMJF was further measured from N2 adsorption–desorption isotherms at 77.35 K using Quantachrome Instruments. At the end, comparison of all mentioned techniques was done for precise estimation of the size.

Preparation of Nanocomposite Films

Nanocomposite films at 3 wt% filler content were prepared by mixing the calculated amount of GJF, DMJF and WMJF with 5 % PLA in chloroform solution using a magnetic stirrer. The stirring was performed at room temperature for 3 h. The composite mixture was further ultrasonicated for 10 min on Bandelin Ultrasonic probe mixer with 50 horn power. The final mixtures then cast on a Teflon sheet. The films were kept at room temperature for 2 days until they were completely dried and then removed from the Teflon sheet. One neat PLA film was also prepared without addition of fillers for comparison purpose.

Testing of Nanocomposite Films

Nanoindentation testing. The nanoindentation experiments were performed using a CSM Instruments Switzerland (NHTX S) with a diamond probe. A diamond Berkovich indenter with a tip radius of 50 nm and a maximum load of 0.50 mN were used for evaluating the Young's modulus and hardness of the films. The loading and unloading rates were fixed at 1 mN/min whereas the approach speed was kept at 2000 nm/min. Subsequently, the images of indentation marks were observed under optical microscope attached to the nanoindenter. Total of 10 samples were used to characterize each material.

Tensile testing. Tensile testing was carried out using a miniature material tester Rheometric Scientific MiniMat 2000 with a 1000 N load cell at a crosshead speed of 10 mm/min. The samples were prepared by cutting strips from the films with a width of 10 mm. The length between the grips was kept at 100 mm. Total of 10 samples were used to characterize each material.

Dynamic mechanical analysis (***DMA***). Dynamic mechanical properties of the nanocomposite films were tested on DMA DX04T RMI instrument, Czech Republic in tensile mode. The measurements were carried out at constant frequency of 1 Hz, strain amplitude of 0.05 %, temperature range of 35–100 °C, heating rate of 5 °C/min and gap distance of 30 mm. The samples were prepared by cutting strips from the films with a width of 10 mm. Four samples were used to characterize each material.

Results and Conclusions

Effect of Milling Condition on Particle Size Reduction of Jute Fibers

Under 1 h of dry milling, jute fibers were pulverised to microparticles with average size of 1480 nm in wider particle size distribution as shown in Table 1, Figs. 1a and 2a. The reason behind multimodal distribution of particles was due to increase in temperature within the mill because of continuous impact of balls. The increased temperature of mill rendered the jute particles to undergo cold welding and deposited a layer on the surface of container and balls as milling progressed. The growth of deposited layer on the milling media changed the impact force of balls on

Table 1 Particle size from DLS measurement for different milling conditions

Sample name	First peak (nm)	Second peak (nm)	Third peak (nm)	Z-average (nm)	Polydispersity index
One hour dry mill	1821	4889	297.5	1480	0.407
One hour wet mill	448.8	2365	0	640.1	0.420

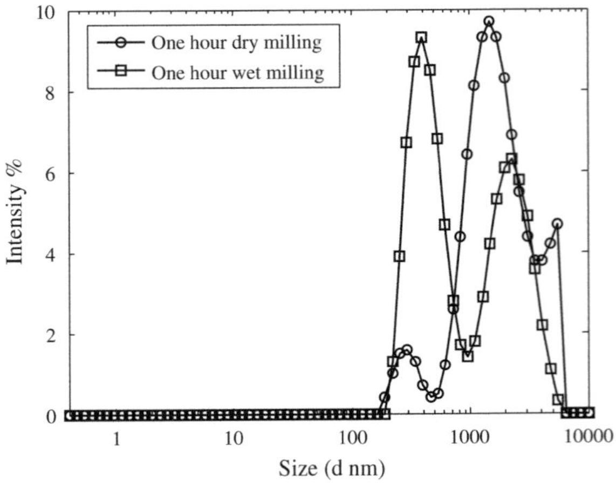

Fig. 1 Particle size distribution of jute particles

the material with least impact on particles at bottom of layer. In case of wet milling, the increase in temperature was slowed down by deionised water which consequently resulted in narrow particle size distribution with significant reduction in average particle size to 640 nm after 1 h of wet milling as shown in Figs. 1b and 2b. This can be attributed to the uniformity in impact action of balls on every individual particle in wet condition.

(a)
(b)

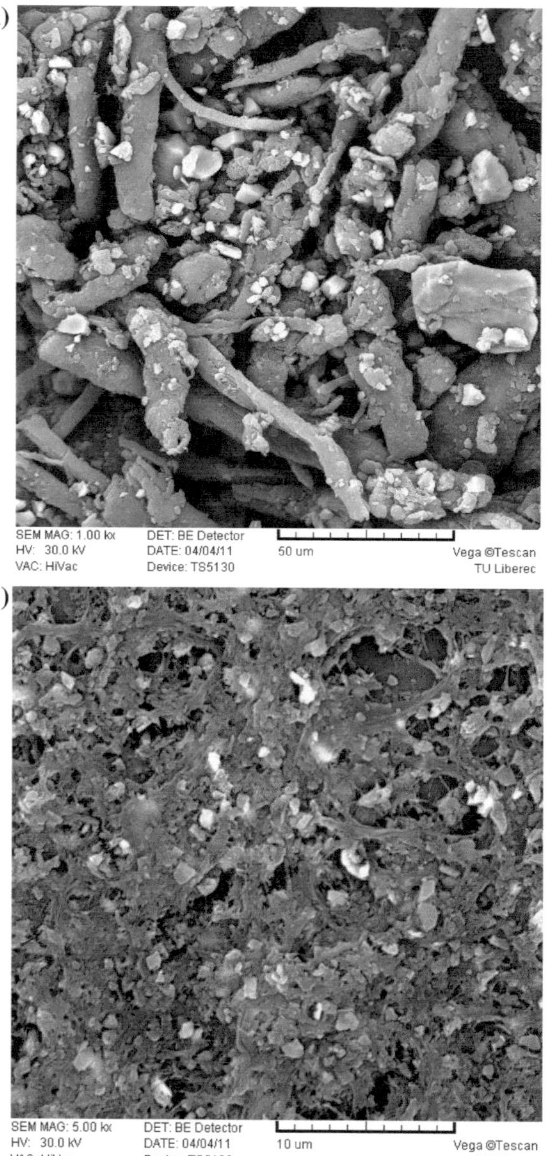

Fig. 2 SEM image of jute particles. **a** One hour dry milling. **b** One hour wet milling

Effect of Wet Milling Time on Particle Size Reduction of Jute Fibers

To further refine the jute particles to smaller size, wet milling was performed for extended duration. The average particle size reached to 443 nm after 3 h of wet milling and the particle size distribution changed slowly from multimodal nature to unimodal nature as shown in Fig. 3 and Table 2. This showed the consistency and homogeneity in milling action on every individual particle as milling continued for longer time. However, the rate of refinement became slower while grinding the smaller particles in addition to the severe damage of milling balls due to direct collision. This could have introduced some inorganic contaminations from mill to the material, so further pulverisation was stopped and jute particles in 500 nm range were used as nano/micro fillers.

The shape and size of jute fibers after 3 h of wet milling was precisely investigated with the help of FESEM image (Fig. 4a) due to its better resolution at nanoscale. The shape of jute particles was seen in the form of nanofibrils with certain aspect ratio. The few jute particles without aspect ratio were considered as agglomerates of hundreds of individual jute nanofibrils. In order to measure the diameter of nanofibrils, NIS Elements BR software was employed and total of 25 observations were made. The probability density function of diameter distribution of 25 readings is shown in Fig. 4b. In this way, the mean diameter and standard deviation was calculated as 46.52 and 13.58 nm respectively.

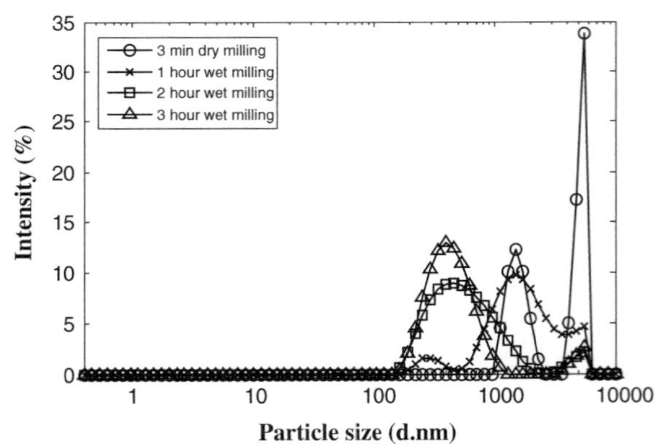

Fig. 3 Effect of extended wet milling time on particle size reduction

Table 2 Particle size from DLS measurement for different wet milling time

Sample name	First peak (nm)	Second peak (nm)	Third peak (nm)	Z-average (nm)	Polydispersity index
One hour	448.8	2365	0	640.1	0.420
Two hour	605.1	4781	0	508.8	0.333
Three hour	449.9	4966	0	443.1	0.299

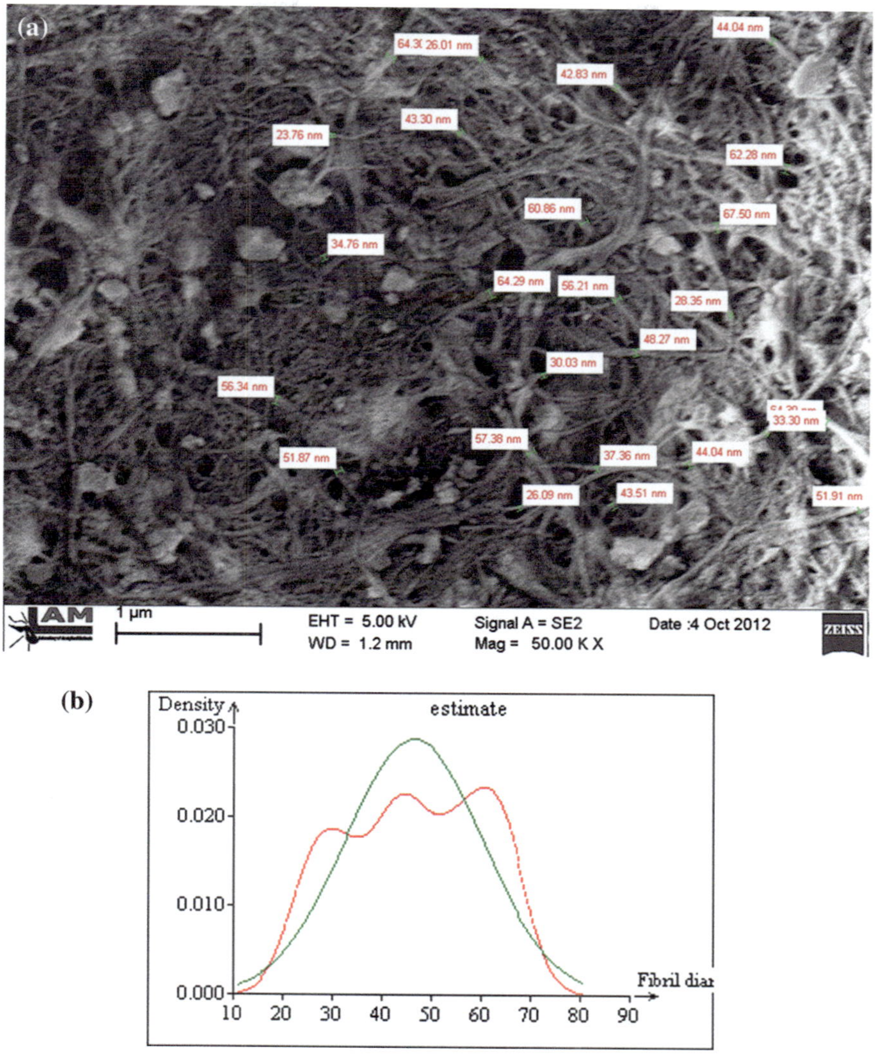

Fig. 4 **a** SEM image of jute nanofibrils. **b** Probability density function of diameter distribution of jute nanofibrils

Comparison of Size Measured on Different Techniques

The size of nanofibrils obtained after dry and wet pulverization of jute fibers is compared on the basis of different techniques of measurements as given in Table 3. The techniques based on DLS and BET measurements were found to overestimate the size of nanofibrils with regard to actual size. The size obtained after image analysis was found much lesser as compared to DLS and BET measurements. The

Table 3 Comparison of jute nanofibril size from different techniques

Sample name	DLS technique (spherical shape) (nm)	BET technique (spherical shape) (nm)	Image analysis (actual shape) (nm)
One hour dry milling	1480	1120	480
One hour wet milling	619	430	85
Three hour wet milling	493	190	46

significant difference in size was attributed to the principles of working of different techniques which were based on certain assumptions. The DLS and BET techniques assume the shape of particles as spherical while calculating the particle size, whereas image analysis measurements are based on actual projected image of particles.

Nanoindentation Measurements

In order to investigate the indentation behaviour, neat PLA and composite PLA films were subjected to the nanoindentation measurement. Figure 5 shows the load-penetration depth curves obtained from neat PLA and after reinforcing with 3 wt% of GJF, DMJF and WMJF in PLA matrix at a maximum load of 0.5 mN. The load-penetration depth behaviour was typical of that for soft materials with a very little signature of an elastic recovery indicating a permanent deformation of the surface beneath the indenter.

Fig. 5 Load-penetration depth profile of neat and composite PLA films

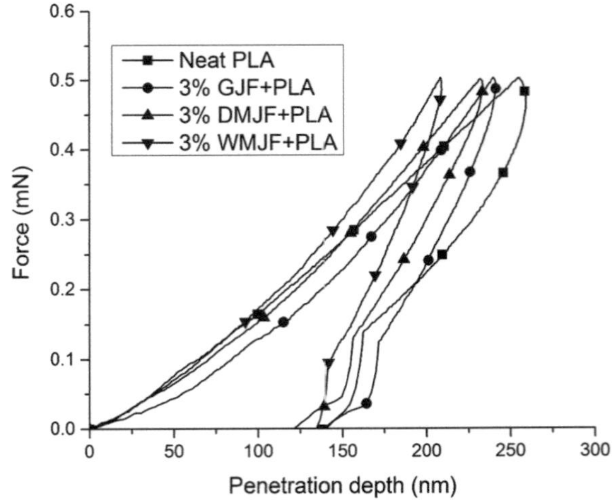

Table 4 Nanoindentation properties of neat and composite PLA films

Sample name	Nanolevel properties	
	Elastic modulus (GPa)	Hardness (GPa)
Neat PLA	4.37 ± 0.45	0.52 ± 0.06
3 % GJF + PLA	4.94 ± 0.41	0.54 ± 0.05
3 % DMJF + PLA	5.48 ± 0.47	0.68 ± 0.08
3 % WMJF + PLA	7.22 ± 0.52	0.70 ± 0.10

(a)

(b)

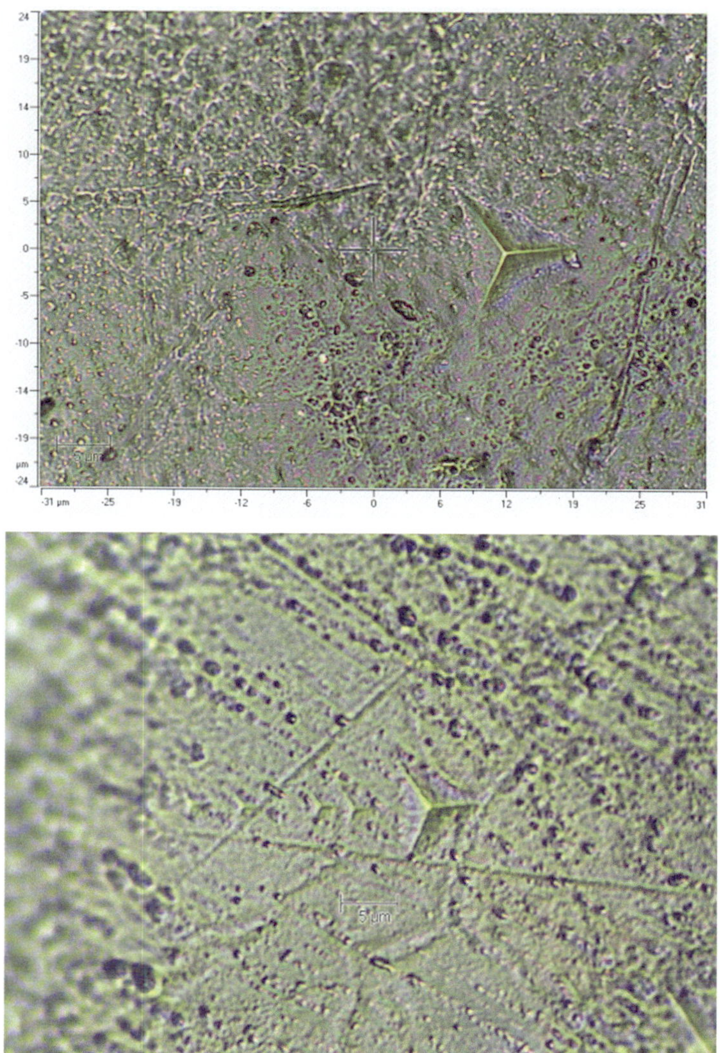

Fig. 6 Indentation mark. **a** Neat PLA film. **b** GJF/PLA film. **c** DMJF/PLA film. **d** WMJF/PLA

(c)

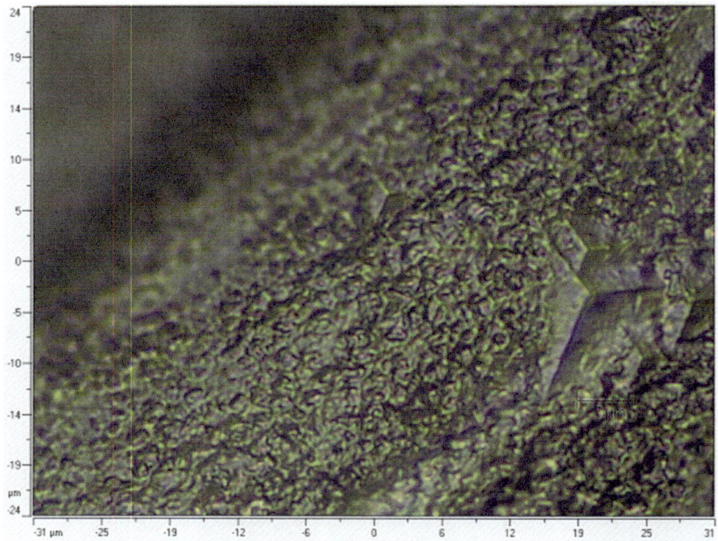

(d)

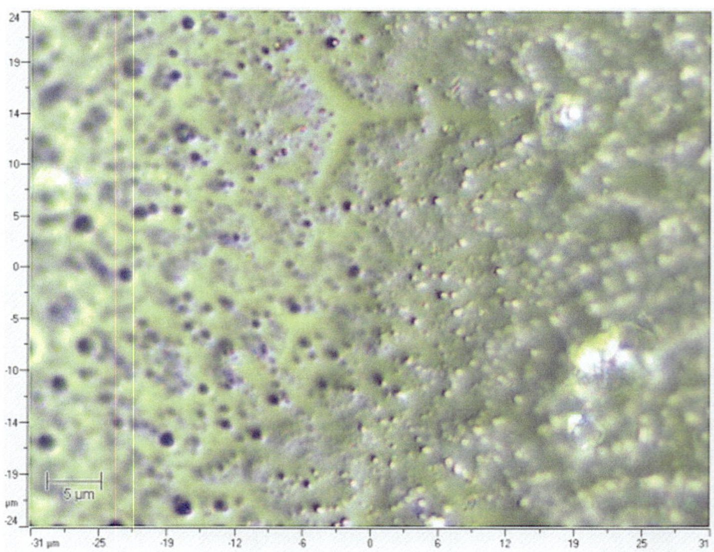

Fig. 6 (continued)

The neat PLA was soft in comparison with the composite films. There was an increase in the measured hardness after addition of fillers (Table 4). The WMJF/PLA films showed maximum hardness of 0.70 GPa compared to 0.68 and 0.54 GPa hardness of DMJF and GJF reinforced composite films, respectively. It was also seen that the elastic recovery (area enclosed between the unloading portion of the load-penetration depth profiles and the depth at maximum load) was maximum in the case of WMJF/PLA composite films. This was attributed mainly due to the effect of higher interaction of nanofibrils and PLA which transmitted the elastic properties of jute nanofibrils to PLA matrix effectively (Materials Today 2006). For the GJF/PLA and DMJF/PLA composite films, the elastic recovery was less as compared to the WMJF/PLA composite films. This might be due to their reduced area of interaction as a result of bigger size of fillers (Schuh 2006). The sharpness of indentation mark shown in Fig. 6 further explained the higher elastic recovery of WMJF/PLA composite films compared to neat PLA and other composite films. The indentation mark was less sharper in case of WMJF/PLA composite films due to faster elastic recovery of deformation (Fig. 6d).

Dynamic Mechanical Analysis

Figure 7a and Table 5 showed that WMJF/PLA composites have higher load bearing capacity than neat PLA due to the transfer of stress from matrix to stiff nanofibrils. The storage modulus of PLA composites at 35 °C increased from 3.09 GPa to the level of 5.92 GPa after the addition of WMJF. The maximum increase in storage modulus for WMJF/PLA composites was attributed to the higher surface area of nanofibrils interacting with the matrix. The deterioration in load bearing capacity of PLA after addition of GJF and DMJF was attributed to the poor bonding of these fillers with PLA due to their bigger particle size. Figure 7b shows that the tan delta peak of PLA was positively shifted only after the addition of WMJF. The maximum shift of 14 °C was reported in case of WMJF/PLA composites due to the higher surface area of nanofibrils interacting with PLA matrix.

Tensile Testing

Figure 8 and Table 6 showed that PLA composite films reinforced with WMJF have significantly higher initial modulus compared to neat PLA film and other composite films. The slight increase in initial modulus was observed in case of GJF and DMJF composite films. The significant increase in initial modulus from 1.04 to 3.14 GPa in case of WMJF was attributed to the increased interaction of WMJF and PLA matrix due to higher surface area of fillers at nanoscale (Baheti 2013, Polymer composites 2013). The least improvements in composite films of GJF and DMJF were attributed to the bigger size of these fillers, which resulted into least interaction with the matrix.

Fig. 7 **a** Storage modulus of neat and composite PLA films. **b** Damping factor of neat and composite PLA film

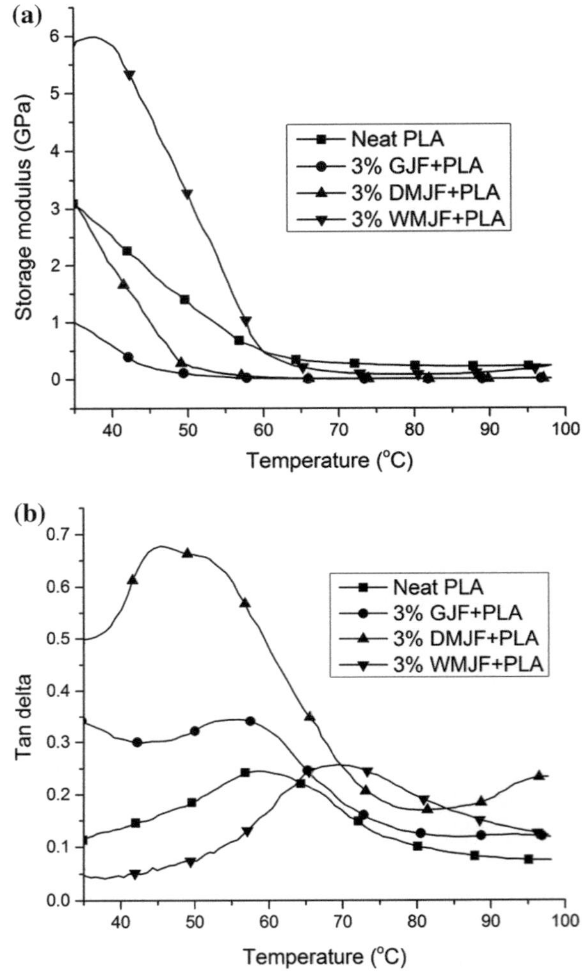

Table 5 Storage modulus of neat and composite PLA films at different temperature

Sample name	T_α	E′ (T_α) (GPa)	E′ (35 °C) (GPa)	E′ (60 °C) (GPa)
Neat PLA	40 ± 1	2.42 ± 0.15	3.09 ± 0.20	0.48 ± 0.02
3 % GJF + PLA	35 ± 1.8	1.01 ± 0.40	1.01 ± 0.38	0.01 ± 0.07
3 % DMJF + PLA	35 ± 1.4	3.15 ± 0.33	3.15 ± 0.30	0.04 ± 0.03
3 % WMJF + PLA	50 ± 1.1	3.27 ± 0.21	5.92 ± 0.23	0.52 ± 0.02

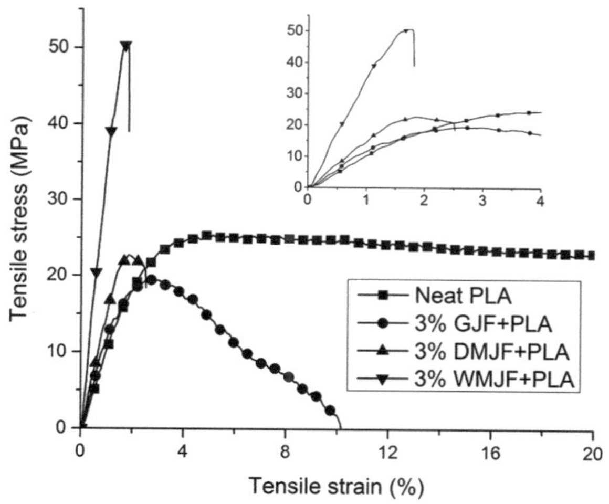

Fig. 8 Stress-strain curve of neat and composite PLA films

Table 6 Tensile properties of neat and composite PLA films

Sample name	Initial modulus (GPa)	Tensile strength (MPa)	Elongation (%)
Neat PLA	1.04 ± 0.03	25.98 ± 0.13	4.84 ± 0.72
3 % GJF + PLA	0.72 ± 0.09	15.14 ± 0.30	3.28 ± 0.48
3 % DMJF + PLA	1.30 ± 0.08	18.41 ± 0.25	2.40 ± 0.43
3 % WMJF + PLA	3.14 ± 0.05	67.90 ± 0.46	1.82 ± 0.32

Conclusions

The goal of present study was to utilize the waste jute fibers in textile industry as a source of cellulose nanofibrils for reinforcement of biodegradable packaging films. The jute nanofibrils were obtained by wet pulverization using high energy planetary ball milling process instead of strong acid hydrolysis due to its simple, economical and environment friendly approach. The extended wet milling for the duration of 3 h resulted into unimodal distribution of jute nanofibrils with diameter below 50 nm. In the subsequent step, composite films at 3 wt% filler content were prepared by mixing the calculated amount of GJF, DMJF and WMJF with 5 % PLA in chloroform solution. The mechanical properties of thin composite films of PLA were studied after reinforcement of nanofibrils, microfibrils and milifibers with the help of nanoindentation technique. The PLA composite films reinforced with jute nanofibrils showed maximum hardness of 0.70 GPa compared to 0.68 and 0.54 GPa hardness of microfibrils and milifibers reinforced composite films, respectively. The enhancement in elastic recovery was observed in case of nanofibril reinforced composite films, which was attributed mainly to the higher surface area of interacting nanofibrils compared to the micro and milifibers. As area of interaction between fibers and matrix increased, the elastic properties of jute fibers transmitted effectively to the PLA matrix. The sharpness of indentation mark also explained the higher elastic recovery of nanofibril reinforced PLA composite films compared to neat PLA and other composite films. The indentation mark was found less sharper in case of nanofibril reinforced PLA composite films, which was attributed to the faster elastic recovery of deformation.

References

W. Wang, Recycling in textiles, (2006) *Woodhead publishing* UK.

R. Horrocks, Recycling textile and plastic waste, (1996) *Woodhead publishing*, UK.

C.W.M. Yuen, Y. F. Cheng, Y. Li, and J. Y. Hu, *Journal of Textile Institute*, **100**,165 (2009).

H. P. S. Khalil, A. H. Bhat, and A. F. Yusra, *Carbohydrate Polymers*, **87**, 963 (2012).

D. Klemm, D. Schumann, F. Kramer, N. Hebler, M. Hornung, H. Schmauder, and S. Marsch, *Advances in Polymer Science,* **205**, 49 (2006).

J. Mussig, and C. Stevens, Industrial Applications of Natural Fibers: Structure, Properties and Technical Applications, (2010) *Wiley.*

A. Dufresne, M. B. Kellerhals, and B. Witholt, *Macromolecules*, **32**, 7396 (1999).

M. F. Rosa, E. S. Medeiros, J. A. Malmonge, K. S. Gregorski, D. F. Wood, L. H. C. Mattoso, G. Glenn, W. J. Orts and S. H. Imam, *Carbohydrate Polymers*, **81**, 83 (2010).

D. Pasquini, E. D. M. Teixeira, A. A. D. S. Curvelo, M. N. Belgacem and A. Dufresne, *Industrial Crops and Products*, **32**, 486 (2010).

R. Zuluaga, J. L. Putaux, J. Cruz, J. Velez, I. Mondragon and P. Ganan, *Carbohydrate Polymers,* **76**, 51 (2009).

R. Li, J. Fei, Y. Cai, Y. Li, J. Feng and J. Yao, *Carbohydrate Polymers*, **76**, 94 (2009).

B. Wang and M. Sain, *Composite Science and Technology*, **67**, 2521 (2007).

A. Alemdar and M. Sain, *Bioresource Technology*, **99**, 1664 (2008).

N. Reddy and Y. Yang, *Polymer*, **46**, 5494 (2005).

D. Y. Liu, X. W. Yuan, D. Bhattacharyya, and A. J. Easteal, *Express Polymer Letters*, **4**, 26 (2010).

P. Satyamurthy, P. Jain, R. Balasubramanya, and N. Vigneshwaran, *Carbohydrate Polymers*, **83**,122 (2011).

W. Li, J. Yue, and S. Liu, *Ultrasonics Sonochemistry*, **19**, 479 (2012).

J. Leitner, B. Hinterstoisser, M. Wastyn, J. Keches, and W. Gindl, *Cellulose*, **14**, 419 (2007).

S. Thomas, and L. A. Pothan, Natural fiber reinforced polymer composites, (2008) *Old City Publishing USA.*

J. Lunt, *Polymer Degradation and Stability*, **59**, 145 (1998).

M. Jonnobi, J. Harun, A. P. Mathew, and K. Oksman, *Composite Science and Technology*, **70**, 1742 (2010).

K. Petersen, P. Nielsen, and M. Olsen, *Starch*, **53**, 356 (2001).

L. Petersson, and K. Oksman, *Composite Science and Technology*, **66**, 2187 (2006).

M. D. Sanchez-Garcia, E. Gimenez, and J. M. Lagaron, *Carbohydrate Polymers*, **71**, 235 (2008).

X. Li, and B. Bhushan, *Materials Characterization*, **48**, 11 (2002).

B. D. Beake, and G. J. Leggett, *Polymer*, **43**, 319 (2002)

A.H.W. Ngan, and B. Tang, *Journal of Materials Research*,**17**, 2604 (2002).

C. A. Schuh, *Materials Today*, **9**, 32 (2006).

V. Baheti, and J. Militky, *Fibers and Polymers*, **14**, 133 (2013).

V. Baheti, J. Militky, and M. Marsalkova, *Polymer Composites*, **34**, 2133 (2013).

Biomedical Applications of Nanocellulose

Catarina Guise and Raul Fangueiro

Abstract Due to remarkable physical properties, especially surface chemistry and excellent biological properties, such as low toxicity, biocompatibility and biodegradability, nanocellulose has gained much attention for its use as biomedical material in the last years. This type of material is widely applied in medical implants, tissue engineering, drug delivery, wound-healing, cardiovascular applications, among others. This paper presents a brief review on the use of nanocellulose in biomedical area. Three different types of nanocellulose, namely cellulose nanocrystals (CNC), cellulose nanofibrils (CNF) and bacterial cellulose (BC) are discussed in terms of their production processes, properties and promising applications, based on the most recent published scientific works. In the same way, possible nanocellulose functional modifications, such as fluorescent modification, with the aim to improve specific properties and behavior are also discussed. Finally, an overview about the future of nanocellulose-based materials in the biomedical field is presented and analyzed.

Keywords Nano cellulose · Classifications · Production · Bio-medical applications

Introduction

In the last years, biomedical area has shown an increased interesting in natural polymers, including collagen, starch, alginate, gelatin, chitosan, elastin, and cellulose (Ivanova et al. 2014). From non-implantable materials, like wound dressings, to implantable materials, as vascular grafts, these natural polymers have been studied. Focus in cellulose, this natural polymer constitutes the most abundant renewable polymer resource available and due to its biodegradability, biocompatibility,

C. Guise (✉) · R. Fangueiro
Center for Textile Science and Technology, University of Minho, Guimarães, Portugal
e-mail: catarinaguise@det.uminho.pt

R. Fangueiro and S. Rana (eds.), *Natural Fibres: Advances in Science and Technology Towards Industrial Applications*, RILEM Bookseries 12,
DOI 10.1007/978-94-017-7515-1_12

exceptional mechanical properties and low cytotoxicity there is a great interest to produce biomedical materials based in cellulose (Lin and Dufresne 2014; Jorfi and Foster 2015).

Researchers have been investigating the processing of cellulose materials, fabrication and design for application in the medical field (Lin and Dufresne 2014; Jorfi and Foster 2015). Cellulose applied at the nanoscale is also being largely explored following the trend of the great expansion of nanotechnology since the last decade. Nanomaterial has its dimensions between 1 and 100 nm, where unique phenomena enable novel applications (Sonia and Sharma 2011).

Nanocellulose is obtained from native cellulose and can be divided in three types: cellulose nanocrystals (CNC), cellulose nanofibrils (CNF) and bacterial cellulose (BC), being this last a promising material for use as implants and scaffolds in tissue engineering (Petersen and Gatenholm 2011; Lin and Dufresne 2014). Both cellulose nanocrystals and cellulose nanofibrils can be extracted from plant cell walls. First type of nanocellulose fibers is prepared from fibers and fibrils via acid hydrolysis. This process degrades the amorphous regions and highly crystalline nanoparticles are obtained. On the other hand, cellulose nanofibrils are obtained by mechanically induced destructuring strategy via high shearing homogenization at high pressure (Xu et al. 2013; Zhang et al. 2013; Salas et al. 2014). Bacterial cellulose is synthesized by bacteria in a pure form, without contaminant molecules. To improve interactions with the human body chemical functionalization of nanocellulose surfaces with bioactive molecules are being studied (Petersen and Gatenholm 2011; Lin and Dufresne 2014).

Based nanocelluose products used for wound healing treatment are nowadays commercially available, including Bioprocess®, XCell®, Biofill® and Nanoskin® (De Olyveira et al. 2011; Petersen and Gatenholm 2011).

Mainly in the last ten years, a wide range of nanomaterials based on cellulose fibers have been developed for application in biomedical area and it is estimated that $97 billion will be impacted for medical and life sciences market (Lin and Dufresne 2014).

From Cellulose to Nanocellulose: Types of Nanocellulose

The interesting of natural fibers in biomedical applications has increased in last years. Chitosan, collagen, alginate, elastin, gelatin, starch and cellulose are examples of natural fibers that have been studied with great success for applications to the biomedical field (Namvar et al. 2014). Low cost, renewable and degradable characteristics are some advantages of interest of these type of fibers (Ticoalu et al. 2010). It is important to note that, besides these characteristics, materials for medical applications must meet several requirements like, for example, biocompatibility to promote an appropriate host response in a specific situation (Cheung et al. 2009).

A particular interest of the researchers lies in cellulose fibers. Cellulose, discovered in 1838 by French scientist Anselme Payen, is the most abundant biopolymer on earth, being estimated that, annually, are produced 7.5 × 10^{10} tons (Dugan et al. 2013; Jorfi and Foster 2015). It is a linear polysaccharide and regular-stereo and its repetition unity is called cellobiose. In turn, cellobiose is formed by anhydroglucopyranose molecules connected by glycosidic linkages (Fig. 1). Cellulose tends to form intramolecular and intermolecular hydrogen bonds due to their strongly interacting hydroxyl groups (Bufalino 2014). The intra- and inter-chain hydrogen bonding network makes cellulose a relatively stable polymer (Moon et al. 2011).

A parallel stacking of multiple cellulose chains forming elementary fibrils that further aggregate into larger microfibrills is promoted during biosynthesis. These fibrils are composed by crystalline regions, where the cellulose chains are arranged in a highly ordered structure, and amorphous regions connecting the crystalline regions (Fig. 2) (Moon et al. 2011). An important property in crystallinity array is packing of the component molecules of individual microfibrils. The sufficiently tightly arrangement prevents penetrations by enzymes and small molecules, such as water (Quiroz-Castañeda and Folch-Mallol 2013). Since microfibrils have amorphous and crystalline regions, native cellulose is paracrystalline and there are

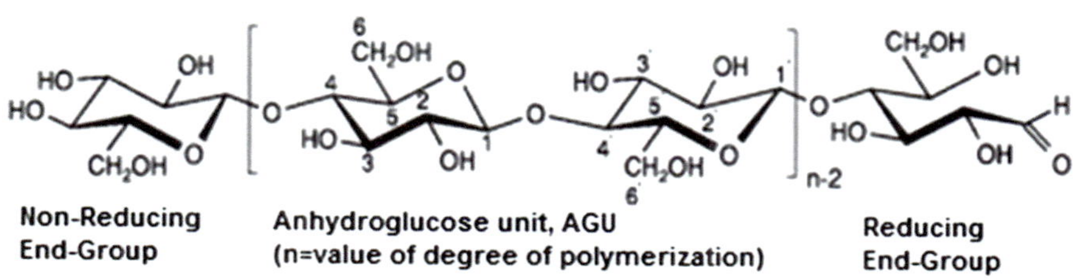

Fig. 1 Molecular structure of cellulose (Klemm et al. 2002)

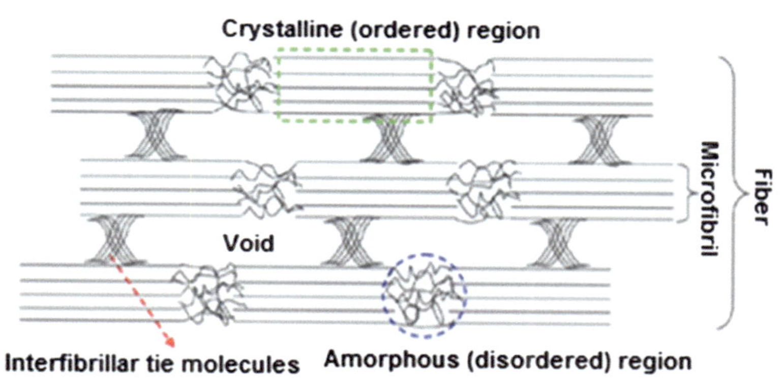

Fig. 2 Schematics of idealized cellulose fiber showing one of the suggested configurations of the crystalline and amorphous regions (Lin et al. 2012)

irregularities in cellulose fibers as twists or voids, which increase their total surface area (Desvaux 2005). For some plant species, amorphous regions can represent up to 50 % of the structure, while in bacterial cellulose and cellulose extracted from some algae, the crystalline domain corresponds to almost 100 % of the fibril, being almost defect-free (Mariano et al. 2014).

Nanocellulose is obtained by native cellulose that can be found in plants, animals, and bacteria and, depending on the cellulosic source and on the processing conditions, nanocellulose can be divided in three types: cellulose nanocrystals, cellulose nanofibrils and bacterial cellulose (Table 1) (Klemm et al. 2011; Lin and Dufresne 2014).

Figure 3 presents the common structures of the three types of nanocellulose obtained by transmission electron microscopy.

CNC and CNF processing system converts the large unit (cm) to the small unit (nm) while biosynthesis of BC involves a processing system from tiny unit (Å) to small unit (nm) (Lin and Dufresne 2014).

Chemically induced destructuring strategy, for example, acid hydrolysis, is usually used to extract CNC from native cellulose. In this process, amorphous regions are removed, preserving the highly-crystalline structure. Regarding to the morphology and dimensions, CNC can be evaluated as elongated rod-like

Table 1 Types, synonyms, sources, formation and average size of nanocellulose materials (Klemm et al. 2011)

Types of nanocellulose	Synonyms	Typical sources	Formation and average size
Cellulose nanocrystals (CNC)	Nanocrystalline cellulose, crystallites, whiskers, rodlike cellulose microcrystals	Wood, cotton, hemp, flax, wheat straw, mulberry bark, ramie, Avicel, tunicin, cellulose from algae and bacteria	– Acid hydrolysis of cellulose from many sources
			– Diameter: 5–70 nm
			– Length: 100–250 nm (from plant celluloses); 100 nm to several micrometers (from celluloses of tunicates, algae, bacteria)
Cellulose nanofibrils (CNF)	Microfibrillated cellulose, nanofibrils and microfibrils, nanofibrillated cellulose	Wood, sugar beet, potato tuber, hemp, flax	– Delamination of wood pulp by mechanical pressure before and/or after chemical or enzymatic treatment
			– Diameter: 5–60 nm
			– Length: several micrometers
Bacterial cellulose (BC)	Bacterial cellulose, microbial cellulose, biocellulose	Low-molecular-weight sugars and alcohols	– Bacterial synthesis
			– Diameter: 20–100 nm
			– Different types of nanofiber networks

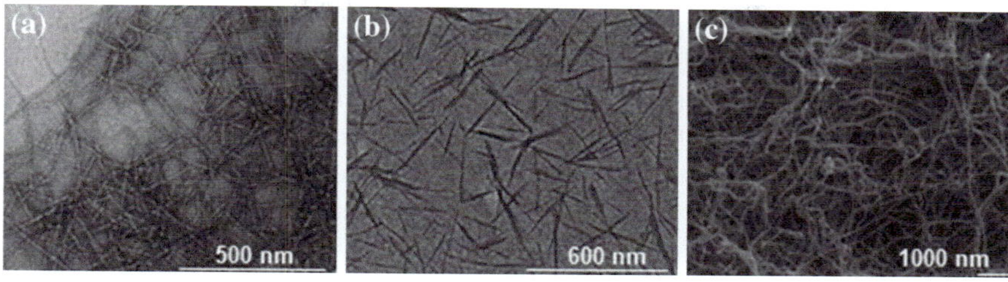

Fig. 3 Typical structures of nanocellulose: **a** cellulose nanofibrilis, **b** cellulose nanocrystals and **c** bacterial cellulose (Klemm et al. 2011)

nanoparticles or needle-like nanoparticles. Therefore, each rod can be considered as a rigid cellulosic crystal with no apparent defect. Values of elastic modulus of CNC are founded in the range between 120 and 170 GPa (Lin et al. 2012).

On the other hand, mechanically induced destructuring strategy is mainly applied on the production of CNF, via high shearing followed by homogenization at high pressure. Individual microfibrils can be successfully delaminated from cellulosic fibers by multiple mechanical shearing actions (Zhang et al. 2013; Lin and Dufresne 2014). Contrary to CNC, these nanofibrils contain both amorphous and crystalline cellulosic regions (Aulin et al. 2009). Figure 4 shows the differences between these two nanocellulose materials.

Fig. 4 Schematic illustrations of cellulose nanofibrils and cellulose nanocrystals (Zhang et al. 2013)

Bacterial cellulose, in general, is synthesized by bacteria, as Acetobacter Xylinum, in a pure form (Esa et al. 2014). This process does not require intensive processing to remove unwanted impurities or contaminants such as lignin, pectin and hemicellulose (Lin et al. 2013) Bacterial cellulose, compared to cellulose extracted from wood, presents better characteristics as higher purity, degree of polymerization, crystallinity, water content and mechanical stability (Rebouillat and Pla 2013). In the literature, many bioreactors have been studied to produce BC on large scale but the highest BC productivity is, only, 0.38 g/(L h) with the aerosol bioreactor (Lee et al. 2014).

There are many research works being undertaking on cellulose materials processing for biomedical applications and nanocellulose has gained attention in this area because of its low cost, sustainability, availability, an appropriate mechanical behavior, biodegradability, biocompatibility and low cytotoxicity (Jorfi and Foster 2015).

Applications in Biomedical Field

Nanocellulose is a biomaterial highly applicable to biomedical industry, specifically in tissue engineering, drug delivery, cartilage replacements, tissue engineering, cardiovascular applications, wound dressings and medical implants.

Regarding to wound dressings based in nanocellulose, specifically, bacterial cellulose, Bioprocess®, Biofill®, XCell® and Nanoskin® are some examples of available products (De Olyveira et al. 2011; Petersen and Gatenholm 2011). The two first products are used in the treatment of burns, skin wounds and ulcers as temporary artificial skin. Gengiflex®, other product available, is applied in treatment of periodontal diseases. The differences between these products are related to the manufacturing technologies used such as initial concentrations of carbon sources, surface to volume ratios and extended times of fermentation. For instance, the fermentation time of Biofill® is only 2 days, while Gengiflex® requires 8 days of fermentation (Brown et al. 2007). On the other hand, Xylos Corporation was the producer of the XCell® family of wound care products, composed by XCell® Cellulose Wound Dressing and XCell® Antimicrobial Wound. These products allow an appropriate hydration and absorption, maintaining the ideal moisture balance required for a good healing process (Thakur 2014).

Czaja et al. (2007) related properties of these bacterial cellulose membranes with properties of an ideal wound dressing material. For example, an ideal wound dressing material is available in various shapes and sizes and bacterial cellulose membranes have the ability to be molded in situ. Similarly, wound dressing material provide physical barrier against bacterial infections and BC membranes are composed by a nanoporous structure that does not allow any external bacteria to penetrate into the wound bed.

Figure 5 shows a highly effective wound dressing material of Nanoskin®, produced by Innovatec's—Biotechnology Research and Development, in the treatment of a diabetic ulcer (De Olyveira et al. 2011).

Another application of bacterial cellulose consists of artificial tubes used as potential replacement of small (<4 mm) or large (>6 mm) size vascular grafts. BActerial SYnthesized Cellulose (BASYC®) is an example. Figure 6 shows BASYC® tubes with different inner diameter, wall thickness and length. The BASYC® tube, marked with an arrow, is sufficient for experimental microsurgical requirements. This tube has an inner diameter of 1 mm, a wall thickness of 7 mm and a length of 5 mm (Klemm et al. 2001).

Due to high mechanical strength in wet state, enormous water retention property, and low roughness of inner tube surface, BASYC® has been successfully used as the artificial blood vessel in rats and pigs for microsurgery. Comparing with conventional synthetic vascular graft materials, these tubes made with bacterial cellulose can be suitable for small diameter vascular conduits and they can restrain the phenomena of thrombus induction and stenosis. Figure 7 shows the recovery of

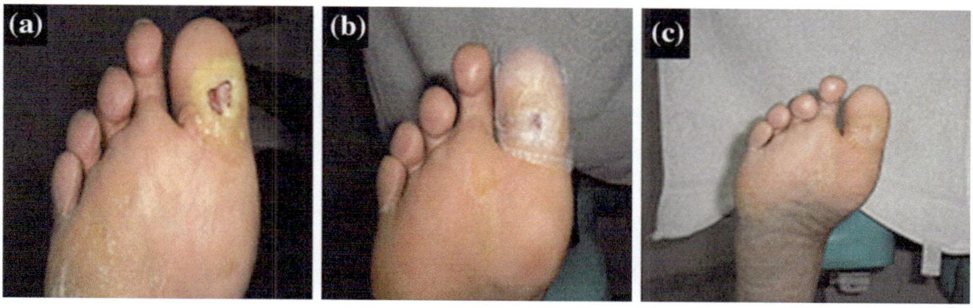

Fig. 5 After 7 days, a diabetic ulcer was completely healed using Nanoskin® (De Olyveira et al. 2011)

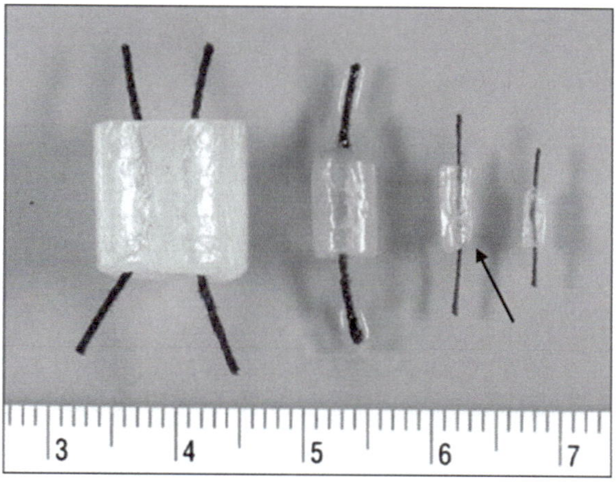

Fig. 6 BASYC® tubes (Klemm et al. 2001)

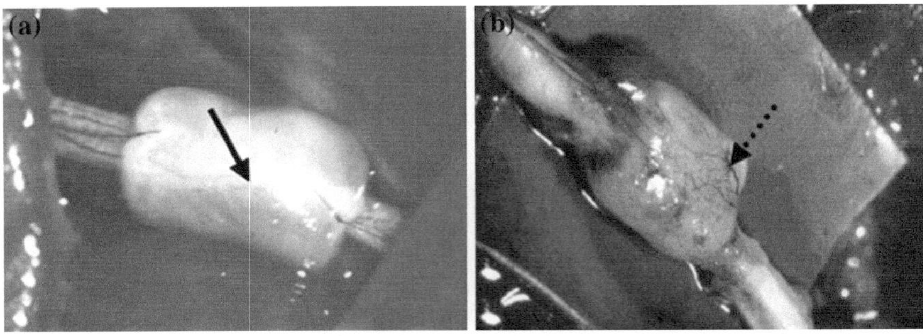

Fig. 7 BASYC® tube in micronerve surgery of a rat: **a** BASYC® tube as protective cover in sciatic nerve after operation; **b** BASYC® tube 10 weeks after operation (Klemm et al. 2001)

static nerve of a rat in 10 weeks after BASYC® placement. Figure 7b shows the BASYC® tube covered with connective tissue and small vessels within (Klemm et al. 2001).

Recently, Azevedo et al. (2013) produced hollow tubes with a small diameter based in cellulose and chitosan blends aiming to mimic human coronary arteries. Both materials are used for biomedical applications, being biocompatible and nontoxic. Firstly, they concluded that the mechanical properties of these membranes could be modulated by changing the ratio between cellulose and chitosan. The ideal proportion to mimic the dimensions of the human coronary artery was 5:5 cellulose/chitosan (Fig. 8).

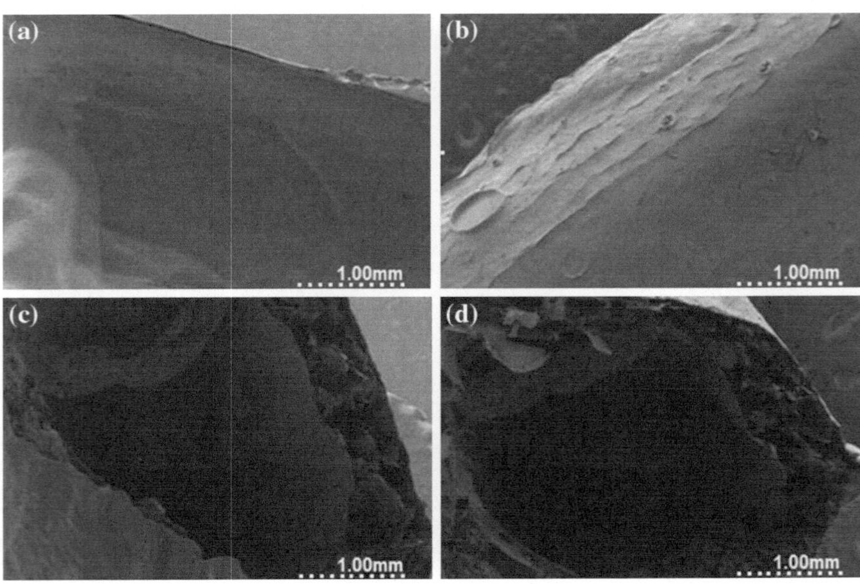

Fig. 8 SEM images of the surfaces **a** inner and **b** outer as well as the cross section (**c**) and **d** of the hollow tubes (Azevedo et al. 2013)

This study showed that the hollow tubes present promising mechanical and cell compatibility properties for further investigation as potential material candidate for coronary artery bypass graft applications.

Nimeskern et al. (2013) studied the mechanical behavior of BC for ear cartilage replacement. Different BC pellicles with different percentage of cellulose were studied to understand the effect on the mechanical properties of the BC samples. Higher percentage of cellulose samples presented better results. Therefore, they concluded that BC is able to reach mechanical properties of relevance for ear cartilage replacement. Also, it was possible to conclude that impatient-specific ear shapes can be produced (Fig. 9).

Eyholzer et al. (2011) investigated the use of a biocomposite hydrogel with carboxymethylated CNF for the replacement of native human nucleus pulposus. About 80 % of the world population will suffer from back pain in their lifetime, and in 75 % of cases this will be a direct consequence of degenerative processes of the disc. The main objective of Eyholzer et al. (2011) was to mimic the swelling and mechanical behavior of the native human nucleus pulposus in the intervertebral disks. A similar behavior between biocomposite hydrogel and native nucleus pulposus was founded in the sample with 0.4 % v/v of carboxymethylated CNF and degrees of substitution of 0.17, such as low strain values after cyclic compression tests, and similar relaxation properties. Another study about that showed, due to its adequate swelling ratio and improved mechanical properties, this biocomposite hydrogel is able to promote a similar behavior of nucleus pulposus (Borges et al. 2011).

Tissue engineering is another area of innovative and exciting potential for nanocellulose materials. The exceptional three-dimensional network formed by

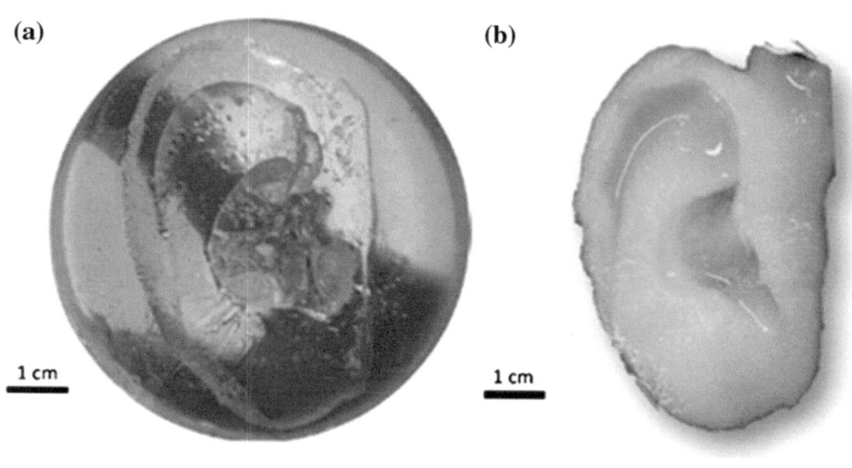

Fig. 9 **a** Negative silicone mold used to guide the bacteria during the bacterial culture to reproduce the large-scale features of the outer ear and **b** 3D BC implant prototype (Nimeskern et al. 2013)

cellulose, mechanical properties to support local forces and biocompatibility are the main advantages of cellulose for bone tissue engineering (Winter 2006; Zander et al. 2014).

Aravamudhan et al. (2013) developed mechanically competent cellulose scaffold materials for bone tissue engineering applications, based in cellulose and collagen. These porous micro-nano structured scaffolds showed mechanical properties in the midrange of human trabecular bone, with compressive modulus equals to 266.75 ± 33.22 MPa and strength equals to 12.15 ± 2.23 MPa.

Another study about scaffolds, developed by Yin et al. (2014), showed that BC has a range of structural and physicochemical properties, particularly, useful for the culture of bacteria. They compared the growth of 14 genera of bacteria on BC substrates with the growth on the commercially available biopolymers (agar, gellan, and xanthan). BC substrates produced rates of bacterial cell growth that typically exceed those on the commercial biopolymers due to the higher rates of nutrient diffusion (Fig. 10). In the same way, it was possible to see, during growth on BC, that the morphology of the cells was not altered.

Various drug carrier systems for biomedical applications have been developed. In recent years, carboxymethylcellulose has been an attractive cellulose material in this area. Weng et al. (2013) reported microspheres based in carboxymethylcellulose and chitosan as a biodegradable embolic agent for arterial embolization applications (Fig. 11).

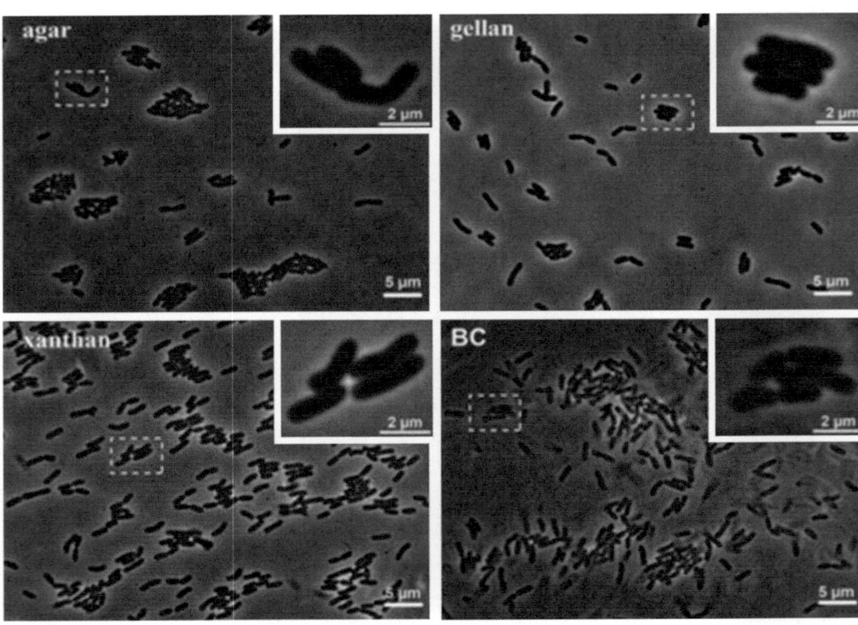

Fig. 10 Microscopy images of cells after growth on each substrate for 4 h (Yin et al. 2014)

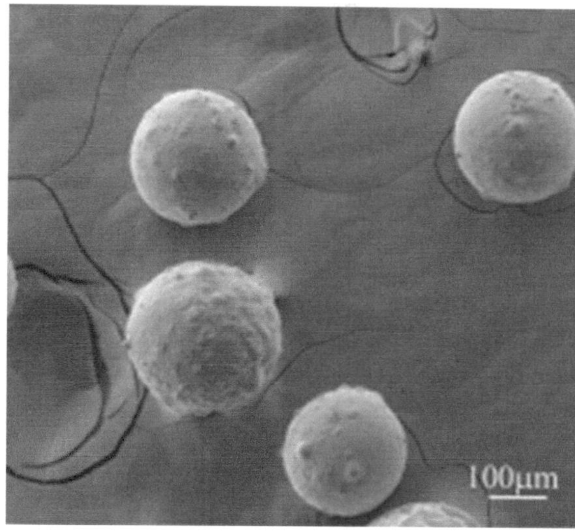

Fig. 11 SEM image of microspheres [adapted from (Weng et al. 2013)]

It was incorporated an anticancer drug for local release, killing cancerous cells. The microspheres were well tolerated by endothelial cells and subject to in vitro enzyme degradation. Regarding in vivo tests, the microspheres were tested in a rabbit renal embolization model and it was revealed that the microspheres were biodegradable with a mild tissue reaction (Weng et al. 2013). Although it is an attractive idea to use nanocellulose as drug deliver, the interactions between nanocellulose and drug molecules and the influence and regulation of drug release are some issues still needing investigation (Kolakovic et al. 2013).

Introduce antibacterial/antimicrobial activity through the functionalization and/or incorporation of antibacterial/antimicrobial agents is an example of the new functions tha can be incorporated to nanocellulose materials (Jorfi and Foster 2015). Nanocellulose-based antimicrobial biomaterials can be divided in two types, according to different types of antimicrobial agents. There are nanomaterials incorporated with inorganic antimicrobial agents, like silver particles and its derivatives, and there are organic antimicrobial agents, as lysozyme (Lin and Dufresne 2014).

Jebali et al. (2013) studied the antimicrobial activity of nanocellulose conjugated with allicin and lysozyme. Antimicrobial activity of nanocellulose without any other component showed to be lower comparing to that of nanocellulose conjugated with allicin and lysozyme. Both set of materials (allicin-conjugated nanocellulose and lysozyme-conjugated nanocellulose) presented good antifungal and antibacterial effects against standard strains of Candida albicans, Aspergillus niger, Staphylococcus aureus and *Escherichia coli.*

Xiong et al. (2013) reported a strategy for synthesis of well dispersed and stable silver nanostructures using cellulose nanocrystals in aqueous solution without any other type of agents. Cellulose nanocrystals nanohybrid materials containing dendritic nanostructured Ag showed better antibacterial activity than that of sphere nanostructured Ag.

Besides several studies about antimicrobial effects of nanocellulose conjugated with antimicrobial agents, several issues are still to explained as the duration of the antimicrobial effect.

The introduction of fluorescent molecules on the nanocellulose surface is other type of modification that can be developed for functionalization. Fluorescent molecules gives the nanocellulose the fluorescent labeling ability to be applied, for example, in bioimaging, as reported by Dong and Roman (2007). The authors developed a method involving a three-step reaction pathway, to covalently attach fluorescein-5′-isothiocyanate molecules to the surface of cellulose nanocrystals. More recently, Abitbol et al. (2013) studied fluorescent labeling and characterization of cellulose nanocrystals with varying charge contents. They used cotton-source cellulose nanocrystals and 5-(4,6-dichlorotriazinyl) aminofluorescein (DTAF). The properties of the fluorescently labeled cellulose nanocrystals were similar to the unlabeled materials, an excellent behavior for studies that require fluorescent cellulose nanocrystals substrates that retain as many of the native characteristics as possible. It was studied the uniformity of cellulose nanocrystals dispersion in

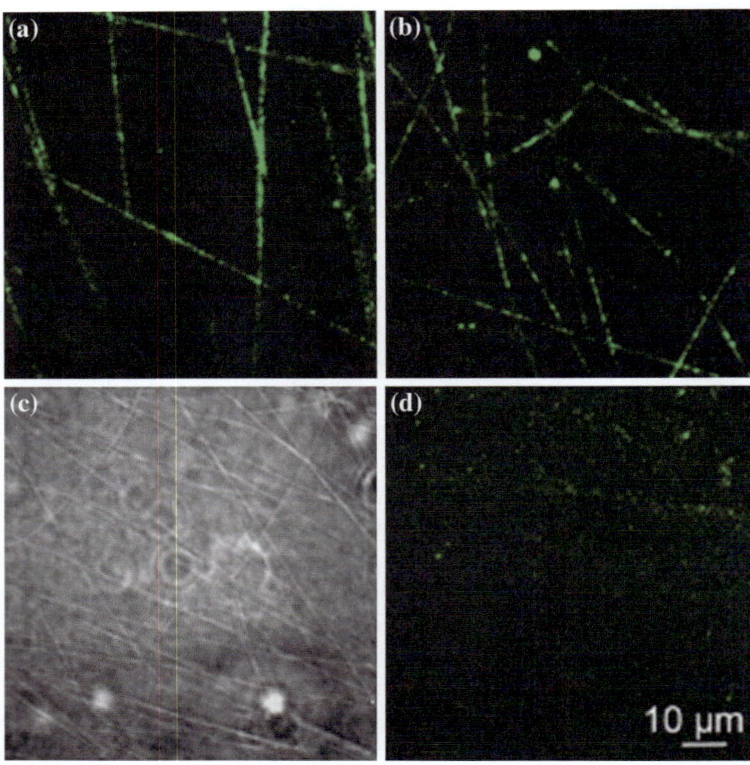

Fig. 12 **a** and **b** show epifluorescence microscopy images of electrospun PVA fibers loaded with DTAF-labeled CNCs, **c** shows bright field image and **d** corresponding epifluorescence image of PVA fibers loaded with unlabeled CNCs

electrospun polyvinyl alcohol fibers (PVA). The images obtained by fluorescence microscopy suggested that the nanoparticles were well dispersed within the polymer matrix (Fig. 12).

Conclusions and Remarks

This paper showed the great potential of several types of nanocellulose based materials in biomedical area. Tissue engineering, drug delivery, cardiovascular applications, wound dressings, and medical implants are just some examples where nanocellulose materials play an important role. These materials are gaining increasing attraction from the scientific community all over the world due to their interesting properties and characteristics for application in several fields, with particular emphasis in biomedical applications. The growing explosion of scientific publications in this field since the beginning of the 21st century is just another indicator of the interest on nanocellulose. Considering the examples presented in this article, it is possible to see the great potential of nanocellulose to create a novel class of biomedical materials, particularly, based on bacterial cellulose. Obviously, nanocellulose is still in its infancy stage being possible to believe that, in a near future, nanocellulose materials may provide the answer for many problems of the materials in biomedical field.

Acknowledgments The authors gratefully acknowledge the funding by Ministério da Ciência, Tecnologia e Ensino Superior, FCT, Portugal under grant SFRH/BD/90324/2012.

References

Abitbol T, Palermo A, Moran-Mirabal JM, Cranston ED (2013) Fluorescent labeling and characterization of cellulose nanocrystals with varying charge contents. Biomacromolecules 14:3278–84. doi: 10.1021/bm400879x

Aravamudhan A, Ramos DM, Nip J, et al (2013) Cellulose and collagen derived micro-nano structured scaffolds for bone tissue engineering. J Biomed Nanotechnol 9:719–31.

Aulin C, Ahok S, Josefsson P, et al (2009) Nanoscale cellulose films with different crystallinities and mesostructures - Their surface properties and interaction with water. Langmuir 25: 7675–7685. doi: 10.1021/la900323n

Azevedo EP, Retarekar R, Raghavan ML, Kumar V (2013) Mechanical properties of cellulose: chitosan blends for potential use as a coronary artery bypass graft. J Biomater Sci Polym Ed 24:239–52. doi: 10.1080/09205063.2012.690273

Borges AC, Eyholzer C, Duc F, et al (2011) Nanofibrillated cellulose composite hydrogel for the replacement of the nucleus pulposus. Acta Biomater 7:3412–3421.

Brown R, Czaja W, Jeschke M, Young DJ (2007) Multiribbon nanocellulose as a matrix for wound healing.

Bufalino L (2014) Filmes de nanocelulose a partir de resíduos madeireiros da Amazônia. Universidade Federal de Lavras

Cheung H, Ho M, Lau K, et al (2009) Natural fibre-reinforced composites for bioengineering and environmental engineering applications. Compos Part B Eng 40:655–663. doi: 10.1016/j.compositesb.2009.04.014

Czaja WK, Young DJ, Kawecki M, Brown RM (2007) The future prospects of microbial cellulose in biomedical applications. Biomacromolecules 8:1–12. doi: 10.1021/bm060620d

De Olyveira GM, Manzine Costa LM, Basmaji P, Xavier Filho L (2011) Bacterial Nanocellulose for Medicine Regenerative. J Nanotechnol Eng Med 2:034001. doi: 10.1115/1.4004181

Desvaux M (2005) Clostridium cellulolyticum: model organism of mesophilic cellulolytic clostridia. FEMS Microbiol Rev 29:741–64. doi: 10.1016/j.femsre.2004.11.003

Dong S, Roman M (2007) Fluorescently labeled cellulose nanocrystals for bioimaging applications. J Am Chem Soc 129:13810–1. doi: 10.1021/ja076196l

Dugan JM, Gough JE, Eichhorn SJ (2013) Bacterial cellulose scaffolds and cellulose nanowhiskers for tissue engineering. Nanomedicine (Lond) 8:287–98. doi: 10.2217/nnm.12.211

Esa F, Tasirin SM, Rahman NA (2014) Overview of Bacterial Cellulose Production and Application. Agric Agric Sci Procedia 2:113–119. doi: 10.1016/j.aaspro.2014.11.017

Eyholzer C, Borges De Couraça a., Duc F, et al (2011) Biocomposite hydrogels with carboxymethylated, nanofibrillated cellulose powder for replacement of the nucleus pulposus. Biomacromolecules 12:1419–1427. doi: 10.1021/bm101131b

Ivanova EP, Bazaka K, Crawford RJ (2014) New Functional Biomaterials for Medicine and Healthcare. New Functional Biomaterials for Medicine and Healthcare. Elsevier, pp 32–70

Jebali A, Hekmatimoghaddam S, Behzadi A, et al (2013) Antimicrobial activity of nanocellulose conjugated with allicin and lysozyme. Cellulose 20:2897–2907. doi: 10.1007/s10570-013-0084-3

Jorfi M, Foster EJ (2015) Recent advances in nanocellulose for biomedical applications. J Appl Polym Sci 41719:1–19. doi: 10.1002/app.41719

Klemm D, Kramer F, Moritz S, et al (2011) Nanocelluloses: A new family of nature-based materials. Angew Chemie - Int Ed 50:5438–5466. doi: 10.1002/anie.201001273

Klemm D, Schmauder H-P, Heinze T (2002) Biopolymers, Polysaccharides II: Polysaccharides from Eukaryotes. In: Vandamme E, De Baets S, Steinbüchel A (eds)Wiley, p 644

Klemm D, Schumann D, Udhardt U, Marsch S (2001) Bacterial synthesized cellulose - artificial blood vessels for microsurgery. Prog Polym Sci 26:1561–1603.

Kolakovic R, Peltonen L, Laukkanen A, et al (2013) Evaluation of drug interactions with nanofibrillar cellulose. Eur J Pharm Biopharm 85:1238–44. doi: 10.1016/j.ejpb.2013.05.015

Lee KY, Buldum G, Mantalaris A, Bismarck A (2014) More than meets the eye in bacterial cellulose: Biosynthesis, bioprocessing, and applications in advanced fiber composites. Macromol Biosci 14:10–32. doi: 10.1002/mabi.201300298

Lin N, Dufresne A (2014) Nanocellulose in biomedicine: Current status and future prospect. Eur Polym J 59:302–325. doi: 10.1016/j.eurpolymj.2014.07.025

Lin N, Huang J, Dufresne A (2012) Preparation, properties and applications of polysaccharide nanocrystals in advanced functional nanomaterials: a review. Nanoscale 4:3274. doi: 10.1039/c2nr30260h

Lin S-P, Calvar IL, Catchmark JM, et al (2013) Biosynthesis, production and applications of bacterial cellulose. Cellulose 20:2191–2219.

Mariano M, El Kissi N, Dufresne A (2014) Cellulose nanocrystals and related nanocomposites: Review of some properties and challenges. J Polym Sci Part B Polym Phys 52:791–806. doi: 10.1002/polb.23490

Moon RJ, Martini A, Nairn J, et al (2011) Cellulose nanomaterials review: structure, properties and nanocomposites. Chem Soc Rev 40:3941–3994.

Namvar et al (2014) Potential use of plants fibres and their composites for biomedical applications. Bioresources Peer-revie:19.

Nimeskern L, Martínez Ávila H, Sundberg J, et al (2013) Mechanical evaluation of bacterial nanocellulose as an implant material for ear cartilage replacement. J Mech Behav Biomed Mater 22:12–21. doi: 10.1016/j.jmbbm.2013.03.005

Petersen N, Gatenholm P (2011) Bacterial cellulose-based materials and medical devices: current state and perspectives. Appl Microbiol Biotechnol 91:1277–1286.

Quiroz-Castañeda RE, Folch-Mallol J (2013) Hydrolysis of Biomass Mediated by Cellulases for the Production of Sugars. In: Chandel AK, Silva SS (eds) Sustainable Degradation of Lignocellulosic Biomass - Techniques, Applications and Commercialization. InTech,

Rebouillat S, Pla F (2013) State of the Art Manufacturing and Engineering of Nanocellulose: A Review of Available Data and Industrial Applications. J Biomater Nanobiotechnol 04: 165–188. doi: 10.4236/jbnb.2013.42022

Salas C, Nypelö T, Rodriguez-Abreu C, et al (2014) Nanocellulose properties and applications in colloids and interfaces. Curr Opin Colloid Interface Sci 19:383–396. doi: 10.1016/j.cocis.2014. 10.003

Sonia T, Sharma C (2011) Chitosan and Its Derivatives for Drug Delivery Perspective. Adv Polym Sci 243:23–54. doi: 10.1007/12

Thakur VK (2014) Nanocellulose Polymer Nanocomposites: Fundamentals and Applications. Wiley

Ticoalu A, Aravinthan T, Cardona F (2010) A review of current development in natural fiber composites for structural and infrastructure applications. South. Reg. Eng. Conf. 1–5.

Weng L, Rostamzadeh P, Nooryshokry N, et al (2013) In vitro and in vivo evaluation of biodegradable embolic microspheres with tunable anticancer drug release. Acta Biomater 9:6823–33. doi: 10.1016/j.actbio.2013.02.017

Winter B (2006) Nano-cellulose in Regenerative Medicine. Conference on Nanotechnology, Atlanta, GA.

Xiong R, Lu C, Zhang W, et al (2013) Facile synthesis of tunable silver nanostructures for antibacterial application using cellulose nanocrystals. Carbohydr Polym 95:214–9. doi: 10. 1016/j.carbpol.2013.02.077

Xu X, Liu F, Jiang L, et al (2013) Cellulose Nanocrystals vs. Cellulose Nano fi brils: A Comparative Study on Their Microstructures and Effects as Polymer Reinforcing Agents. ACS Appl Mater 5:2999–3009.

Yin N, Santos TMA, Auer GK, et al (2014) Bacterial cellulose as a substrate for microbial cell culture. Appl Environ Microbiol 80:1926–32. doi: 10.1128/AEM.03452-13

Zander NE, Dong H, Steele J, Grant JT (2014) Metal cation cross-linked nanocellulose hydrogels as tissue engineering substrates. ACS Appl Mater Interfaces 6:18502–10. doi: 10.1021/ am506007z

Zhang Y, Nypelö T, Salas C, et al (2013) Cellulose Nanofibrils: From Strong Materials to Bioactive Surfaces. J Renew Mater 1:195–211. doi: 10.7569/JRM.2013.634115

Part III
Natural Fibre Reinforced Polymeric Composites

A Finite Element Analysis to Validate the Rule-of-Mixtures for the Prediction of the Young's Modulus of Composites with Non-circular Anisotropic Fibres

Amandeep Singh Virk, John Summerscales and Wayne Hall

Abstract This paper considers the rule-of-mixtures in the context of the tensile modulus of unidirectional fibre reinforced polymer (FRP) composites made with fibres of irregular cross-section, having anisotropic mechanical properties. A finite element model is used to generate data for the determination of the tensile modulus of the FRP composite. A range of degrees of anisotropy are considered. The error in the predicted modulus is found to be small for irregular fibres.

Keywords Fibres · Elasticity · Finite element analysis (FEA) · Natural fibres

Introduction

The elastic modulus of a composite parallel to the fibre direction is generally predicted using the Rule of Mixture (RoM) (Daniel and Ishai 2005; Jones 1998; Summerscales et al. 2010a, 2013), Eq. 1.

$$E_c = V_f E_f + V_m E_m \tag{1}$$

where E_c is the composite modulus in fibre direction, E_f and E_m are the fibre and matrix modulus respectively and V_f and V_m are the fibre and matrix volume fraction.

A.S. Virk · W. Hall
Griffith School of Engineering, Griffith University, Gold Coast Campus, Gold Coast, QLD 4222, Australia

J. Summerscales (✉)
Advanced Composites Manufacturing Centre, School of Marine Science and Engineering, Reynolds Building, University of Plymouth, Plymouth PL4 8AA, Plymouth, UK
e-mail: j.summerscales@plymouth.ac.uk

R. Fangueiro and S. Rana (eds.), *Natural Fibres: Advances in Science and Technology Towards Industrial Applications*, RILEM Bookseries 12,
DOI 10.1007/978-94-017-7515-1_13

However, the RoM has a number of underlying assumptions:

(a) a perfect bond between the fibres and matrix,
(b) the fibres are parallel, continuous, homogenous, linear elastic and regularly spaced in the composite, have uniform cross-section along the length, and
(c) the matrix is homogenous, linear elastic and void free.

Despite all these assumptions, the RoM has been successfully used to predict the modulus of synthetic fibre reinforced composites (Hyer and Waas 2000; Hull and Clyne 1996), where the fibres have a regular cross-section shape (generally circular) and variation between fibre cross-sectional areas is low. However, the Cross-Sectional Area (CSA) of natural fibres is irregular (Virk et al. 2010a), exhibits greater variation in fibre-to-fibre cross-sectional area (Virk et al. 2010a) and can vary along the length of the fibre. Moreover, natural fibres have a high degree of anisotropy in their mechanical properties (Thomason 2009) and large variability in their mechanical properties (Summerscales et al. 2010a, 2011; Virk et al. 2009a). These variations are in addition to the above assumptions made to simplify the RoM model. A Finite Element Analysis (FEA) was carried out to predict the composite modulus of a unidirectional natural fibre (jute) reinforced polymer composite and the applicability of the RoM to the natural fibre composite was assessed. The deviation between FEA and RoM predictions was quantified.

Finite Element Analysis (FEA)

A unidirectional natural fibre composite structure with a fibre volume fraction of 20 % was modelled using finite element software to predict the modulus of the composite structure. The fibre cross-section shape and area were determined from micrographs of jute fibres (Fig. 1). The sample preparation used to obtain the micrographs is given in Virk et al. (2010a). The variation in the fibre cross-sectional area along the fibre length has been disregarded as only a small length (100 μm) of fibre was simulated in FEA (to reduce FEA model size). The fibres were assumed to be arranged in uniform square packing in the matrix as the simulated fibre volume fraction was low.

Typical experimental tensile stress-strain curves for the jute fibre are shown in Fig. 2 (Virk et al. 2009a). The fibres show elastic behaviour up to failure, therefore the fibres were modelled as linear elastic. The composite was assumed to be simulated below the glass transition temperature of the matrix therefore, the behaviour of the matrix is also assumed to be linear elastic. The material properties for the jute fibres [corrected for true fibre area (Summerscales et al. 2013; Virk et al. 2012)] and the epoxy matrix used for the FEA are given in Table 1 (Anon 2014; Thomason 2009; Virk et al. 2009a, b, 2011, 2012). The jute fibres are modelled as orthotropic with transverse isotropy. The material orientation direction for the anisotropic natural fibre FEA model is shown in Fig. 3. Direction '1' oriented parallel to the

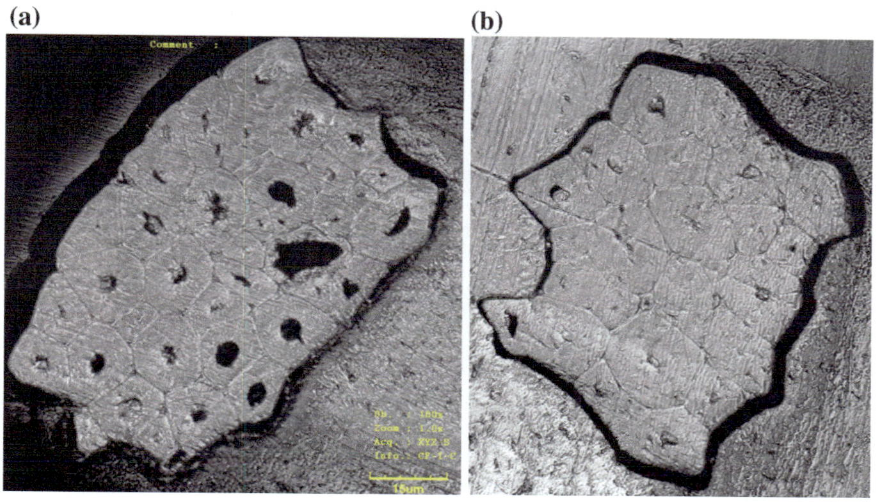

Fig. 1 Typical jute fibre cross-sections

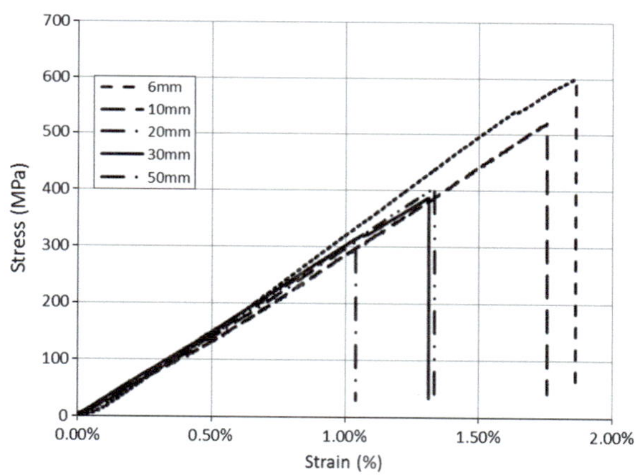

Fig. 2 Typical jute fibre stress-strain curves

Table 1 Material properties for FEA model (MPa)

Material	E_1	E_2	E_3	ν_{12}	ν_{13}	ν_{23}	G_{12}	G_{13}	G_{23}
Jute	39,618	5500	5500	0.11	0.11	0.35	7124.1	7124.1	2037
Matrix	2650	–	–	0.35	–	–	–	–	–

Anon (2014), Thomason (2009), Virk et al. (2009a, b, 2011, 2012)

fibre principal axis (along the global Z axis) denotes the direction of E_1 fibre modulus. Fibre properties E_2 and E_3 are assumed radial symmetric, and hence numerically equal and are orientated along the radial (2 and 3) directions (global X and Y axes) respectively.

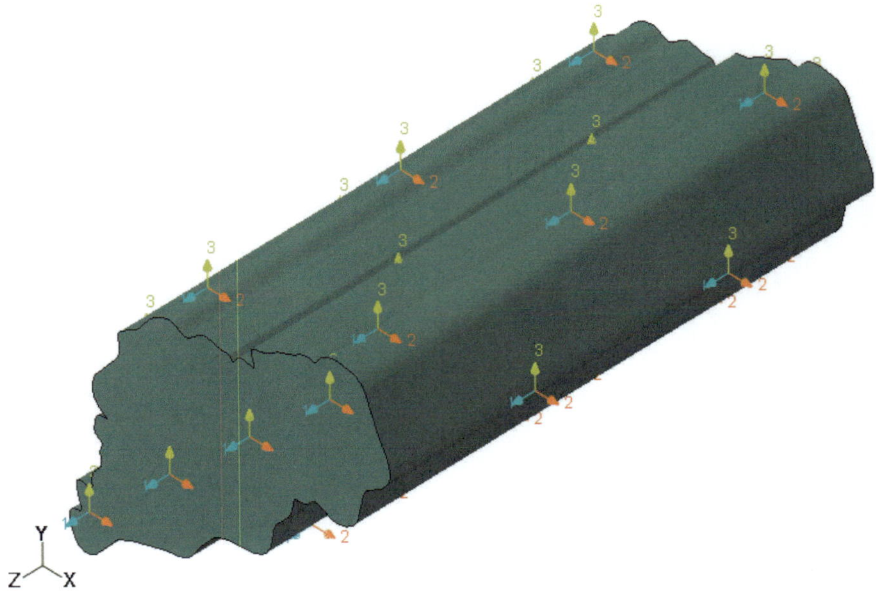

Fig. 3 Material orientation direction for the jute fibres

The interface between the fibre and the matrix was modelled using tie constraint i.e. the bond between the fibres and the matrix was assumed to be perfect. The composite representative volume was constrained using symmetrical boundary conditions along the direction normal to the faces of the cube. On each of the orthogonal axes (−X, −Y and −Z), symmetrical constraints were applied to only one face of the representative composite cube shown in Fig. 4. The composite was

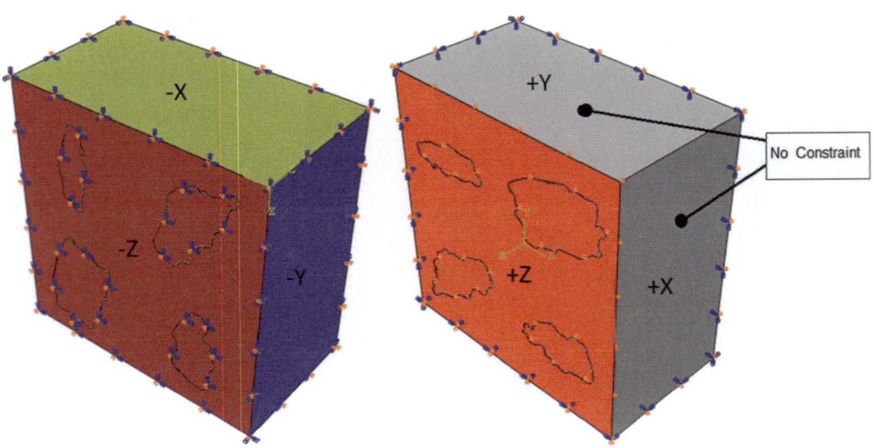

Fig. 4 Boundary condition definition for FEA model

loaded along the +Z axis (parallel to the fibre axis) with the prescribed displacement, the load was applied to the face opposite to the symmetric constrained face (−Z).

The fibres and the matrix were modelled using a mesh of 3D stress elements (Abaqus®C3D8R: an 8-node linear brick, reduced integration, hourglass control). In each case the mesh used ~25,000 elements for the matrix and ~4000 elements for each fibre as shown in Fig. 5.

An implicit solver was used to obtain the linear solution for the FE model. The computation time was 10 min on a Quad core 1.6 GHz computer. The reaction forces and the corresponding extension in the structure were recorded for each equilibrium iteration. These were used to calculate the average stress and strain respectively using the original composite area and fibre length. The effective modulus of the structure was calculated from the average stress and strain values generated from the forces and displacements in the FEA analysis

The FEA predicted the composite stiffness for a 20 % fibre volume fraction irregular fibre jute/epoxy composite to be 10.12 GPa. The corresponding RoM prediction (Virk et al. 2012) was 10.04 GPa. The discrepancy between the FEA and RoM predictions is 0.72 %. Thus it can be concluded that the RoM can be safely applied to predict the elastic modulus of composite reinforced by anisotropic natural fibres with irregular CSA.

To provide a baseline comparison, equivalent models were run with (a) circular cross-section fibres with equal diameters, and (b) circular cross-section fibres with varying fibre diameters (Fig. 6). The FEA derived composite modulus for circular cross-section fibres with equal and varying diameter was 10.02 and 10.01 GPa respectively. The small difference can be due to the circle being approximated by

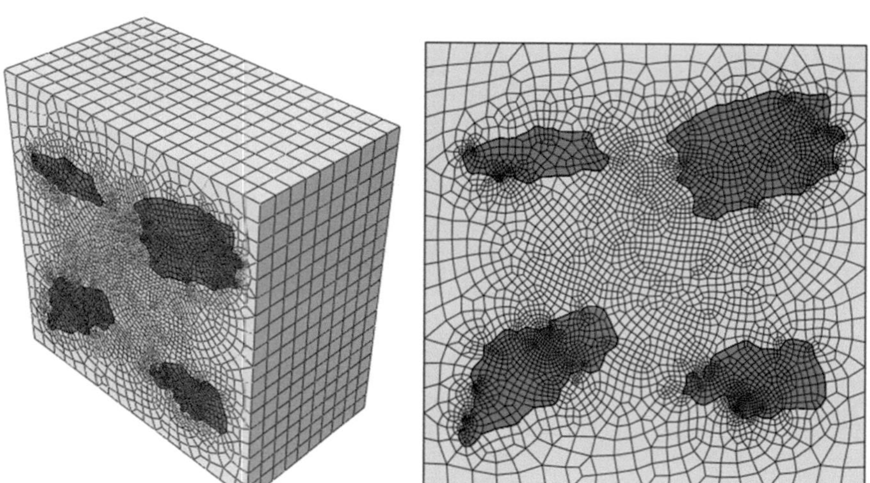

Fig. 5 FEA mesh for the natural fibre composite model

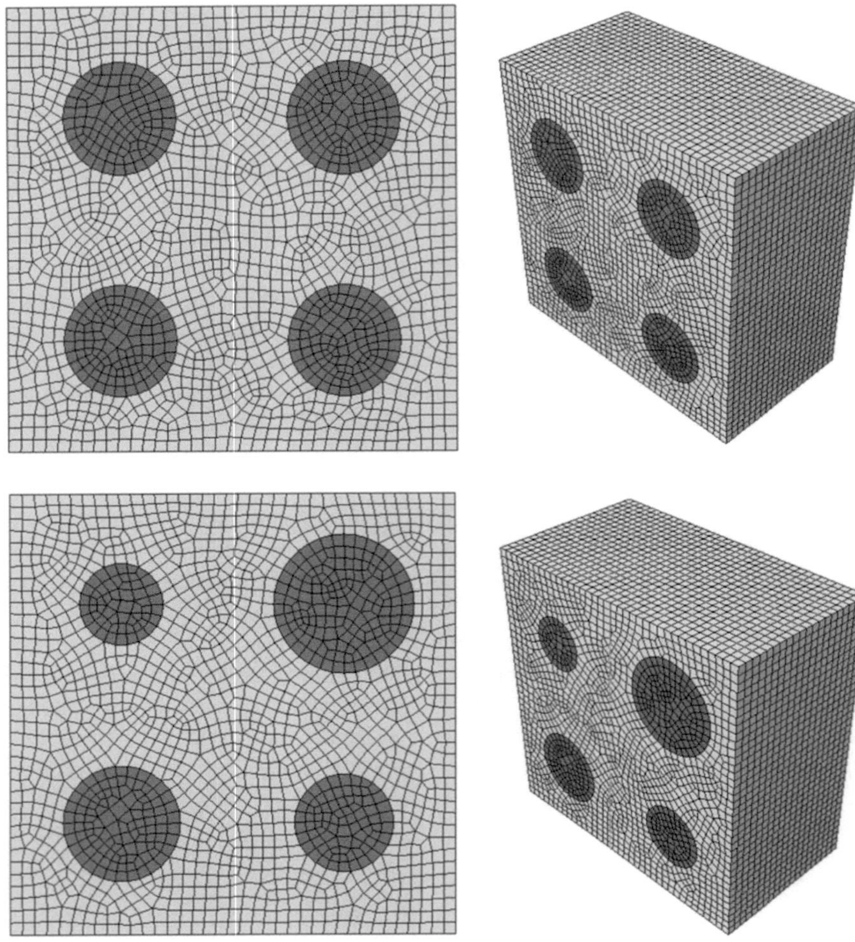

Fig. 6 FEA mesh for the circular fibre composite models

flat sided elements. Comparing FEA prediction to the RoM estimate, an error of −0.23 and −0.29 % was calculated for circular fibres with equal or varying diameters respectively.

Poisson's Ratio

It is not easy to accurately measure the Poisson's ratios of natural fibres. Therefore, to assess the effect of the Poisson's ratio on the predicted composite modulus a FEA study was carried out where the fibre Poisson's ratio, $\mathbf{v_{12}}$ was assumed to vary between 0.05 and 0.95. The first subscript '1' indicates the stimulus and the second subscript '2' response. The fibre material properties were varied such that the material stability requirements specified by the conditions of symmetry of compliances (Eqs. 2–3) and the Lemprière (1968) criteria for the limits on Poisson's ratios in orthotropic materials (Eqs. 4–5) were satisfied.

$$\nu_{ij}E_j = \nu_{ji}E_i \tag{2}$$

$$|\nu_{ij}| < (E_i/E_j)^{1/2} \tag{3}$$

$$E_1, E_2, E_3, G_{12}, G_{13}, G_{23} > 0 \tag{4}$$

$$1 - \nu_{12}\nu_{21} - \nu_{23}\nu_{32} - \nu_{31}\nu_{13} - 2\nu_{21}\nu_{32}\nu_{13} > 0 \tag{5}$$

The shear modulus of the fibre was predicted using Huber (1923) equation,

$$G_{ij} = \frac{\sqrt{E_i E_j}}{2\left[1 + \sqrt{\nu_{ij}\nu_{ji}}\right]} \tag{6}$$

The bulk modulus of the fibre was calculated using Eq. 7 (Summerscales 2000),

$$K_f = \frac{\sqrt[3]{E_1 E_2 E_3}}{3\left[1 - 2\sqrt[3]{\nu_{12}\nu_{31}\nu_{23}}\right]} \tag{7}$$

The analytical Eqs. 8 and 9 were used to predict the upper and lower bounds on the elastic modulus in the fibre direction in a unidirectional composite (Zweben 1994).

$$E_c(U) = V_f E_f + (1 - V_f)E_m + \frac{4V_f(1 - V_f)(\nu_f - \nu_m)^2}{V_f/K_m + (1 - V_f)/K_f + 1/G_f} \tag{8}$$

$$E_c(L) = V_f E_f + (1 - V_f)E_m + \frac{4V_f(1 - V_f)(\nu_f - \nu_m)^2}{V_f/K_m + (1 - V_f)/K_f + 1/G_m} \tag{9}$$

where E_c is the composite modulus in the fibre direction, U and L are upper and lower bound respectively, V_f is the fibre volume fraction, ν_f is the axial Poisson's ratio of the fibre (ν_{12}), ν_{m} is the Poisson's ratio of the matrix, E is the axial modulus, G is the shear modulus, K is the bulk modulus and subscript f and m indicate the fibre and matrix respectively.

The fibre material properties used for the FEA study and Eqs. 8 and 9 are given in Table 2. The FEA models used for this study were same as detailed above i.e. irregular fibre cross-section and two equivalent models with round fibres, with (a) constant and (b) varying fibre diameter.

Table 2 Fibre properties for FEA model (MPa)

E_1	$E_2 = E_3$	$\nu_{12} = \nu_{13}$	$\nu_{21} = \nu_{31}$	ν_{23}	$G_{12} = G_{13}$	G_{23}	K_f
39,618	5500	0.05	0.007	0.35	7246	2037	3930
39,618	5500	0.11	0.015	0.35	7090	2037	4253
39,618	5500	0.15	0.021	0.35	6990	2037	4459
39,618	5500	0.25	0.035	0.35	6752	2037	4984
39,618	5500	0.35	0.049	0.35	6529	2037	5554
39,618	5500	0.45	0.062	0.35	6321	2037	6196
39,618	5500	0.55	0.076	0.35	6125	2037	6941
39,618	5500	0.65	0.090	0.35	5942	2037	7827
39,618	5500	0.75	0.104	0.35	5769	2037	8906
39,618	5500	0.85	0.118	0.35	5605	2037	10,259
39,618	5500	0.95	0.132	0.35	5451	2037	12,013

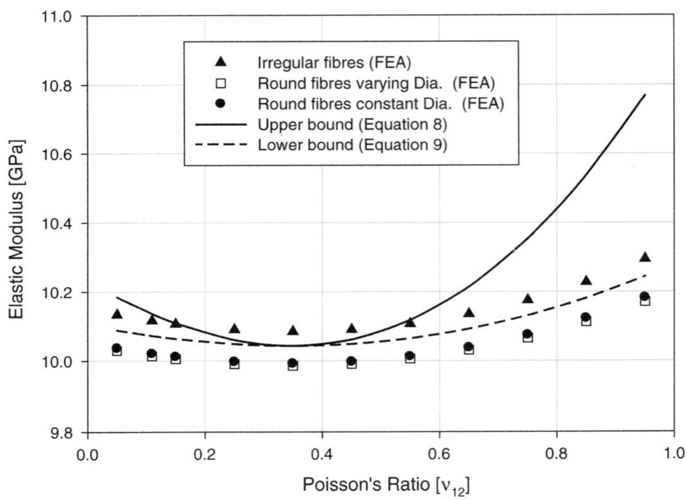

Fig. 7 Predicted composite modulus

The composite modulus predicted for different Poisson's ratio using FEA and Eqs. 8 and 9 at 20 % fibre volume fraction is shown in Fig. 7. For all the cases it was observed that with the change in the Poisson's ratio (from 0.05 to 0.95) the predicted minimum modulus occurs when the fibre axial (ν_{12}) and the matrix Poisson's ratios are numerically equivalent. Note that the right hand term in Eqs. 8 and 9 goes to zero under this condition. The predicted composite modulus increases either side of this minimum. The FEA model predicts slightly higher modulus for the irregular fibres than for the circular fibres with varying or constant diameter. The difference between FEA predicted modulus for circular fibres with varying or constant diameter was small. The lower bound of the modulus predicted by Eq. 9

Table 3 Comparison of the errors arising in each of the FEA models

Poisson's ratio	Round fibres of constant diameter		Round fibres of varying diameter		Irregular cross-section fibres	
	Composite modulus (GPa)	Error (%)	Composite modulus (GPa)	Error (%)	Composite modulus (GPa)	Error (%)
0.05	10.04	−0.07	10.03	−0.14	10.13	0.90
0.11	10.02	−0.23	10.01	−0.29	10.12	0.72
0.15	10.01	−0.31	10.01	−0.38	10.11	0.63
0.25	10.00	−0.46	9.99	−0.52	10.09	0.47
0.35	9.99	−0.51	9.99	−0.57	10.08	0.41
0.45	10.00	−0.46	9.99	−0.52	10.09	0.47
0.55	10.01	−0.31	10.01	−0.37	10.11	0.63
0.65	10.04	−0.05	10.03	−0.13	10.14	0.91
0.75	10.07	0.31	10.07	0.23	10.18	1.31
0.85	10.12	0.78	10.11	0.69	10.23	1.83
0.95	10.18	1.38	10.17	1.26	10.29	2.49
Max	10.18	1.38	10.17	1.26	10.29	2.49
Mean	10.04	0.01	10.04	−0.07	10.14	0.98
Min	9.99	−0.51	9.99	−0.57	10.08	0.41

was able to capture the trend in FEA predicted modulus, but the upper bound (Eq. 9) only predicted the minimum and deviated strongly either side of this value

The discrepancy between the FEA and standard RoM (Eq. 1) predictions are given in Table 3. The difference between the two prediction methods (standard RoM and FEA) is less than 2.5 %. Thus it can be concluded that the variation in the Poisson's ratio has only small (potentially insignificant) effect on the predicted composite modulus.

Conclusions

The finite element analysis reported above shows that the RoM can be safely applied to predict the elastic modulus of composites reinforced by anisotropic natural fibres of non-circular cross-section. Further, variation in the anisotropy by changing Poisson's ratio only has minimal effect on the predicted composite modulus.

Acknowledgements ASV is grateful to the University of Plymouth for a scholarship to pursue his doctorate.

References

Anon. (2014), SR 8100 epoxy system for Injection and Infusion, < http://www.sicomin.com/datasheets/product-pdf94.pdf> (accessed 04 June 2015).

Daniel I M and Ishai O (2005), Engineering Mechanics of Composite Materials, Oxford University Press. ISBN 019515097X.

Jones R M (1998), Mechanics of Composite Materials, CRC Press. ISBN 156032712X.

Huber M T (1923), The theory of crosswise reinforced ferroconcrete slabs and its application to various important constructional problems involving rectangular slabs, Der Bauingenieur, 4, 354-360 and 392-395.

Hull D and Clyne T W (1996), An Introduction to Composite Materials, Cambridge University Press. ISBN 0521388554.

Hyer M W and Waas A M (2000), Micromechanics of Linear Elastic Continuous Fiber Composites. Chapter 12 in Kelly A and Zweben C (editors), Volume 1: Fibre Reinforcements and General Theory of Composites, Elsevier Science, 345-375,

Lemprière B M (1968), Poisson's ratio in orthotropic materials, AIAA Journal, 6, 2226-2227.

Summerscales J (2000), The bulk modulus of carbon fibres, Journal of Materials Science Letters, 2000, 19(1), 15-16.

Summerscales J, Dissanayake N P J, Virk A S and Hall W (2010a), A review of bast fibres and their composites. Part 2-Composites, Composites Part A: Applied Science and Manufacturing, 41, 1336-1344.

Summerscales J, Dissanayake N P J, Virk A S and Hall W (2010b), A review of bast fibres and their composites. Part 1 - Fibres as reinforcements, Composites Part A: Applied Science and Manufacturing, 41, 1329-1335.

Summerscales J, Hall W, Virk A S (2011), A fibre diameter distribution factor (FDDF) for natural fibre composites, Journal of Materials Science, 2011, 46 (17), 5876-5880.

Summerscales J, Virk A and Hall W (2013), A review of bast fibres and their composites: Part 3 - Modelling, Composites Part A: Applied Science and Manufacturing, 44, 132-139.

Thomason J L (2009), Dependence of interfacial strength on the anisotropic fiber properties of jute reinforced composites, Society of Plastic Engineers, 31, 1525-1534.

Virk A S, Hall W and Summerscales J (2009a), The tensile properties of jute fibres, Materials Science and Technology, 25, 1289-1295.

Virk A S, Hall W and Summerscales J (2009b), Multiple data set (MDS) weak-link scaling analysis of jute fibres, Composites Part A: Applied Science and Manufacturing, 40, 1764-1771.

Virk A S, Hall W and Summerscales J (2010a), Physical characterisation of jute technical fibres: fibre dimensions, Journal of Natural Fibres, 7, 216-228.

Virk A S, Hall W and Summerscales J (2010b), Failure strain as the key design criterion for fracture of natural fibre composites, Composites Science and Technology, 70, 995-999.

Virk A S, Hall W and Summerscales J (2011), "Modelling tensile properties of jute fibres", Materials Science and Technology, 27, 458-460.

Virk A S, Hall W and Summerscales J (2012), "Modulus and strength prediction for natural fibre composites", Materials Science and Technology, 28, 864-871.

Zweben C (1994), Fibrous Composites: Thermomechanical Properties,. In: A Kelly, Concise Encyclopaedia of Composite Materials, Pergamon, 98–106.

Effects of Water Ageing on the Mechanical Properties of Flax and Glass Fibre Composites: Degradation and Reversibility

Guilherme Apolinario, Patrick Ienny, Stéphane Corn, Romain Léger, Anne Bergeret and Jean-Marc Haudin

Abstract Mechanical properties of flax-fibre reinforced composites (FFRC) are strongly affected by water ageing compared to glass-fibre reinforced composites (GFRC). This study highlights the influence of water absorption during immersion at 30 °C on mechanical properties of unsaturated polyester reinforced composites. Flax-fibre composites showed a Fickian absorption behaviour and a water uptake 15 times higher than that of glass-fibre composites. GFRC's tensile modulus and maximum stress were slightly affected by water uptake while FFRC's tensile modulus decreased by 37 %, and ultimate stress increased by 34 %. A silane-based chemical treatment (1 % compared to flax fibres) was applied onto flax fibres. Water uptake was slightly reduced by 9 % while tensile modulus at saturation was enhanced by 22 % on treated FFRC compared to untreated ones. Moreover, the complete recovery of the tensile modulus after desiccation suggests that ageing was mainly reversible: fibre and matrix plasticizing phenomena occurred during immersion at 30 °C. No damage was noticed but composites' initial properties changed with the action of water: further crosslinking of matrix and release of fibre's small cell-walls components into water were observed. Finally, the drying conditions influenced the return to the initial state before ageing insofar the flax fibres partially lost their initial humidity.

Keywords Flax fibers · Water ageing · Composites · Mechanical properties · Morphology

G. Apolinario (✉) · P. Ienny · S. Corn · R. Léger · A. Bergeret
Centre des Matériaux des Mines d'Alès (C2MA), Ecole des Mines d'Alès, Alès, France
e-mail: guilherme.apolinario@mines-ales.fr

J.-M. Haudin
CEntre de Mise En Forme des matériaux (CEMEF), Mines ParisTech, Paris, Sophia-Antipolis, France

R. Fangueiro and S. Rana (eds.), *Natural Fibres: Advances in Science and Technology Towards Industrial Applications*, RILEM Bookseries 12,
DOI 10.1007/978-94-017-7515-1_14

Introduction

According to the French FIN (Nautical Industries Federation) around 95 % of boats, which reach their end-of-life in 2015, are manufactured with glass-fibre reinforced composites (GFRC) mainly unsatured polyester based composites (Le Duigou et al. 2014). The application of GFRC to ship structures includes hull, decks and structural bulkheads but also framing parts and non-structural bulkheads (Smith 1990; Grabovac and Turley 1993). Marine applications of composites require materials with weak influence of water on diffusivity and mechanical properties (Davies et al. 1996; Camino et al. 1997; Gellert and Turley 1999; Bergeret et al. 2001). New environmental legislation and public pressure have promoted the search for bio-based materials to substitute these conventional non-renewable composite components (Le Duigou et al. 2009). Plant fibres, in particular flax, present some advantages comparing with glass fibres, such as low density, specific stiffness and biodegradability (Bledzki and Gassan 1999), and using natural fibres as alternatives to glass fibres is relatively recent (Mussig 2010; Baley et al. 2006). The feasibility of this substitution is the subject of numerous studies (Bledzki and Gassan 1999; Pillin et al. 2011; Shah 2014; Faruk et al. 2012) since natural fibres present a strong sensitivity to water which leads to composite degradation (Le Duigou et al. 2009; Scida et al. 2013; Dhakal et al. 2007; Assarar et al. 2010).

Concerning GFRC, Davies et al. (1996) have demonstrated that these composites showed a Fickian diffusion behaviour during immersion at room temperature and that the weight gain was not affected by post-cure treatment. Davies et al. reported also non-Fickian behaviour during immersion at 50 °C attributed to water transport through free edges along the fibre/matrix interface (wicking mechanism). Fraga et al. (2003) highlighted the plasticizing effect of water in polymer composites and the removing of extractibles during immersion. The resin plasticization (physical modification) was displayed by the glass transition temperature changes but chemical modification (hydrolysis of chain molecules) was not observed neither at 40 °C, nor at 80 °C. Thus the main reason of the observed weight decrease should be the extraction of monomers or oligomers.

In recent years, biocomposites degradation studies have been performed to characterize long-term influence of water on diffusivity and mechanical properties. Assarar et al. (2010) showed that water ageing of flax-fibre composites reinforced composites (FFRC) at room temperature exhibited a Fickian behaviour and a saturated weight gain 12 times higher than that of GFRC. Water ageing of FFRC degraded the Young's modulus (−39 %) and the ultimate tensile stress (−15 %), but improved the ultimate tensile strain (+63 %). Concerning GFRC, both Young's modulus and maximum strain were slightly affected by water (saturated weight gain around 1 %). Acoustic emission technique suggested that matrix cracking was the dominant damage mechanism in GFRC and FFRC. However, the 25 % ultimate stress decrease in GFRC indicated that other damage mechanisms were taking place.

Chemical treatments can be used for modifying the surface characteristics of natural fibres. Xie et al. (2010) showed that a γ-methacryloxypropyltrimethoxysilane

(MPS) treatment can offer a good affinity between natural fibres and polyester matrix. Moreover, polysiloxane formed a monolayer on the fibre surface, and then was adsorbed by the hydroxyl groups of fibres. Alix et al. (2011) showed that weight gain can be slightly reduced by around 1 % by treating flax-fibre unsaturated polyester composites with MPS. The silane treatment impact on water sorption of composites could be interpreted by an interface effect, which consists in improving the fibre/matrix interface by a crosslinking reaction between methacrylate groups of silane agent and the unsaturated functions of polyester chains.

Materials and Methods

Materials

Resin and Fibres

A dicyclopentadiene isophthalic unsaturated polyester resin Enydyne® (Cray Valley—Rouvroy, France) was used. The resin was polymerized with 1.8 % w/w of methyl ethyl ketone peroxide Luperox K1S® (Arkema—Colombes, France).

Unidirectional glass fibres were supplied by Chomarat (Le Cheylard, France) for a weight of 416 g/m^2 (Weft: 408 g/m^2/Warp: 8 g/m^2). Unidirectional flax fibres were manufactured by Fibre Recherche Developpement (Troyes, France) for a weight of 390 g/m^2 (Weft: 360 g/m^2/Warp: 30 g/m^2).

Chemical Treatment

Glass fibre fabrics were pre-treated by the manufacturer with a silane based sizing. A chemical treatment (1.0 wt% γ-methacryloxypropyltrimethoxysilane or MPS compared to flax fibres) was set up to flax fibres by immersing the fabrics into an ethanol and water (60/40 wt%) solution containing MPS. Silane was hydrolysed during 3 h at pH 4. Flax fibre fabrics were thereafter dried in an oven at 105 °C during 2 h and stored at 23 °C and 50 % H.R. before processing.

Vacuum Infusion Processing of Composite Sheets

Composite sheets (300 mm × 250 mm) consisting of 6 layers of glass fabrics for GFRC and 4 layers of flax fabrics for FFRC were vacuum infused with the resin. A vacuum pump was used to maintain impregnated fibres under a constant pressure of 100 mbar during 24 h (curing at room temperature). Subsequently, a post-curing step took place in an oven at 60 °C for 24 h to obtain a polyester crosslinking rate of 88 %. The thickness of GFRC (respectively FFRC) ranged from 1.8 to 2.1 mm (respectively 2.7–3.4 mm). Samples were cut to size of 250 mm × 25 mm using a diamond cutter. Composites of 54 and 32 % vol. of fibres were fabricated for GFRC and FFRC respectively. All specimens were stored in a climatic room, at 23 °C and 50 % H.R., before testing.

Methods

Water Ageing Conditions

Six samples were immersed into water for more than 6 months at 30 °C, period during which relative weight uptake and tensile modulus were followed respectively by gravimetric and vibration analysis. Ultimate tensile stress and strain were determined by uniaxial tensile tests and the composites fracture surfaces analysed by scanning electron microscopy.

Water Uptake, Volume and Density Variations

Ageing tests were performed in a controlled temperature bath with specimens immersed for different periods and taken away from the bath for weight measurements. Composite samples were carefully wiped to remove water from surface, weighed and replaced into the water baths. The water uptake W was determined according to Eq. (1):

$$W\,(\%) = ((w_t - w_o)/w_o) \times 100 \tag{1}$$

where w_t (resp. w_0) is the mass of sample at time t (resp. at time t = 0 h).

Composites absorbed water until saturation, and exhibited a diffusion kinetics that can be described by the Fick's law, whose expression for a plane sheet of thickness h is developed in Eq. (2):

$$\frac{W}{W_m} = 1 - \frac{8}{\pi^2}\sum_{n=0}^{\infty}\frac{1}{(2n+1)^2}\exp\left(\frac{-(2n+1)^2\pi^2 Dt}{h^2}\right) \tag{2}$$

where W is the water uptake at time t, W_m is the maximum water uptake, at equilibrium state, D is the diffusion coefficient and h is the composite thickness.

The water uptake varies linearly according to the square root of immersion time in the initial part of the Fick's curve (i.e., $\frac{W}{W_m} < 0.6$), thus leading to the assessment of the diffusion coefficient D from Eq. (3):

$$D = \left(\frac{k}{4W_m}\right)^2 \pi \tag{3}$$

where k is the slope of the linear part of the curve $W = f\left(\sqrt{t}/h\right)$.

Composites Drying

After saturation was reached, samples were desiccated at 30 °C/2 % RH until weight stabilisation. Then, mechanical properties were determined for dried samples to assess the impact of reversible (physical) and irreversible (chemical) ageing.

Viscoelastic Properties Measured by Vibration Analysis

The viscoelastic parameters (elastic modulus and loss factor) of polyester matrix, FFRC and GFRC were determined using vibration analysis. On a general point of view, vibration techniques aim to study the dynamic behaviour of a structure thanks to its natural vibration modes (Corn et al. 2012). This technique is non-destructive and easy to implement, therefore it is well suited for monitoring materials during ageing. For this study, the set-up simulated free boundary conditions by supporting the sample with soft suspensions (Gibson 2000). The sample, a 250 mm × 25 mm beam, was excited in free vibration by an impulse hammer and its response was monitored by an accelerometer. An additional steel mass (86 g) was clamped at each end of the beam in order to bring down the natural frequency f_1 of its first "traction-compression" mode inside the measurable frequency range (100–9000 Hz). The resonant frequencies of the sample are given by the series of peaks in the frequency response. The software MODAN®, developed by the FEMTO-ST (Franche-Comté University, Besançon, France), was used to extract natural frequencies and damping ratios from the frequency response function (by using a curve fitting technique). Then, the expression of the elastic modulus E was given in Eq. (4):

$$E = \frac{\rho L^2 f_1^2}{\left(\frac{\beta_1}{\pi}\right)^2} \tag{4}$$

with β_1 being the lowest positive root of Eq. (5):

$$\beta_1 \cdot \tan \beta_1 = \frac{m}{M} \tag{5}$$

where M is the total added mass (172 g), L and m being respectively the length and the mass of the part of the sample located between these two clamped masses, and ρ its mass density.

Tensile quasi-static measurements were carried out to ensure that dynamic elastic modulus was close to Young's modulus for these materials.

Uniaxial Tensile Tests

Ultimate tensile properties were assessed with a MTS testing machine (model Criterion C45.105) equipped with a 100 kN capacity load cell. Samples were loaded at a constant crosshead displacement rate of 1 mm/min at room temperature until breaking according to the standard ISO EN 2747. The reproducibility was evaluated on 5 samples.

Scanning Electron Microscopy

An Environmental Scanning Electron Microscope (FEI Quanta 200 ESEM) was used to observe biocomposites fracture surfaces obtained from tensile tests before and after immersion into water. Composite samples were observed at different scales: low magnification allows a global fracture view (particularly the revelling zone), and high magnification revealed the interface between matrix and fibre and the fibres failure mode.

Differential Scanning Calorimetry

Thermograms were recorded using a Diamond DSC—PerkinElmer. Polyester resin samples of approximately 20 mg were analysed in aluminium pans submitted to temperature ramps from 25 °C up to 220 °C at a heating rate of 10 °C/min. The cross-linking degree (X) was estimated from post-curing peak using Eq. (6):

$$X = \frac{\Delta H_{total} - \Delta H_{peak}}{\Delta H_{total}} \times 100 \tag{6}$$

with ΔH_{total} the total reaction enthalpy and ΔH_{peak} the residual enthalpy of the polymerized resin.

Results and Discussion

Figure 1 describes the water uptake of GFRC and FFRC compared to polyester resin as a function of ageing time. It can be observed that flax fibres tended to accelerate diffusion process while glass fibre acted like a barrier to water absorption in composite (compared to non-reinforced polyester). Moreover, the water uptake obeyed a Fickian behaviour with a saturation weight gain 15 times higher for FFRC than for GFRC. Assarar et al. (2010) showed a saturated weight gain 12 times higher for flax/epoxy composites than for glass/epoxy composites (samples immersed into a water bath at room temperature). Finally it was shown that the flax

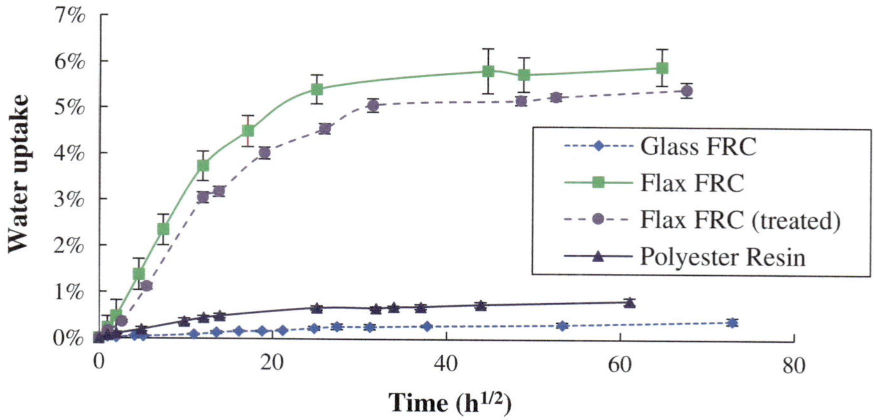

Fig. 1 Water uptake as a function of ageing time for polyester resin, glass and flax-fibre composites (untreated and silane treated flax fibre) immersed in water at 30 °C

fibre treatment slowed down water diffusion in composite and leaded to a slight reduction of 9% on weight gain compared to untreated FFRC.

Table 1 shows the values of the modulus (E_0) and damping (a_0) before ageing, while Fig. 2 displays the relative modulus evolution of GFRC and FFRC compared

Table 1 GFRC, FFRC and polyester resin modulus and damping before ageing

References	Initial	
	E_0 (GPa)	a_0 (%)
Polyester resin	3.25 ± 0.64	1.25 ± 0.18
GFRC	34.88 ± 1.16	0.24 ± 0.01
FFRC	20.26 ± 1.16	1.00 ± 0.06
FFRC (treated)	20.44 ± 0.74	0.99 ± 0.04

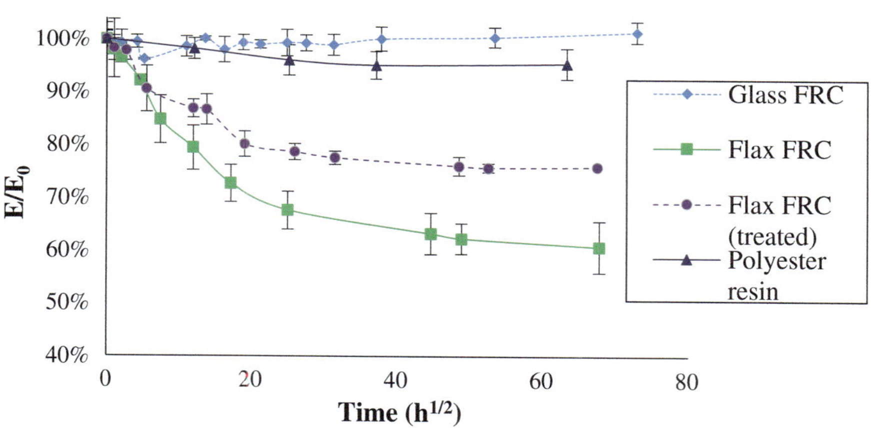

Fig. 2 Relative modulus evolution as a function of ageing time for polyester resin, glass and flax-fibre composites (untreated and silane treated flax fibre) immersed in water at 30 °C

to polyester resin as a function of ageing time. The relative modulus value is the modulus of the aged sample divided by the modulus of the unaged one (E_0).

On one hand no change in modulus was observed for GFRC while a slight decrease was obtained for polyester resin. On the other hand a 37 % decrease with ageing time was depicted for FFRC which was lowered (−22 %) when flax fibre fabrics were treated. It is well known that water diffusion induces both plasticization that leads to a decrease in mechanical properties and post-crosslinking (because of higher macromolecular chain mobility) that leads to higher mechanical properties (Azwa et al. 2013). These two phenomena were in competition during ageing (Fig. 3). The polyester resin post-crosslinking had been verified by differential scanning calorimetry: the crosslinking rate increased from 88 % before immersion to 96 % at saturation (Fig. 3a). Nevertheless a decrease in modulus was shown, thus denoting that plasticization was predominant. The plasticizing effect for matrix was shown on Fig. 3b. For GFRC both phenomena induced equivalent effects. For FFRC the water uptake was so important that only plasticization was observed. The fibre plasticizing effect can be highlighted by monitoring the damping ratio of the materials during ageing.

Figure 4 corresponds to the damping of the polyester resin, GFRC and FFRC as a function of ageing time. The increase of FFRC's damping during immersion confirmed that viscoelastic properties changed with water uptake. This increase was quite less important for treated flax fibre fabrics. This change was due to the flax fibre plasticization in the presence of water (Joffe et al. 2003). GFRC's damping remained constant along the time confirming the hydrophobic behaviour of glass fibres. The polyester plasticization in water was also observed by following the

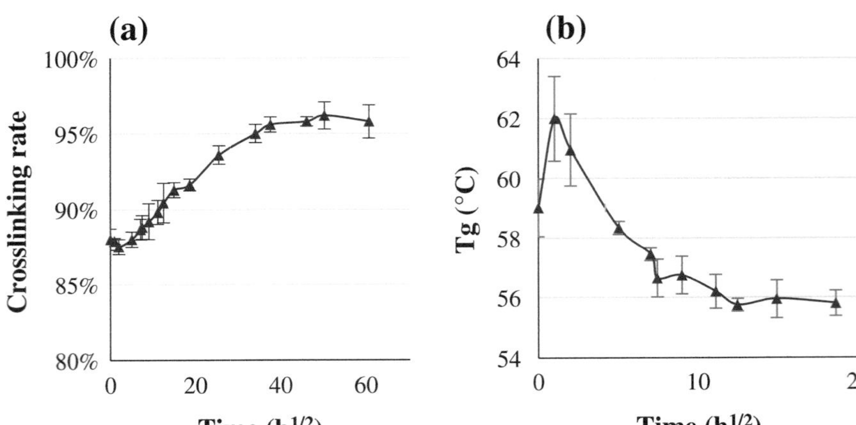

Fig. 3 Crosslinking rate (**a**) and T_g evolution (**b**) as a function of ageing time for polyester resin immersed in water at 30 °C

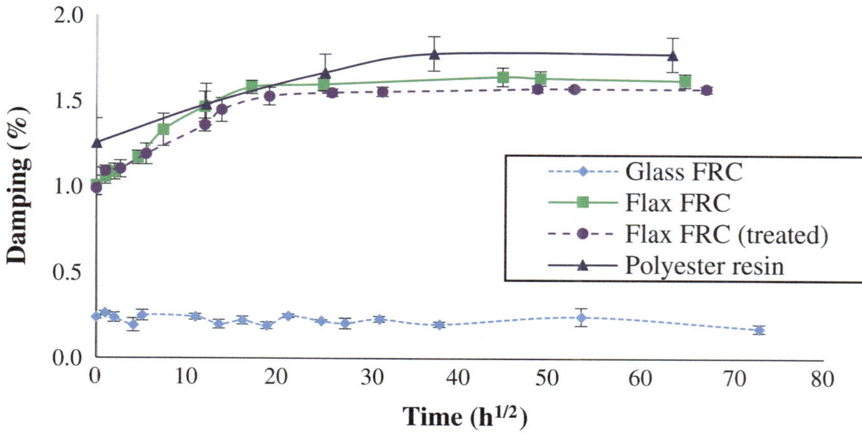

Fig. 4 Damping ratio as a function of ageing time for polyester resin, glass and flax-fibre composites (untreated and silane treated flax fibre) immersed in water at 30 °C

glass transition temperature during ageing (by differential scanning calorimetry): polyester's T_g decreased from 60 to 55 °C for 1600 h in water (Fig. 3b) which is in agreement with other work (Fayolle and Verdu 2005).

When saturation was reached, ultimate tensile tests were performed on composites and the fracture surfaces analysed by ESEM (Fig. 5). The presence of unbounded zones (Fig. 5a) after water immersion suggested that debonding took place at the interface between matrix and fibres. This debonding can lead to fibre slippage, which was highlighted by a longer revelling length (Fig. 5b). Finally, the fibre ductile fracture behaviour after immersion (Fig. 5c) confirmed that fibre plasticization was related to the presence of water within composites.

Figure 6 exhibits the mechanical properties of composites at initial state (before immersion), at saturation, and after drying. The results confirmed that GFRC modulus was not affected by the ageing performed in this work. The slight decrease in ultimate stress at saturation could be explained by interfacial debonding induced by differential swelling between fibres and matrix (Gautier et al. 1999). Concerning FFRC, both mechanical properties were affected by water sorption: modulus decreased by 37 % and ultimate stress increased for about 34 %. A similar evolution for stress was reported in other studies (Dhakal et al. 2007). Dhakal et al. attributed the increase in wet ultimate tensile stress (compared to dry samples), to the fact that water caused swelling of the fibres and could fill the gap between fibre and matrix, which could lead to an increase in ultimate properties. In this work, as shown in Fig. 5, no gaps were observed before immersion and the increase of ultimate stress was probably due to an improved frictional effect caused by fibres swelling (Le Duigou et al. 2012). When flax fibres were treated, the decrease in

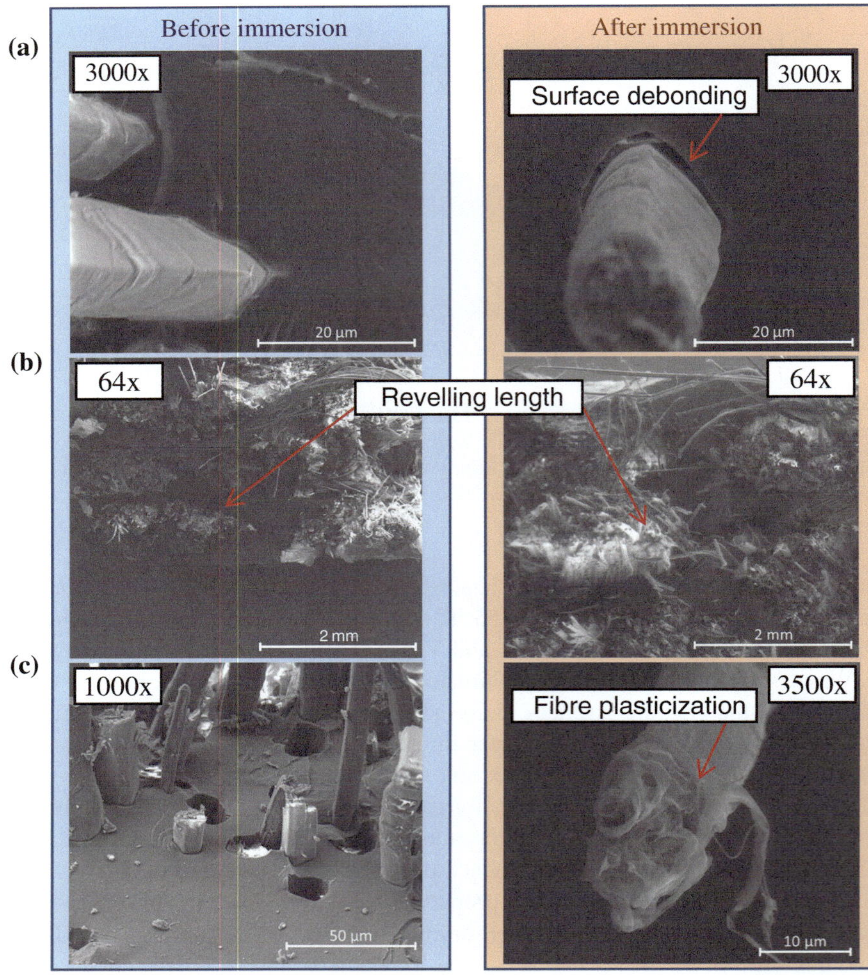

Fig. 5 Fracture surface analysis of untreated FFRC before ageing and after ageing at saturation (immersion in water at 30 °C)

Young's modulus and the increase in ultimate stress were less important. These results are in agreement with the expected behaviour for a composite which absorbs less water than the untreated ones.

After desiccation, FFRC tensile modulus had increased compared to initial modulus. Two phenomena could explain that result: (i) further matrix crosslinking (observed through DSC analysis) (ii) lower moisture content compared to initial state. Indeed, composites reached a dried mass 2 % lower than initial state before immersion (see Fig. 7). The weight loss after desiccation could be explained by two factors: (i) the fibres get drier than at the initial state and (ii) the small cell-walls

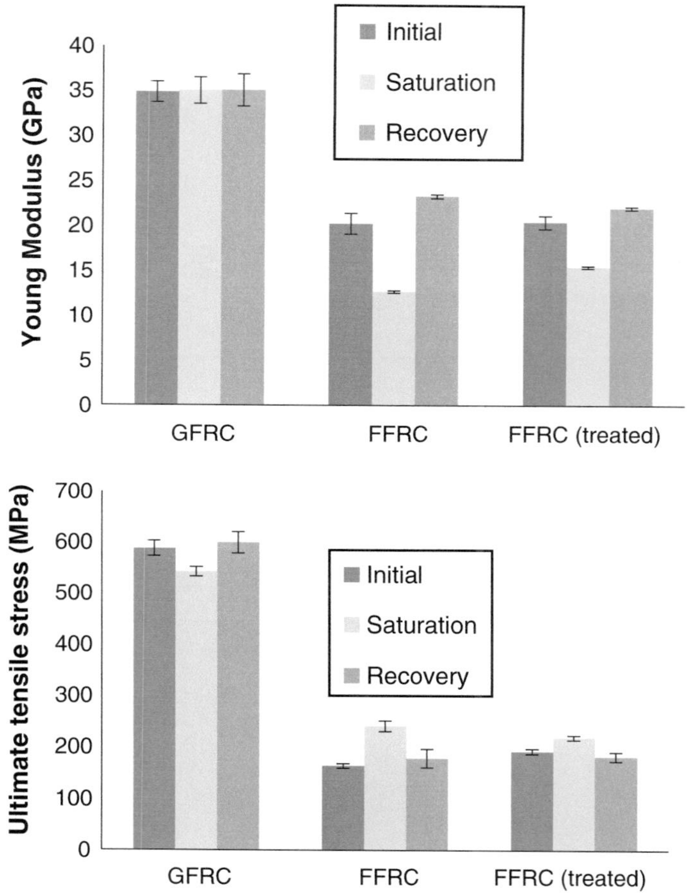

Fig. 6 Mechanical properties (Young's modulus and ultimate tensile stress) of glass and flax-fibre composites (untreated and silane treated flax fibre) before immersion, at saturation and after drying

components were removed during immersion which was observed by scanning electron microscope analysis (see Fig. 8). Furthermore, glass-fibre and treated flax-fibre composites showed ultimate tensile stress that had not changed compared to initial values. Pavlidou and Papaspyrides (2003) reported that fibre-resin bond was reversible on redrying wet specimens because silane coupling agent bonds were easily hydrolysed in the presence of water. In the case of untreated flax fibre composites, interlaminar shear tests are being carried out in order to verify the interfacial adhesion between fibre and matrix after drying.

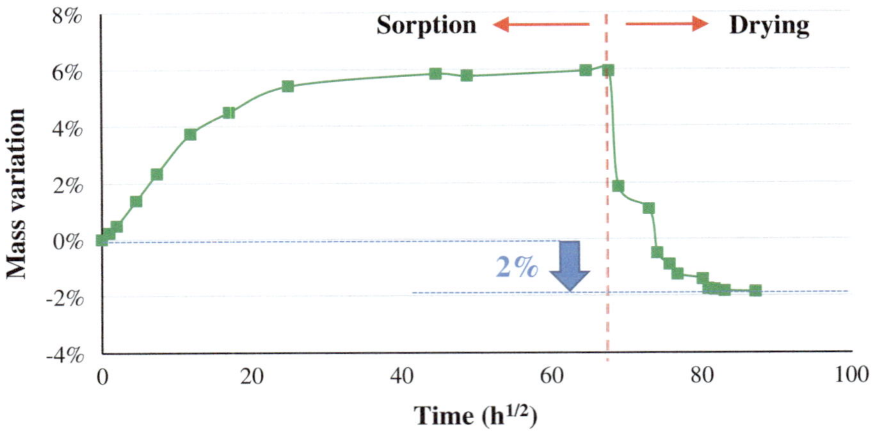

Fig. 7 Mass variation of flax fibre reinforced composites during sorption and drying at 30 °C

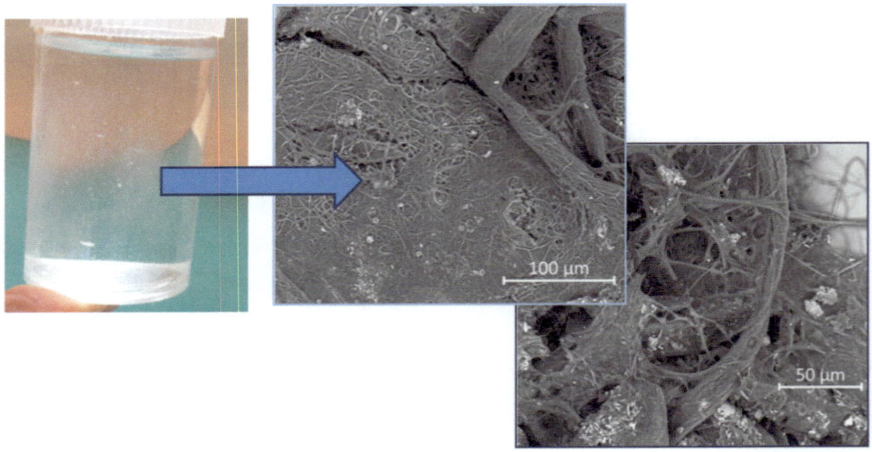

Fig. 8 Small cell-walls components removed during immersion observed by analysing the ageing water using an electronic microscope

Conclusions

This study showed that composite's water uptake in immersion at 30 °C was mainly controlled by natural fibres and reversible phenomena like fibre and matrix plasticizing took place were observed. The composite's stiffness decreased and damping increased during ageing, highlighting the plasticizing effect of flax fibres. Thus the biocomposite's ultimate tensile stress increased at saturation. No damage was noticed but irreversible phenomena such as further matrix crosslinking and release of small cell-walls components took place during ageing. It was noted that the modifications of biocomposites during immersion were complex to study: physico-chemical plus mechanical modifications are important factors that need to be taken into account simultaneously.

References

Alix, S., Lebrun, L., Morvan, C., Marais, S., 2011. Study of water behaviour of chemically treated flax fibres-based composites: A way to approach the hydric interface. Compos. Sci. Technol. 71, 893–899.

Assarar, M., Scida, D., Mahi, A. El, Poilâne, C., Ayad, R., 2010. Influence of water ageing on mechanical properties and damage events of two reinforced composite materials: Flax–fibres and glass–fibres. Mater. Des. 32, 788–795.

Azwa, Z.N., Yousif, B.F., Manalo, A.C., Karunasena, W., 2013. A review on the degradability of polymeric composites based on natural fibres. Mater. Des. 47, 424–442.

Baley, C., Perrot, Y., Busnel, F., Guezenoc, H., Davies, P., 2006. Transverse tensile behaviour of unidirectional plies reinforced with flax fibres. Mater. Lett. 60, 2984–2987.

Bergeret, A., Pires, I., Foulc, M.P., Abadie, B., Ferry, L., Crespy, A., 2001. The hygrothermal behaviour of glass-fibre-reinforced thermoplastic composites: a prediction of the composite lifetime. Polym. Test. 20, 753–763.

Bledzki, A.K., Gassan, J., 1999. Composites reinforced with cellulose based fibres. Prog. Polym. Sci. 24, 221–274.

Camino, G., Luda, M.P., Polishchuk, A.Y., Revellino, M., Blancon, R., Merlec, G., 1997. Kinetic aspects of water sorption in polyester-resin/Glass-fibre composites. Compos. Sci. Technol. 51, 1469–1482.

Corn, S., Dupuy, J., Ienny, P., Daridon, L., 2012. Vibration analysis techniques for detecting filler-matrix decohesion in composites. Rev. des Compos. des matériaux avancés 77–90.

Davies, P., Pomiès, F., Carlsson, L., 1996. Influence of Water and Accelerated Aging on the Shear Fracture Properties of Glass/ Epoxy Composite. Appl. Compos. Mater. 3, 71–87.

Dhakal, H., Zhang, Z., Richardson, M., 2007. Effect of water absorption on the mechanical properties of hemp fibre reinforced unsaturated polyester composites. Compos. Sci. Technol. 67, 1674–1683.

Faruk, O., Bledzki, A.K., Fink, H.-P., Sain, M., 2012. Biocomposites reinforced with natural fibers: 2000–2010. Prog. Polym. Sci. 37, 1552–1596.

Fayolle, B., Verdu, J., 2005. Vieillissement physique des matériaux polymères. Tech. L'Ingénieur 1–19.

Fraga, A., Alvarez, V., Vazquez, A., 2003. Relationship between dynamic mechanical properties and water absorption of unsaturated polyester and vinyl ester glass fiber composites. J. Compos. Mater. 37, 1553–1574.

Gautier, L., Mortaigne, B., Bellenger, V., 1999. Interface damage study of hydrothermally aged glass-fibre-reinforced polyester composites. Compos. Sci. Technol. 59, 2329–2337.

Gellert, E.P., Turley, D.M., 1999. Seawater immersion ageing of glass-fibre reinforced polymer laminates for marine applications. Compos. Part A Appl. Sci. Manuf. 30, 1259–1265.

Gibson, R.F., 2000. Modal vibration response measurements for characterization of composite materials and structures. Compos. Sci. Technol. 60, 2769–2780.

Grabovac, I., Turley, D., 1993. Present and future composite materials for marine applications. In: Miravete A, E. (Ed.), Proceedings of the Ninth International Conference on Composite Materials (ICCM-9). Miravete A, editor, Madrid, p. 89.

Joffe, R., Andersons, J., Wallström, L., 2003. Strength and adhesion characteristics of elementary flax fibres with different surface treatments. Compos. Part A Appl. Sci. Manuf. 34, 603–612.

Le Duigou, A., Bourmaud, A., Balnois, E., Davies, P., Baley, C., 2012. Improving the interfacial properties between flax fibres and PLLA by a water fibre treatment and drying cycle. Ind. Crops Prod. 39, 31–39.
Le Duigou, A., Bourmaud, A., Davies, P., Baley, C., 2014. Long term immersion in natural seawater of Flax/PLA biocomposite. Ocean Eng. 90, 140–148.
Le Duigou, A., Davies, P., Baley, C., 2009. Seawater ageing of flax/poly(lactic acid) biocomposites. Polym. Degrad. Stab. 94, 1151–1162.
Mussig, J., 2010. Industrial Applications of Natural Fibres. John Wiley & Sons, Ltd, Chichester, UK.
Pavlidou, S., Papaspyrides, C.., 2003. The effect of hygrothermal history on water sorption and interlaminar shear strength of glass/polyester composites with different interfacial strength. Compos. Part A Appl. Sci. Manuf. 34, 1117–1124.
Pillin, I., Kervoelen, A., Bourmaud, A., Goimard, J., Montrelay, N., Baley, C., 2011. Could oleaginous flax fibers be used as reinforcement for polymers? Ind. Crops Prod. 34, 1556–1563.
Scida, D., Assarar, M., Poilâne, C., Ayad, R., 2013. Influence of hygrothermal ageing on the damage mechanisms of flax-fibre reinforced epoxy composite. Compos. Part B Eng. 48, 51–58.
Shah, D.U., 2014. Natural fibre composites: Comprehensive Ashby-type materials selection charts. Mater. Des. 62, 21–31.
Smith, C.S., 1990. Design of marine structures in composite materials. Elsevier Science Publishers, London.
Xie, Y., Hill, C. a. S., Xiao, Z., Militz, H., Mai, C., 2010. Silane coupling agents used for natural fiber/polymer composites: A review. Compos. Part A Appl. Sci. Manuf. 41, 806–819.

Processing of Wet Preserved Natural Fibers with Injection Molding Compounding (IMC)

Hans-Jörg Gusovius, Gesine Wallot, Stefan Schierl, Roman Rinberg, Tobias Hartmann, Lothar Kroll and Ivonne Jahn

Abstract The paper shows selected results of a comprehensive national network activity related to the integrated value chain for new whole plant based fibrous raw materials. Therefore, both the supply of raw materials and the processing with the innovative one step technology injection molding compounding (IMC)—are optimized. As a result the fiber length in the product is longer compared to the standard two-step process with pelletizing. The new technology helps in improving the manufacturing of fiber reinforced composites by considering the integrated value chain.

Keywords Plant fibers · Injection molding compounding · One-step technology · Mechanical properties

List of Abbreviations

°C	Temperature in grade centigrade
ATB	Leibniz-Institute for Agricultural Engineering Potsdam-Bornim
BD/SM	Butadiene to styrene-monomer ratio
cf.	Compare
D	Diameter
DM	Dry matter content

H.-J. Gusovius (✉) · G. Wallot
Department of Post Harvesting Technologies, Leibniz-Institute for Agricultural Engineering Potsdam-Bornim, Potsdam, Germany
e-mail: hjgusovius@atb-potsdam.de

S. Schierl
KraussMaffei Technologies GmbH, Munich, Germany

R. Rinberg · T. Hartmann · L. Kroll
Faculty of Mechanical Engineering, Department of Lightweight Structures and Polymer Technology, Technical University Chemnitz, Chemnitz, Germany

I. Jahn
Fraunhofer Institute for Mechanics of Materials IWM, Halle, Germany

R. Fangueiro and S. Rana (eds.), *Natural Fibres: Advances in Science and Technology Towards Industrial Applications*, RILEM Bookseries 12,
DOI 10.1007/978-94-017-7515-1_15

FENAFA	Ganzheitliche Bereitstellungs-, Verarbeitungs- und FErtigungsstrategien von NAturFAserrohstoffen
Fig.	Figure
IMC	Injection molding compounder
IWM	Fraunhofer Institute for Mechanics of Materials IWM
KM	KraussMaffei
L/D	Length diameter ratio
μm	Micrometer
mm	Millimeter
rpm	Revolution in rounds per minute
R&D	Research and development
SEBS	Styrene-ethylene/butylene-styrene block copolymer
temp.	Temperature
TPE	Thermoplastic elastomer
WPC	Wood polymer composite
wt%	Weight percent

Introduction

Growing, processing and utilization of agricultural based fiber resources has a long tradition both in textile and industrial applications (Mohanty et al. 2005; Hänninen and Hughes 2010; Shahzad 2012; Faruk et al. 2012). Recently an increasing amount of such raw materials is produced and specifically used in industrial applications within the European Union (Carus et al. 2014). Even though only 6 % (15,000 t) respectively 2 % (2000 t) of forest or agricultural based raw materials are processed by injection molding in the plastics industry. There is a growing interest for natural fiber reinforced materials in several fields such as construction and automotive. Thus, several bottlenecks need to be solved for the utilization of natural fibers by such technologies.

For more than 5 years 15 industrial companies and scientific institutions have been working in the natural fiber network FENAFA under the leadership of Chemnitz University of Technology. The network focused on new supply chain strategies and processing methods for domestic fiber plants. This more efficient and less risky use of fibers should attract more farmers to increase the cultivation of these plants. Starting point of this cooperation network have been raw material supply procedures with significantly reduced efforts and costs. FENAFA focused its work on three primary fields: First, in the agricultural technology sector, field procedures were to be improved in order to provide competitive raw materials at reduced price and in high quality. A second area focused on the development of processes to manufacture natural fiber based components reinforced with nonwoven semi-finished products. The third core activity has been the utilization of

"TechnoFlax", wet preserved hemp, and wood fibers in injection molding processes. Along the entire processing chain, mass market products such as portable cases, toolboxes and biodegradable toys for children have been produced with the developed compounds reinforced by regionally grown fibers.

Novel Supply Chain for Fibrous Raw Materials

The conventional production of natural fibers from materials such as hemp or flax is based on field drying and retting of straw. Under the condition of usual harvest time, weather conditions are often problematical for the harvest of dry straw (Gusovius et al. 2000; Hoffmann et al. 2013; Liu et al. 2015).

A weather-independent post harvest technique is under investigation at the ATB (Idler et al. 2011). The harvest of fiber crop (e.g. hemp) by means of a chopper followed by an immediate anaerobic storage is favourable for the farmer because available machines can be operated. Additional steps are the same as for ensiling of fodder. Thus, a raw material on a homogeneous quality level is available for further processing and utilization throughout the following period until the next growing season and harvest.

Additionally, one of a few advantages of this novel technology is the processing of the whole plant material to final products such as fibers for plastic reinforcement or fiber boards on an economic viable level (Gusovius et al. 2010; Wallot et al. 2012). A pilot scale plant with a raw material input of up to 1 t h^{-1} (300 kg h^{-1} DM) was established at the ATB in order to investigate the processing and further utilization of the resulting fiber material in several industrial applications (Wallot et al. 2012). In recent years the main focus of R&D at ATB has been the specification of the storage process and the processing steps for the manufacturing of fiber boards (Hoffmann et al. 2009; Pecenka et al. 2009).

In contrast to that significant adoptions of harvest, preservation and fiber generation have been actual key aspects in order to serve products useful for plastics reinforcement. Prior studies have already shown that the reinforcement potential of natural fibers is highly dependent on the fiber fineness as well the aspect-ratio of fiber elements (Ruth et al. 2002; Stark et al. 2003; Specht 2007; Schirp and Stender 2010).

Specific investigations were carried out within the 5 year duration of the FENEFA network in order to improve the quality of the raw material beginning with an optimized preservation. Identical varieties (Fedora 17 and Santhica 27) were grown in each of the growing seasons at the experimental field of ATB in Potsdam-Bornim. Dependent on plant maturity and weather the yearly harvests have been realized by means of a forage chopper at particle size of approx. 10–20 mm between end of August and end of September. Subsequently the resulting plant material was properly densified and stored under anaerobic conditions in preserving bins or model silos (Fig. 1).

Fig. 1 Anaerobic storage in model silo bins (*left*) and chopped plant material (*right*)

Regularly, small sub-samples were taken from the stock for the biochemical analysis of preservation progress and resulting metabolism products of microorganisms. There is a substantial importance of such investigations because the desired stabilization of the preservation is a result of microbial activities and the conversion of carbohydrates into lactic acid. Furthermore the content of this and other organic acids as well as alcohols is influencing the organoleptic properties of the plant material. In particular the main focus of these experiments has been to determine the effect of different silage additives in order to reduce the content of odour relevant substances.

The results show that the addition of silage aid agents has a considerable influence on the content of odour relevant metabolism products in the raw material (Figs. 2, 3). Thus, the effect of propionic acid and as well in combination with a pectinase enzyme (Natuzym) results in a reduction of the content of both alcohols and the majority of organic acids. Organoleptic observations during the removal from the stock confirmed these measurements as the mentioned samples gave the most pleasant impressions.

Fig. 2 Content of metabolism products in wet preserved hemp after 6 weeks within the storage period 2012–2013

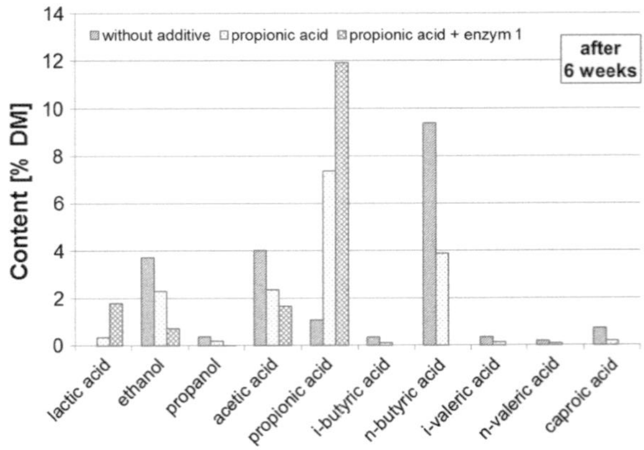

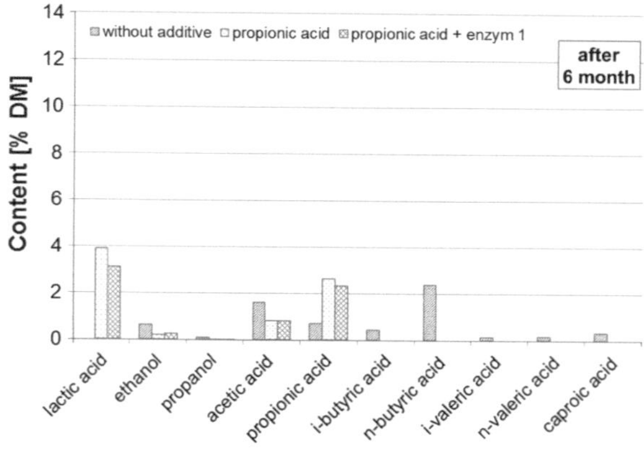

Fig. 3 Content of metabolism products in wet preserved hemp after 6 month within the storage period 2012–2013

Further analysis of the preserved plant materials were carried out to assess the stock composition and the change of the pectin content particularly. From previous investigations it is known that the microbiological activities in the course of wet preservation are comparable to those known as retting. The soluble part of the fiber gluing middle lamella is degraded by enzyme activities originating from anaerobic bacteria's (Fig. 4, right).

The content of hemicelluloses as an indicator of pectin's is noticeable reduced by up to 80 % independent of the type and variation of wet preservation (Fig. 4, left).

It can be summarized that the procedure of wet preservation is simple and low cost possibility to supply fiber crop raw materials for further processing and utilization. The use of silage aid additives can support the improvement of the odor characteristics but seems to be insignificant for a suitable reduction of fiber gluing substances. The latter is an important precondition in order to enable a purposeful fiber processing into an intermediate applicable for plastic reinforcement (Fig. 5).

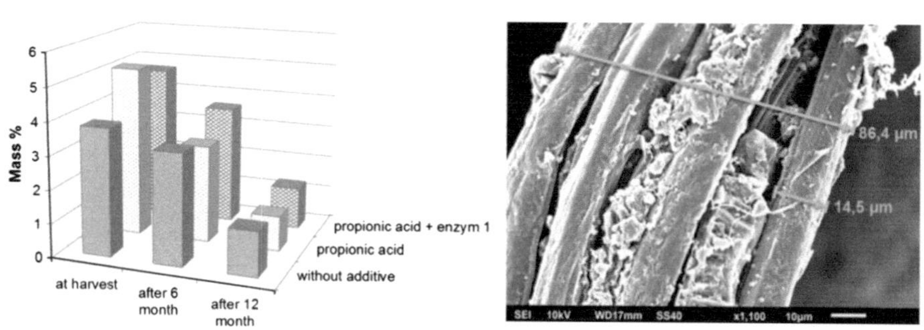

Fig. 4 Content of hemicelluloses in wet preserved hemp within storage period 2011–2012 (*left*) and REM-view of a fiber bundle after 12 month wet preservation (*right*)

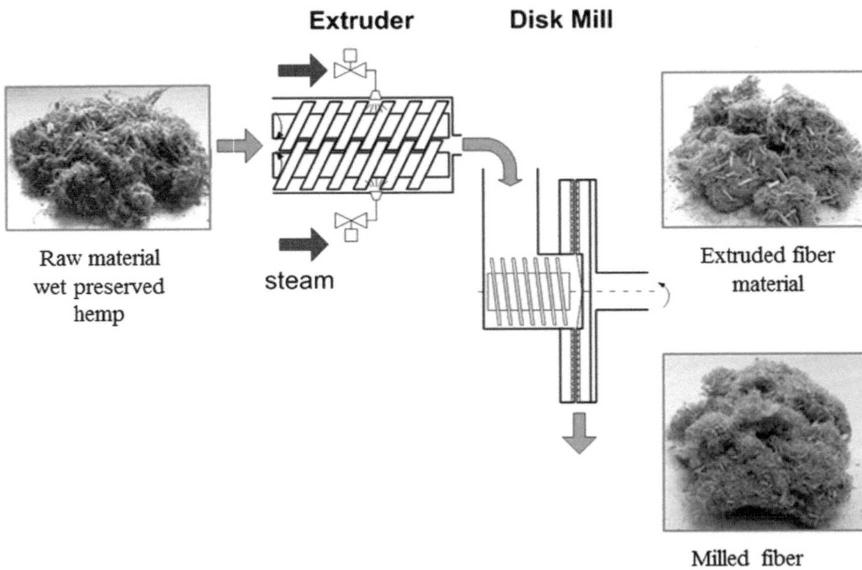

Fig. 5 Process sketch of fiber processing as part of ATB pilot plant

Raw material processing is carried out by means of a defibration extruder and a disc mill (Fig. 5). At both stages the resulting fiber quality can be influenced by the moisture content of the feedstock as well as operating parameters of both of the devices.

The processing of raw material with a lower dry matter content at the extruder results in preferred geometric fiber characteristics. A larger outlet gap of the extruder results in a similar effect (Table 1).

Further experiments were focused on the reduction of under and over sized particles in particular dust as well as remaining shive pieces. The former can already be influenced by a modification of the harvest process. Separate cutting and handling of the plant tops (flowers, inflorescence, seeds, leaves) and exclusive

Table 1 Particle size characteristics according to selected operating parameters of the extruder and dry matter content of wet preserved hemp

Test series		Mesh analysis	Image analysis by "FibreShape" (length weighted)		
Revolution/outlet gap (rpm/mm)	Dry matter content (%)	x_{50}-value (µm)	Aspect-ratio (–)	x_{50}-length (µm)	x_{50}-width (µm)
40/16	30	888	9.6	559	72
40/16	47	765	8.8	457	65
40/23	30	1055	10.6	702	80
40/23	47	761	9.5	530	70

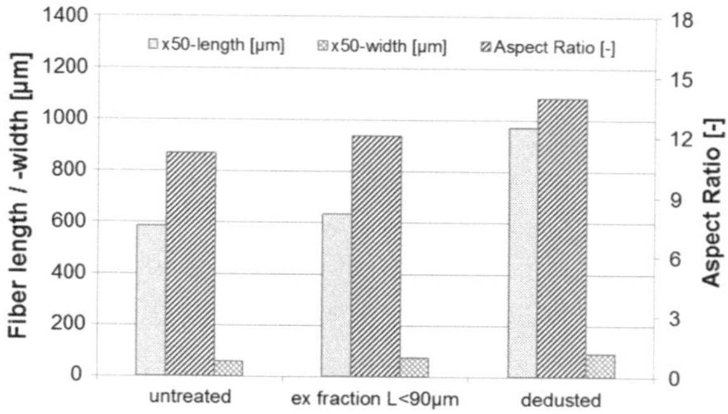

Fig. 6 Length weighted parameters of particle size distribution of a fiber material originated from processing of wet preserved hemp and different sizing treatments

chopping and storage of the remaining stem can result in a reduced dust content in the fiber fraction after processing. Within the described processing experiments simple sieves and a sieve drum separator (condenser) were used to improve particle size parameters of the intermediate (Fig. 6).

The resulting fiber material is characterized by an improved particle size distribution particularly regarding the fiber length as well as aspect-ratio which increased from 11 to about 14.

All experiments on harvest, storage and processing were always scheduled and carried out in order to supply intermediate fibrous materials to the cooperating partners in the network. Samples of different quantities were used to investigate if favorable fiber characteristics resulted in a good processability and composite properties as well.

Processing with IMC

The confection of customised solutions for raw materials with special combinations of their properties demands technological innovations with a high cost effectiveness. Therefore, the process of injection molding compounding (IMC) was investigated within the FENAFA network to enable a simplified processing of native fibrous raw materials into composites. The main goal was the realization of a consistent process chain from the raw material to the final product by a substantial improvement of material and energy efficiency with single step processing in comparison to conventional procedures. The IMC direct compounding provides specific advantages for the processing of heat-sensitive materials like natural fibers. The only singular thermal and mechanical treatment leads to better composite characteristics compared to the traditional two-step-processing. The preceding pelletizing is not applied and higher fiber length can be realized in the final product (Widmayer 2012).

Main advantages of the injection molding compounding process are the obvious reduction of the costs of the component manufacturing and the flexibility of the material combination. The continuous mode of an extruder is combined with the discontinuous function of an injection molding machine in order to produce parts consisting of different raw materials in one process step. The one time heating of the material leads to a tapering of the energy consumption as well as to a lower thermic strain of temperature sensitive materials like composites reinforced with natural fibers. Furthermore, there is no need for granulation and drying of granulate before further processing.

The Injection Molding Compounder (IMC) consists of a fusing and homogenization unit, a twin-screw extruder and an injection system (KraussMaffei 2013). These parts are connected by a diverter pipe. A material storage used as buffer is, in Y-setting, connected between. The material supply is made gravimetrical with an underfeed extruder to get a material throughput nondependent from the rotational speed.

Granulate, powder and additives are added over dosing units. Short-, intermediate-, and endless fibers are added downstream into the extruder. In the closely intermeshing, co-rotating twin-screw extruder the material (polymer, matrix, additive, fiber) is homogenized and conveyed isothermal over the diverter pipe in the injection unit, respectively during the injection process into the material storage. For the start-up there is a diverter valve between extruder and material storage, which is open to derive the melt into a container until the melt quality is in tolerance. Therefore, in normal operation the valve is closed. The material storage buffers the melt supplied from the twin-screw extruder during the injection process and holding pressure phase. This is necessary, to combine the continuous extrusion process with the discontinuous injection molding process. Additionally, the melt buffer applies an adjustable pressure on the melt to ensure a constant part quality. The composite is dosed dependent on throughput into the injection piston. Due to the dosing process the material storage is completely emptied. Configurable dynamic pressure works against the retreat of the injection piston. When the adjustable shot volume for the next injection cycle is reached the diverter valve is closed. This causes a separating of plasticize and injection unit while the injection and holding pressure phase. The shut-off nozzle valve is opened and the injection process starts. After completion of the holding pressure phase the cooling period starts followed by the mold opening time, the remove time and the closing time. Simultaneously the dosing phase starts by closing shut off nozzle valve and open the transfer valve. As soon as the expected threshold value is met the cycle starts again.

Initially different compositions of materials and methods were developed and proofed in a two-step process (1. Compounding, 2. Injection Molding) within the network activities of FENAFA. In order to scale-up and verify the practical feasibility these material systems were processed on an Injection Molding Compounder of KraussMaffei (KM 1300-14000 IMC) available in the Pilot Plant of the Fraunhofer Institute for Mechanics of Materials (IWM) (Fig. 7).

Fig. 7 Injection molding compounder at Fraunhofer Pilot Plant Center Schkopau (Michel 2015)

For the scale-up of the process a demonstrator part from the logistics sector (transport and storage box) was used. The trials are carried out without preceding pelletizing the flax and hemp based compounds.

The production process consists of the steps compounding, injection molding, automatic removal of the molded part from the tool and the drop off on a conveyor with a robot.

In the preliminary stage, the screw geometry of the compounding extruder (D = 84 mm, L/D = 37), the mold, the dosing device and process parameters (e.g. temperature of the cylinder, rotational speed of the screw, mass throughput) were adapted to the configuration of the KM 1300-14000 IMC due to knowledge found ex ante. The mixing areas consist of kneading and tooth—elements to get a homogenous fiber filled melt.

An innovative gravimetrical universal dosing unit developed during this project by the TU Chemnitz was used for the dosing and feed of hemp fibers, which have very poor flow properties.

Processing, Sampling and Characterization of Products

The demonstration of feasibility of the R&D results was performed by producing transport and storage boxes (shot weight approximately 1100 g) with the industrial sized injection molding compounder KM 1300-14000 IMC (Fig. 10).

All compounds consist of matrix polymer polypropylene (BE170MO from Borealis) with 30 wt% of hemp or flax and 3 wt% of a maleated polypropylene as coupling agent (Scona TPPP8112 GA from BYK Kometra). In Fig. 8, the coloring of the first trails is shown. In addition to a strong odor of burned fibers which occurred during processing, a dark coloring in comparison to the wood-filled part was noted. The odor and the darkening were caused by degradation of the agro-based natural fibers (flax and hemp). The frictional heat resulted in high melt

Fig. 8 Comparison of colouring of different fiber types (wood *left*; flax *middle*; hemp *right*)

temperatures. Thus, in comparison to the wood-filled material (60 wt% of wood powder), the process parameters for the flax and hemp filled compounds had to be adapted.

Compared to wood fibers, natural fibers like hemp and flax are more sensitive in terms of thermal stability. A special run-up regime of the IMC for natural fibers as well the reduction of the barrel temperature below 160 °C (instead of 175 °C in Fig. 8) resolved that problem. Due to this, the main energy input for melting and compounding was caused exclusively by frictional heat. Furthermore, the throughput was adapted by decreasing the rotational speed of the screw from 200 to 400 rpm to 60 rpm.

In Fig. 9 the result of that processing adaption is shown. There is a significant difference between 160 and 175 °C, which is a remarkable progress in processing of preserved natural fibers.

The mechanical testing of the transport and storage boxes (cf. Fig. 9) with flax and hemp fiber-filling were performed at test specimens cut by waterjet. The specimens for tensile testing (according to DIN EN ISO 527-1 and 527-4) were obtained from the long side of the box and those for Charpy impact testing (according to DIN EN ISO 179-2) from the short side. Figure 10 provides an overview of results of mechanical testing of the transport and storage boxes. All samples are reinforced by a natural fiber content of 30 wt%.

Fig. 9 Comparison of coloring with different barrel heat of 160–175 °C

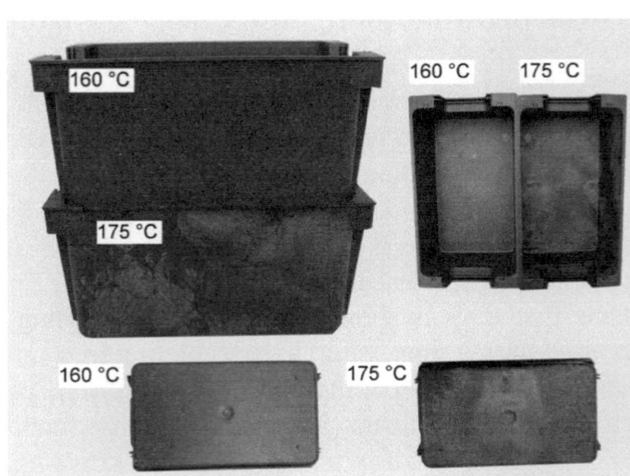

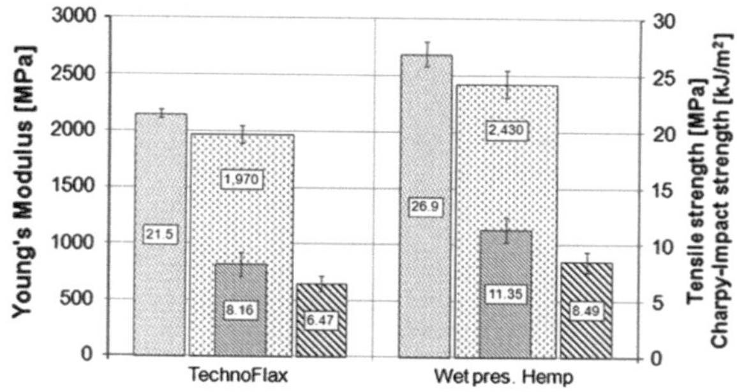

Fig. 10 Mechanical characteristics of test specimens cut out of natural fiber-reinforced boxes (30 wt% fiber content, flax *left*: hemp *right*)

Overall, the results of the hemp-filled (cf. Fig. 10, right) sample exceed the ones of flax-reinforced (cf. Fig. 10, left) sample in all tested mechanical properties, but the general properties are still less than those of common wood polymer composites (WPCs). An addition of a thermoplastic elastomer (TPE) as impact modifier was performed in the second step of experiments in order to improve especially the impact properties. Within the studies two types of TPEs were tested, TPE-1 Exxon Mobile Chemical's propylene-based Vistamaxx™ 6202 and styrene-based TPE-2 Kripoflex which is actually a dry-blend of 65 wt% linear high molecular weight styrene-ethylene/butylene-styrene block copolymer (SEBS with butadiene to styrene-monomer ratio BD/SM of 67/33) and 35 wt% of a low molecular weight oily component. TPE-2 showed especially excellent performance in context of processing WPC due to both mechanical properties and processability (Hartmann et al. 2014). In Fig. 11 the results of impact strength tests are shown in comparison to the unmodified composites. The addition of 20 wt% TPE-2 increased Charpy impact strength at room temperature by more than 100 % regarding the mean values or 50 % by taking the standard deviations into account. The addition of 20 wt% of TPE-1 leads to an increase of 70 % of Charpy impact strength regarding the mean

Fig. 11 Comparison of unmodified and modified natural fiber-reinforced composites (with 30 wt% flax and 20 wt% impact modifier)

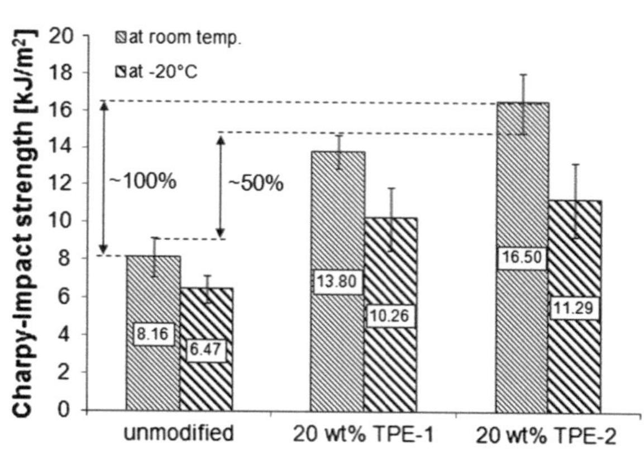

Table 2 Processing parameters of modified and unmodified composites (with 30 wt% flax)

Impact modificator	Torque of extruder—compounding (%)	Melt pressure—injection molding (bar)
Unmodified	39	500
20 wt% TPE-1	32	525
20 wt% TPE-2	27	480

values. The tests at −20 °C show an equal enhancement of impact properties. Moreover, the acceptable advancements for impact properties at the same time Young's modulus shows a decrease of almost 40 % for TPE-1 and 50 % for TPE-2. The tensile strength was also reduced by 25 % on TPE-1 and 32 % on TPE-2 addition. In summary, the mechanical properties still have a huge potential for optimization and will be a topic of further studies.

During the monitoring of the processing parameters for melt pressure and torque, another important fact was observed. The parameters for the torque of extruding section under stable processing conditions and the maximum melt pressure in injection unit are listed in Table 2.

The data show that an addition of TPE-1 and TPE-2 had opposing effects on the material behavior during processing compared to the unmodified sample. While an addition of TPE-1 leads to a small torque reduction of 7 % combined with an disadvantageously increase of melt pressure due to stress of the fiber content, an addition of TPE-2 reduces both, the torque by 12 % and the melt pressure. Thus, adding TPE-2 enables sensitive processing.

Summary

The supply chain for fibrous raw materials is optimized in order to supply particle morphology adapted to the processing requirements. Subsequently, an optimization of the process parameters and the screw configuration of the IMC are realized. Thus, the arrangement of kneading and mixing elements as well as the screw speed is investigated. Finally, downstream and upstream feeding of the fibers is examined. The results reveal that an optimal configuration is characterized by high friction and low heat stress on the material. Higher screw speed leads to lower variations of the resultant values which probably exist due to a better homogenization of fibers in the melt.

Products such as transport and storage boxes are produced on an industrial sized injection molding compounder KM 1300-14000 IMC to demonstrate that the R&D results are not only achieved under lab conditions but also on industrial scale machines. Using an innovative dosing system, Polypropylene, 3 % coupling agent and whole plant materials from hemp and flax at 30 wt% are applied.

The results indicate that there are substantial differences on the mechanical properties of different fibrous raw materials to be used as a reinforcing agent. The supply and processing of raw materials on an industrial and an economic viable scale and respective characteristics of the products is important both in practice and research.

Acknowledgments The authors gratefully acknowledge the funding by German Ministry of Food and Agriculture (BMEL) and its funding body Agency of Renewable Raw Materials (FNR) under grants 22023808, 22025608, 22025708 and 22025508.

References

Carus M, Dammer L, Scholz L, Essel R, Breitmayer E, Eder A, Korte H. European and Global Markets, 2012 and Future Trends. bioplastics magazine, 2014, Vol. 9, p. 42-44

Faruk O, Bledzki A, Fink H P, Sain M. Biocomposites reinforced with natural fibers: 2000-2010. Prog Polym Sci 2012, 37:1552-1596

Gusovius H-J, Prochnow A, Streßmann U, Hahn J. The weather-determined process risk of harvesting fibre hemp. in: Proceedings of the International Conference on Agricultural Engineering, Part 2, 2000, Warwick, p. 46-48

Gusovius H-J, Pecenka R, Wallot G, Rinberg R, Nendel W. An alternative processing technology for agricultural raw materials to supply fibrous materials for reinforced composites. Online-Proceedings of the BioComp 2010 – 10th Pacific Rim Bio-Based Composites Symposium, 5-8th October 2010, Banff, Canada

Hänninen T, Hughes M. Historical, Contemporary and Future Applications. In Industrial applications of natural fibres: structure, properties and technical applications/ edited by Jörg Müssig. 2010, Chichester: John Wiley, p. 385-395

Hartmann T, Bürgermeister S, Rinberg R, Kroll L. TPE-modification of wood plastic compounds for advanced rheological and impact properties. 58. Internationales Wissenschaftliches Kolloquium (IWK); Ilmenau, 2014

Hoffmann T, Pecenka R, Fürll C. Mathematical model of the binder application for the production of fibre boards. Journal of Biobased Materials and Bioenergy. 3, 2009, p. 265-268 Online: http://dx.doi.org/10.1166/jbmb.2009.1033

Hoffmann T, Pecenka R, Schemel H, Gusovius H. Process-technological evaluation of harvesting hemp in winter. Journal of Natural Fibers. 10 (2), 2013, p. 159-175 Online: http://dx.doi.org/10.1080/15440478.2013.783451

Idler Ch, Pecenka R, Fürll Ch, Gusovius H-J. Wet Processing of Hemp: An Overview. Journal of Natural Fibers, 2011, Vol. 8, Iss. 2, p. 59-80

KraussMaffei: KraussMaffei injection molding compounder supports lightweight construction. Press release July 24th, 2013, http://www.kraussmaffeigroup.com/en/corporate-press-releases/d/KraussMaffei_injection_molding_compounder_lightweight_construction.html (access June 9th, 2015)

Liu M, Fernando D, Daniel G, Madsen B, Meyer A S, Ale M T, Thygesen A. Effect of harvest time and field retting duration on the chemical composition, morphology and mechanical properties of hemp fibers. Industrial Crops and Products, 2015,Vol 69, p. 29-39

Michel P. Fraunhofer-Institut für Werkstoffmechanik Leichtbau-Anwendungen, Hafenhinterland Konferenz, Halle, March 11th, 2015

Mohanty A K, Misra M, Drzal L T. Natural fibers, biopolymers, and biocomposites. Boca Raton, FL: CRC Press, 2005, ISBN-13: 978-0849317415

Pecenka R, Fürll C, Idler C, Grundmann P. Fibre boards and composites from wet preserved hemp. International Journal of Materials and Product Technology. 36 (1-4), 2009, p. 208-220 Online: http://dx.doi.org/10.1504/IJMPT.2009.027832

Ruth J, Fritz H G, Brükle E, Zimmet R. Innovative Direktverarbeitung von Naturfasern, KU Aufbereitung 92 (2), 2002, p. 29-34

Schirp A, Stender J. Properties of extruded wood-plastic composites based on refiner wood fibres (TMP fibres) and hemp fibres, Eur. J. Wood Prod., 68 (2), 2010, p. 219-231

Shahzad A. Hemp fiber and its composites—a review. Journal of Composite Materials, vol. 46, 2012, p. 973–986

Specht K. Holz- und hanffaserverstärktes Polypropylen in der Spritzgieß¬verarbeitung, Dissertation, 2007, Universität Kassel

Stark N M, Rowlands R E. Effects of wood fiber characteristics on mechanical properties of wood/polypropylene composites, Wood and Fiber Science, 35 (2), 2003, p. 167-174

Wallot G, Gusovius H-J, Pecenka R, Schierl S, Rinberg R, Nendel W. Developments in the use of fibres from wet-preserved hemp for composite production. CIGR Ejournal, 2012, 14 No 4, p. 218-223

Widmayer S. Generating and Processing of natural fiber composites on an injection molding machine. 9th WPC, Natural Fibre and other innovative Composites Congress and Exhibition; Stuttgart, 2012

Fluorination as an Effective Way to Reduce Natural Fibers Hydrophilicity

K. Charlet, F. Saulnier, D. Gautier, M. Pouzet, M. Dubois and A. Béakou

Abstract In order to get composite materials with high mechanical properties, the quality of the interface between the fibres and the matrix has to be good enough to enable the load transfer. In the case of wood polymer composites, made of hydrophilic wood particles and of a generally non-polar polymer, the lack of natural compatibility between the constituents hinders the load transfer. Aiming at decreasing the gap of polarity between wood fibres and polymer matrices, fluorination has been applied to wood. This treatment is known to be very efficient to make more hydrophobic materials without requiring solvent or high temperature. After the optimization of the treatment parameters so as to get a high level of fluorine grafting without burning the particles, the hygroscopic and thermal behaviors of the fluorinated wood flour have been evaluated and compared to the non-treated flour. For that purpose, several analyses were carried out: FT-IR spectroscopy, ^{19}F solid-state NMR spectroscopy, SEM, contact angle measurements, TGA. The fluorine based treatment was shown to decrease notably the capacity of the wood particles to absorb water without damaging their surfaces. Lastly, at the composite scale, the wood fluorination was shown to strongly reduce its hydrophilicity and to largely enhance its tensile and flexural properties. This is directly linked with the improvement of the compatibility between the treated (and thus, less hydrophilic) wood particles and the polymer matrix, as also proved by X-ray tomography.

Keywords Wood · Fluorination · Composite · Hydrophilicity

K. Charlet (✉) · F. Saulnier · D. Gautier · M. Pouzet · A. Béakou
Institut Pascal, UMR 6602, UBP/CNRS/IFMA, Clermont-Ferrand, France
e-mail: karine.charlet@ifma.fr

D. Gautier · M. Pouzet · M. Dubois
Institut de Chimie de Clermont-Ferrand, UMR CNRS 6296 UBP/CNRS/ENSCCF, Clermont-Ferrand, France

R. Fangueiro and S. Rana (eds.), *Natural Fibres: Advances in Science and Technology Towards Industrial Applications*, RILEM Bookseries 12,
DOI 10.1007/978-94-017-7515-1_16

Introduction

Wood fibres, among other natural fibres, are increasingly used in replacement of usual aramid, carbon or glass reinforcements in composites, in order to reduce the amount of fossil carbon produced every day. The most common application fields of wood polymer composites are building, automotive, furniture and packaging industries. These composites give added value to a relatively abundant and cheap raw material, i.e. co-products of wood transformation. Nevertheless, their use is restricted to non-structural applications because of the lack of compatibility between the hydrophilic wood reinforcements and the hydrophobic polymer matrix, which generates porosity and low ability to transfer load at the interface, leading to poor composite mechanical properties. Therefore, a pre-treatment of the wood reinforcements is often necessary before processing these composites. During the last few years, various treatments have been developed to decrease wood's hydrophilic character and make it more compatible with polymer matrices (Guyonnet 1999; Podgorski et al. 2000; Ichazo et al. 2001; Lu et al. 2005; Tserki et al. 2005; Karmarkar et al. 2007; Nachtigall et al. 2007; Dominkovics et al. 2007; Dányádi et al. 2010; Ayrilmis et al. 2011; Acda et al. 2012; Islam et al. 2012; Zhang 2014). However, none of them is really convincing at the industrial scale in the case of wood based composites since they are either environmentally harmful, hardly efficient, highly energy-consuming or very costly.

Since few years, direct fluorination has been used to treat the surface of polymers, in order to enhance their barrier property against multiatomic molecules, e.g. hydrocarbons, and improve their chemical resistance to solvent attack (Kharitonov et al. 2005; Kharitonov 2008). It is known to decrease the permeability of materials, mainly polymers. The application of direct fluorination to synthetic fibres, such as aramid fibres, was shown to have a clear impact on material surface energy (Maity et al. 2007). In particular, when applied to carbon fibres, direct fluorination led to a noticeable increase in fibre hydrophobicity (Bismarck et al. 1997; Dubois et al. 2005; Ho et al. 2007; Guérin et al. 2012). The application of direct fluorination to lignocellulosic products has been studied in the past with the intention of making Kraft paper waterproof (Sapieha et al. 1990). Then the surface fluorination of paper was developed using the CF_4 radio-frequency plasma technique (Sahin et al. 2002). In both studies, the published results reported a significant increase in the paper contact angle (i.e. a decrease in the hydrophilic character) after fluorination.

These findings suggest that direct fluorination could be applied in the treatment of wood fibres used as reinforcements for polymer matrices. To the authors' best knowledge, the use of fluorine chemistry to enhance the properties of wood based composites has never been studied before. This work focuses on the direct fluorination of wood flour in order to reduce its hydrophilic character and then improve its compatibility with unsaturated polyester. Changes in the structure of the wood and the impact on its properties are observed and discussed, in order to verify the validity of this method to produce good quality composites. The main difficulty and, at the same time, the main advantage, is the high reactivity of molecular fluorine F2 towards wood. With such a reactive sample, fluorination, i.e. the formation of C–F covalent bonds, competes with wood decomposition; when the fluorination is not controlled the wood may react with fluorine gas to form volatile perfluorinated groups such as CF_4 or C_2F_6. In other words, the wood may burn in the fluorine atmosphere as it does in air at high temperature.

The first objective was therefore the perfect control of the fluorination conditions in order to treat only the surface of the wood. Once the fluorination parameters were optimised, both the non-treated and fluorinated wood flours were characterized by Fourier transform infrared (FT-IR) spectroscopy, ^{19}F solid-state nuclear magnetic resonance (^{19}F NMR) spectroscopy, scanning electron microscopy (SEM), thermo-gravimetric analysis (TGA), and hygroscopic analysis in order to highlight the effects of the treatment.

The second objective was to evaluate the influence of the wood flour treatment on the composite properties. Thus, after the composite processing, the amounts of pores within the treated and non-treated wood composites were revealed by X-ray tomography. Then, tensile and flexural tests were performed on unsaturated polyester reinforced with non-treated or fluorinated wood flours. Contact angle measurements were used to estimate the change in the composite surface energy due to fluorination. Finally, the effect of humidity was investigated by testing composites that had been stored several days at normal or high relative humidity.

Materials and Methods

Materials

The wood flour under study was a mix of spruce and Douglas species obtained from sawmill co-products in Auvergne, France. Its density was measured by the solvent method using xylene and toluene and was found equal to 1.41 ± 0.17. The flour was sifted so that its grading was smaller than 250 μm. The polymer used to process the derived composites was unsaturated polyester Norsodyne G703, from Cray Valley, whose density was given equal to 1.17.

Direct Fluorination of the Wood Flour

About 5 g of wood flour were placed on a tray in a nickel reactor. The gas inlet was located to the left of the sample. The sample was first outgassed under a dynamic vacuum (−960 mbar) at 150 °C for 2 h in order to remove all traces of water. Actually, water molecules may be adsorbed on the surface of the wood flour and react with fluorine gas (pure F_2) to form HF molecules, which may act as a catalyst for fluorination, leading to the inhomogeneity of the treatment. The reaction oven was divided into three parts: it was set at 42 °C on the left, 55 °C in the middle and

70 °C on the right. Such a temperature gradient is necessary to avoid a high fluorination rate near to the gas inlet of the oven. This variation allows the control of the fluorination process and leads to a homogenous treatment. A partial vacuum was applied to the closed reactor (−20 mbar) and F_2 gas was injected in addition to N_2 gas to reach 1 atm. The total pressure inside the closed reactor was maintained constant for 3 h, and five additions of fluorine were performed in order to compensate for the consumption of molecular fluorine during the reaction. Finally, after 2 h under a flow of N_2 gas at 150 °C to remove all the F_2 molecules, and cooling to ambient temperature for 11 h, the flour was heated again to 150 °C with a flow of N_2 gas for 1 h to eliminate traces of adsorbed F_2, HF, CF_4 and C_2F_6 molecules.

FT-IR and NMR Spectroscopy

Samples of fluorinated wood flour were taken at different locations in the reactor in order to check the homogeneity of the treatment using FT-IR and ^{19}F NMR analyses. FT-IR spectroscopy was performed in attenuated total reflectance (ATR) mode using a SHIMADZU FT-IR 8300 spectrometer. The spectra were recorded in the mid infrared region (4000–400 cm^{-1}), and 64 co-added scans were collected for each sample. ^{19}F NMR experiments were carried out using a Bruker Avance spectrometer, with working frequencies of 282.2 MHz. For magic angle spinning (MAS), a Bruker probe operating with 2.5 mm rotors was used. A simple sequence was performed with a single $\pi/2$ pulse length of 4.0 μs. 128 scans were recorded. The MAS spinning speed was set to 34 kHz, high enough to significantly average and weaken the ^{19}F–^{19}F homonuclear coupling, which is the main interaction. ^{19}F chemical shifts were externally referenced to $CFCl_3$.

TGA

The thermogravimetric analysis was performed under a nitrogen atmosphere of 0.9 bars using a PerkinElmer TGA 4000 analyser. About 10 mg of wood flour was heated from 35 to 450 °C with a heating rate of 5 °C/min. Then the sample was cooled to 35 °C with a cooling rate of 50 °C/min.

SEM Analysis

The surface morphology of the wood particles before and after fluorination was observed by SEM analysis, performed with a Phenom FEI microscope.

X-ray Tomography

Wood polyester composites were analyzed by X-ray tomography in order to determine the amount of pores present within the volume of the sample and to compare these amounts according to the applied treatment. The device used, a compact SkyScan 1174, enables the 3D reconstruction of the composite with a resolution of 30 μm under 50 kV, which means that the micro-porosity cannot be visualized. Therefore, this may give an idea of the potential improvement of the fibre/matrix compatibility. The dimensions of the observed composites were 10 mm × 10 mm × 1.76 mm.

Hygroscopic Testing of the Wood Flour

Humidity-controlled chambers were created following the NF EN ISO 483 standard, so as to study the hygroscopic behaviour of the wood flour under different relative humidity (RH) conditions (9, 33, 59, 75 and 98 % RH). Hermetic jars were filled with diverse saturated salt solutions, which are listed in Table 1. A plastic tube and an iron grid were used to maintain the sample 3 cm above the solution and thus prevent direct contact with water. The wood flour was placed in plastic caps and dried in an oven at 90 °C for 48 h to attain the anhydrous state. Then each sample was conditioned in the humidity-controlled chambers until its weight was stabilized.

Hygroscopic characterization was carried out by measuring the weight variation of the wood samples throughout the conditioning procedure for each relative humidity. Finally, the water content of the wood was evaluated using the following formula, where m is the sample weight at the considered time and m_0 the weight at the anhydrous state:

$$x(t) = (m(t) - m_0)/m_0 \quad (1)$$

For each relative humidity value, five samples of non-treated and fluorinated wood flour were analysed.

Table 1 Saturated salt solutions and corresponding relative humidity at 20 °C

Saturated aqueous salt solutions	RH (%)
Potassium hydroxide	9
Magnesium chloride hexahydrate	33
Sodium bromide	59
Sodium chloride	75
Potassium sulphate	98

Wood-Polyester Composite Processing

Wood polyester composites with a reinforcement weight fraction of 45 % (corresponding to a volume fraction of 40 %) were processed by hot compression molding, using a SATIM hot press. The wood polyester mixture was poured into an aluminium mold of 100 mm × 100 mm × 2 mm covered with 0.12 mm thick PTFE sheets on each internal face (aimed at easing the final demolding). Then, the closed mold was placed in the press and kept under a pressure of 60 kN and a temperature of 80 °C for 2 h, so as to ensure the resin cross-linking. The cooling to ambient temperature was performed using air under pressure. Treated and non-treated wood flour composites were processed in exactly the same way, without any pre-curing of the materials. Samples for the mechanical tests were taken from the plates thus obtained; their width was 10 mm and their thickness was 1.76 mm.

Composite Mechanical Characterization

Tensile tests The tensile tests were performed on a Zwick/Roell Zmart Pro test machine, with a prescribed crosshead displacement rate of 1 mm/min and a gage length of 50 mm, until the rupture of the sample. The initial length of the samples was 70 mm, adapted from the standard NF EN ISO 527-3. The tests were conducted at room temperature (about 23 °C) and relative humidity (about 40 %). The longitudinal stress (σ) and strain (ε) of the sample were calculated from the measured force (F), the sample elongation (ΔL), the sample section S_0 (supposed to be constant), and the gage length L_0 as follows:

$$\sigma = F/S_0 \tag{2}$$

$$\varepsilon = \Delta L/L_0 \tag{3}$$

Flexural tests Three point bending tests were performed on an Instron 5543 test machine equipped with a 500 N load cell, at room temperature (about 23 °C) and relative humidity (about 40 %). The length of the sample was 35 mm and the distance between the supporting pins was set to 28 mm, following the standard NF EN ISO 178. The samples were tested at a crosshead displacement rate of 1 mm/min with the load applied at mid-span. The flexural strain (ε) and stress (σ) at the middle of the sample were calculated from the deflection (Y, corresponding to the crosshead displacement) and the force (F). Considering the material as homogeneous, we had:

$$\varepsilon = 6hY/L_0^2 \tag{4}$$

$$\sigma = 3FL_0/2bh^2 \tag{5}$$

where L_0 is the span, b and h respectively the width and the thickness of the samples.

Hygroscopic Behavior of the Composites

Contact angle measurements Contact angle measurements were carried out on an Easy Drop Krüss device. A water microdroplet of 5 μL was deposited on the composite surface. Once the droplet was stabilized, the picture was frozen and analysed using the Krüss DSA software that evaluates directly the contact angle. For each composite, five microdroplets were deposited and an average contact angle was calculated.

Composite water absorption After curing 24 h at 80 °C to remove water, 35 mm long composite samples were stored in a climatic chamber at a constant relative humidity (80 %) and a constant temperature (20 °C). The weight of the samples (4 of each type of composites) was measured every day at the same moment of the day until equilibrium. Then, the water content of the sample was estimated as:

$$W(t) = (m(t) - m_0)/m_0 \tag{6}$$

with m(t) the mass of the moist sample at time t at equilibrium and m_0 its dry mass. After 18 days, their weight being almost constant, the samples were rapidly tested to get their flexural properties, following the same protocol as described in section 'Composite Mechanical Characterization'. At the time of the tests, the ambient temperature was 21 °C and the relative humidity 44 %.

Results and Discussion

Covalent Grafting of Fluorine on Wood

The ATR FT-IR spectra of the wood flour before and after direct fluorination are given in Fig. 1. The absorbance peaks in the non-treated wood flour spectrum correspond to the main bonds of the wood components (Müller et al. 2003; Popescu et al. 2013; Schwanninger et al. 2004). The broad peak around 3320 cm^{-1} is due to the –OH hydroxyl groups in cellulose whereas the peak around 2900 cm^{-1} is ascribed to the asymmetric and symmetric mode of hydrocarbon (C–H) stretching. The C=O stretching vibration of carbonyl, carboxyl and acetyl groups appears at 1740 cm^{-1}. The band at 1640 cm^{-1} is due to C=O stretch of aromatic ketones. The band at 1600 cm^{-1} corresponds to the C=C stretching of aromatic skeletal in lignin.

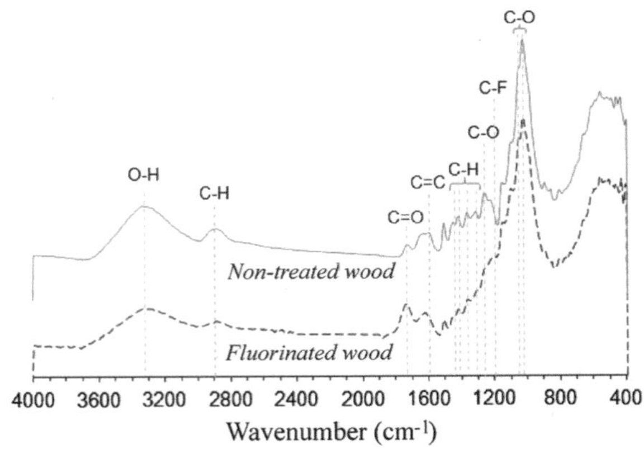

Fig. 1 FT-IR spectra of wood flours

C–O stretching in lignin is observed at 1260 cm^{-1} and the large peak at 1020 cm^{-1} is attributed to C–O stretching vibration.

It is well known that wood hydrophilicity is essentially due to the hydroxyl groups (–OH) of both the cellulose and hemicelluloses. Indeed, these hydroxyl groups have the property of absorbing water molecules, forming hydrogen bonds. The FT-IR spectroscopy analysis of fluorinated wood flour reveals a significant decrease in the number of hydroxyl bonds (3320 cm^{-1}). As there are fewer sites for hydrogen bonding, the fluorinated wood particles should adsorb less water. In other words, the decrease in the number of –OH groups in the wood flour was one of the first proofs of the weakening of its hydrophilicity.

Fluorination also induces the substitution of certain specific atoms in the treated material by fluorine atoms. Hydrogen atoms in hydrocarbon groups on the wood surface are expected to be replaced by fluorine, because of both the weakness of C–H bonds and the reactivity of molecular fluorine. In fact, the substitution of C–H bonds (2900 cm^{-1}) by carbon-fluorine (C–F) bonds (1180 cm^{-1}) was observed after the fluorination of the wood flour. This latter band corresponds to covalent C–F bonds (Kharitonov 2008). Even with weak intensity, its presence underlines the formation of C–F bonds onto the wood components. Moreover, the fact that samples taken at various distances from the gas inlet of the reactor showed similar spectra confirms that the temperature gradient ensured the homogeneity of the fluorine grafting onto the wood.

The ^{19}F NMR measurements were used to confirm both the covalent grafting of fluorine atoms and the nature of the fluorinated groups, i.e. –CF, $–CF_2$ or $–CF_3$ (Isbester et al. 1997; Aimi et al. 2004; Zhang et al. 2010). An example of the ^{19}F MAS NMR spectrum of fluorinated wood flour is presented in Fig. 2. It is typical of the samples, whatever their location in the reactor. The resonances are attributable to the different groups, namely –CF, $–CF_2$ or $–CF_3$. The broad signal at −135 ppm versus $CFCl_3$ matches $–CF_2$ groups, while the signal at −182 ppm corresponds to C–F bonds and the signals at −74 and −83 ppm are both attributed to $–CF_3$ groups with different neighbours. Finally, the signal at −207 ppm is due to an aromatic –CF bond or –CHF bond. Bearing out the evidence of IR spectroscopy, NMR

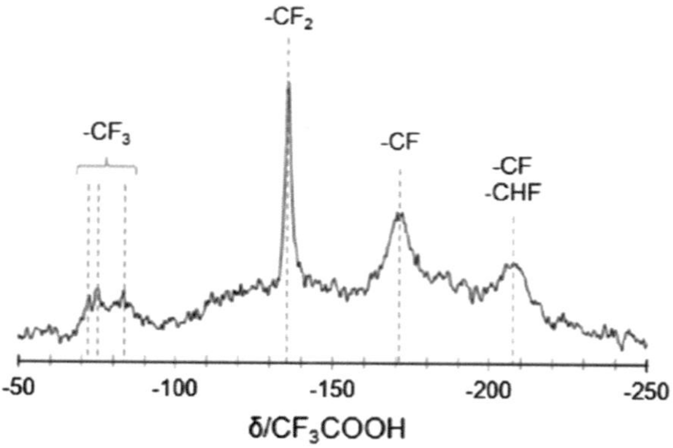

Fig. 2 ^{19}F NMR spectrum of fluorinated wood flour

unambiguously highlights the presence of both fluorinated and perfluorinated groups on the wood surface.

The electronegativity of the fluorine atom, the highest in the periodic table, confers a strong polarity to the carbon-fluorine bonds, which creates a polar hydrophobicity of the fluorinated compounds. For this reason, the presence of fluorine atoms on the wood surface should imply a decrease in the hydrophilic character of the resulting samples. This decrease has to be confirmed by hygroscopic characterisation.

Decrease of Wood Hydrophilicity

Through the adsorption effect, the water content of the wood moves towards an equilibrium state according to the ambient relative humidity. Thus, when the wood flour was conditioned in a humidity-controlled chamber after drying, the water content increased to match the conditions of the enclosure. Continuous measurements of the sample weight allowed the evolution of the water content for each relative humidity value to be monitored, from the anhydrous state to the equilibrium point. The adsorption curves of untreated and fluorinated wood flours are given in Fig. 3. The hygroscopic analysis of the non-treated wood flour was performed both to validate the method and to make a reference to check the effect of direct fluorination on the hygroscopic behaviour of wood flour. The experimental results corresponded with previous studies found in the literature in terms of water content (Stamm et al. 1964). For example, the water content of wood after equilibrium is known to be about 11 % for a relative humidity of 60 % at 20 °C, and the wood flour studied here contained 10.5 % of water under 59 % relative humidity at the same temperature.

After fluorination, an increase in the adsorption rate was observed. For instance, in the case of a relative humidity equal to 75 %, the water content of fluorinated wood flour was stabilized in 5 h, compared to 24 h for the non-treated wood flour.

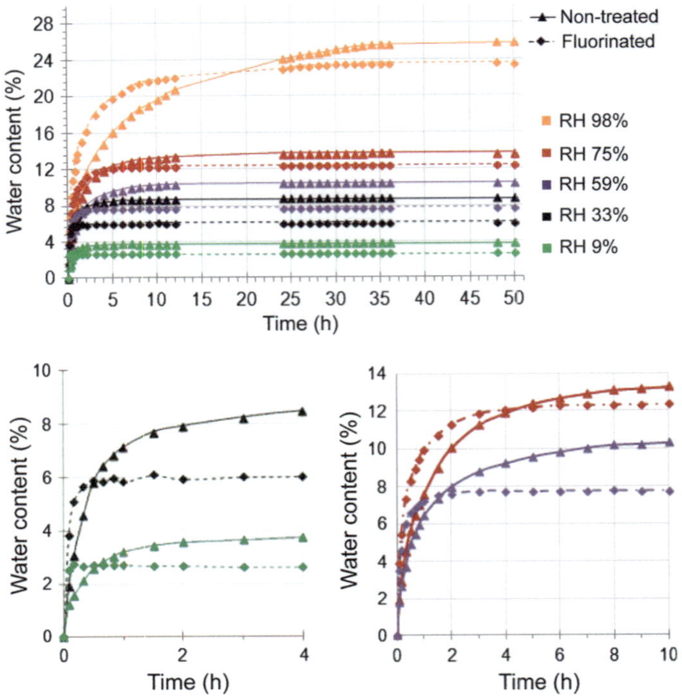

Fig. 3 Adsorption curves of the wood flour before and after fluorination for various relative humidity conditions (*top* global curves; *bottom* zoom on short times)

Hygroscopic analysis also underlined the decrease in the equilibrium water content of the wood flour after fluorination. The water content of both non-treated and fluorinated wood flours under the different relative humidity conditions are listed in Table 2. An average decrease of 2 points in the water content of the wood flour after treatment can be observed, representing a drop of 10–30 % in the water content.

This analysis shows that direct fluorination efficiently lowers the water content in wood flour. This decrease can be explained by the reduction in the number of sites for hydrogen bonding. In addition, the fluorinated wood flour reaches its water

Table 2 Water content of the wood flours under various humidity conditions

Relative humidity (%)	Water content in the non-treated wood (%)	Water content in the fluorinated wood (%)	Loss of water content (%)
9	3.7 ± 0.2	2.7 ± 0.3	30
33	8.7 ± 0.3	6.1 ± 0.5	30
59	10.5 ± 0.4	7.8 ± 0.2	26
75	13.8 ± 0.3	12.4 ± 0.2	10
98	25.8 ± 0.8	23.6 ± 0.2	9

equilibrium point faster than the non-treated wood flour. The presence of electronegative fluorine in the constitution of the fluorinated wood increases the attractiveness of its surface and accelerates its reaction with water.

Preservation of the Structure of Wood

Excessive fluorination may cause important damage to a material surface because of the formation of volatile species such as CF_4 and C_2F_6. In order to verify that the treatment does not damage the wood texture, a SEM analysis was performed. Examples of micrographs of wood particles (with or without fluorination) are given in Fig. 4. The tracheid physical structure appears preserved after direct fluorination; even the wood pits were maintained intact.

Analysis of the Thermal Behavior of Wood

TGA was performed on raw and fluorinated wood flours. The curves obtained are presented in Fig. 5. The first weight loss in the curves corresponds to water desorption, which occurred before 100 °C due to the pressure and the N_2 atmosphere. The weight loss for the fluorinated sample between 140 and 220 °C is due to the decomposition of carbon-fluorine bonds. It corresponds to about 5 % of the sample weight. Wood degradation was noticed from 220 °C, in agreement with recent studies on wood thermal degradation (Jeske et al. 2012; Poletto et al. 2012). The TGA data underline that the treatment is effective up to 140 °C. Since the unsaturated polyester resin is usually polymerized at 80 °C, the direct fluorination of wood flour appears to be an efficient treatment for wood flour used as a polyester composite reinforcement.

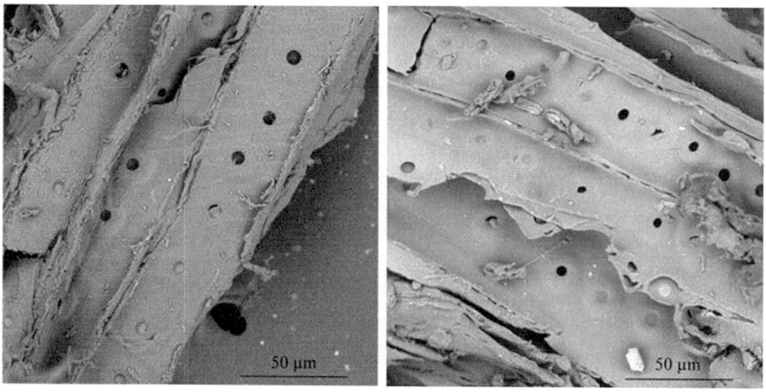

Fig. 4 SEM micrographs of an untreated (*left*) or a fluorinated (*right*) wood particle

Fig. 5 TGA curves of wood flours

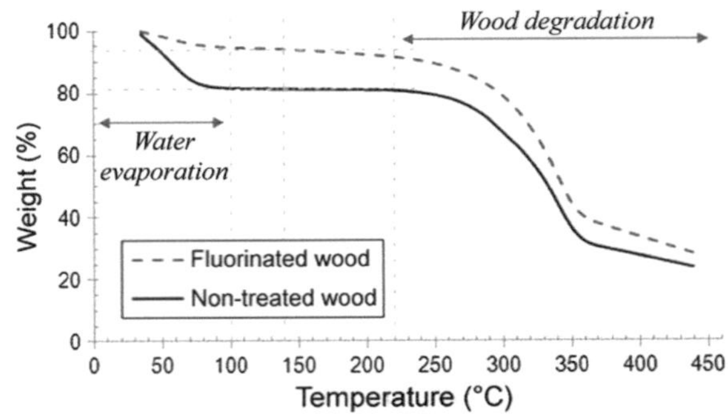

Improvement of the Composite Health

X-ray tomography enabled us to see inside the composite samples. Due to the negligible differences in densities between wood (about 1.4) and polyester (about 1.2), the particles were not visible. Nevertheless, we managed to distinguish the pores, which are air bubbles entangled in the composite during its processing. Figure 6 shows examples of scans obtained for the two kinds of composites.

The difference between the two composites made with non-treated wood or with fluorinated wood is obvious: the latter exhibited much fewer pores than the former. Since the composites were processed in exactly the same way, this observation can only be attributed to the fluorination. It may enhance the compatibility between wood particles and polyester, allowing less volume to be available for air and hindering the formation of bubbles when the counter mold is pressed onto the materials. This tends to prove that the composite's quality is improved by the pre-treatment of the wood. Better properties are thus expected from the fluorinated wood composite.

Improvement of the Composite Mechanical Properties

The tensile stress-strain curves of the composites with non-treated and fluorinated wood flours are given in the Fig. 7 and Table 3. A slight increase in the composite

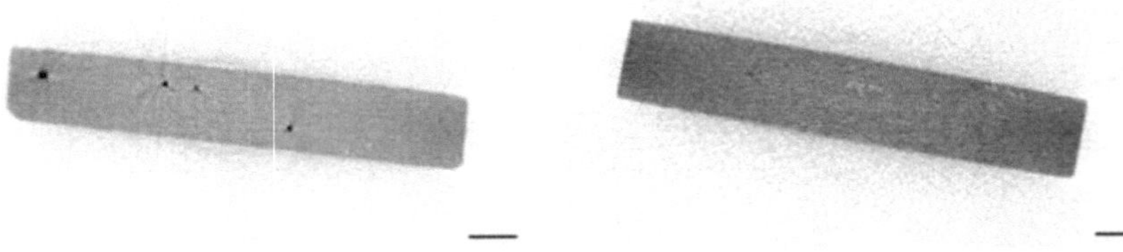

Fig. 6 X-ray tomography images of the cross-sections of two composites made of non-treated (*left*) or fluorinated (*right*) wood flours (scale bars = 1 mm)

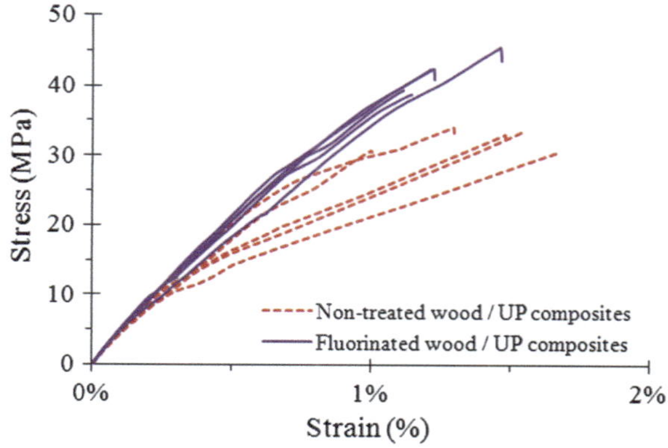

Fig. 7 Tensile stress-strain curves of composites with non-treated or fluorinated wood flours

Table 3 Tensile properties of the composites with non-treated or fluorinated wood flours

Composites	E (GPa)	σ_R (MPa)	A_R (%)
Non-treated wood/UP	4.4 ± 0.2	32.4 ± 1.5	1.4 ± 0.3
Fluorinated wood/UP	4.8 ± 0.2	41.7 ± 2.4	1.3 ± 0.1

stiffness (E) was observed with the direct fluorination of the reinforcements (+8 %), as shown in the higher initial slope of the curve. The mean tensile strength (σ_R) was also improved by the treatment (+29 %). Nevertheless, the elongation at break (A_R) was slightly decreased (−6 %) but this reduction is negligible due to the high standard deviations observed for this property (10–20 %).

We can deduce from the enhancement of the tensile properties that the fibre/matrix interface is of better quality when the wood flour has been previously fluorinated. Compared with other treatments, the direct fluorination developed here enables an improvement of the composite tensile properties comparable to treatments using maleic anhydride or silane. Actually, wood-polypropylene composites showed increases in their elastic modulus of about 6 % and in their tensile strength of maximum 20 % after a maleic anhydride treatment (Garcia et al. 2005), whereas the addition of 1 wt% of silane to a 30 wt% wood reinforced polypropylene improved the composite tensile strength by 12 % (Nachtigall et al. 2007).

The effect of the fluorination on the flexural mechanical properties of the composites was also studied. The stress-strain curves obtained for both types of composites are presented in Fig. 8 and the corresponding mean values are given in Table 4. As for the tensile tests, the composites appeared to be stiffer after the direct fluorination of the reinforcements (+25 %). For both types of composite, failure occurred for similar values of strain. The strength was also increased by the wood treatment (+27 %). Compared with the data available in the literature on the effect of other treatments on the composite flexural properties, fluorination appears as the most efficient. Actually, the addition of 4 % of bamboo charcoal to wood polymer composites was found to increase strength to a similar level (+32 %) but the

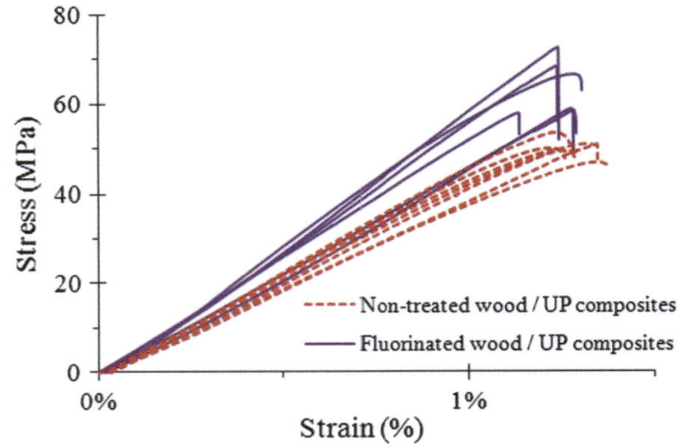

Fig. 8 Flexural stress-strain curves of composites with non-treated or fluorinated wood flours

Table 4 Flexural properties of the composites with non-treated or fluorinated wood flours

Composites	E (GPa)	σ_R (MPa)	A_R (%)
Non-treated wood/UP	4.5 ± 0.3	50.1 ± 1.8	1.4 ± 0.1
Fluorinated wood/UP	5.6 ± 0.6	63.8 ± 5.7	1.3 ± 0.1

increase in the flexural modulus was relatively low (+9 %) (Li et al. 2014). Also, an alkali treatment applied on different tropical wood polymer composites improved the strength of up to 16 % and the modulus of 13 % maximum (Islam et al. 2012).

To conclude on this part, the fluorination of the wood flour tends to improve both flexural and tensile stiffness and strength of the derived composites, which could be explained by a better load transfer due to an improvement in the compatibility between the wood reinforcements and the polyester resin. This is in agreement with the decrease of the amount of pores after the wood treatment, as presented in section 'Improvement of the Composite Health'.

Improvement of the Composite Hygroscopic Behavior

Decrease of the composite surface energy An overview of the pictures obtained for the measurement of contact angles between water and the wood polymer composite surfaces is given in Fig. 9. It is to be noted that these images were taken when the droplets were stabilized. The contact angle increases with the treatment of the wood reinforcements by about 10 %, from 75.9° ± 13.1° to 83.6° ± 8.6°. The relatively large standard deviations can be explained by the presence of random surface micropores influencing the measurements. Despite this scattering of data, the fluorination of wood flour is likely to enhance the surface properties of the composite against water.

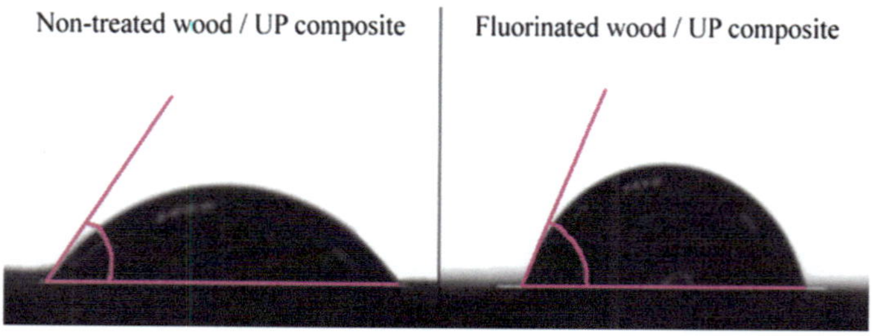

Fig. 9 Examples of contact angle images for the non-treated (*left*) or fluorinated (*right*) wood composites

Decrease in the composite water absorption The composites that were stored under a constant and relatively high humidity (80 %) at 20 °C were weighed every day until their weight stabilized. The results of these measurements are given in Fig. 10. It appears first that the fluorinated wood composite exhibits a lower water content at the hygroscopic equilibrium (4.2 %) compared with the non-treated wood composite (5.5 %), which means that the former absorbs less water than the latter. This is in agreement with the values found for the fluorinated wood flour itself which, after storage under a relative humidity of 85 %, presented a water content about 30 % below the one of non-treated wood flour (Saulnier et al. 2013). This drop of water content (−23 %) has already been observed for immersed silane- or MAPP-treated wood composites, but at a lower level (−10 and −15 % respectively) (Ichazo et al. 2001). Also, it was reported that the addition of 4 % of bamboo charcoal in a wood polymer composite decreased the equilibrium water content of 6 % (Li et al. 2014). Benzylation was proved to be very efficient to reduce the composite water uptake but at the expense of its mechanical properties (Dányádi et al. 2010). Secondly, this equilibrium is reached faster in the case of treated wood (14 days as against 16 days). Yet again, this follows the tendency of the fluorinated wood flour which, after storage at a relative humidity of 75 %, reached weight

Fig. 10 Evolution of the water content of fluorinated or non-treated wood composites

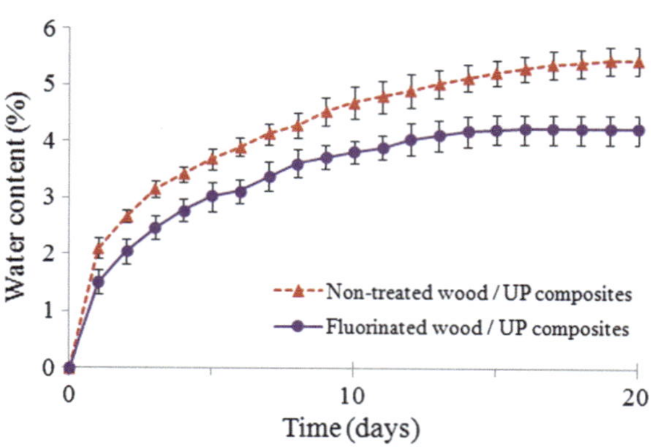

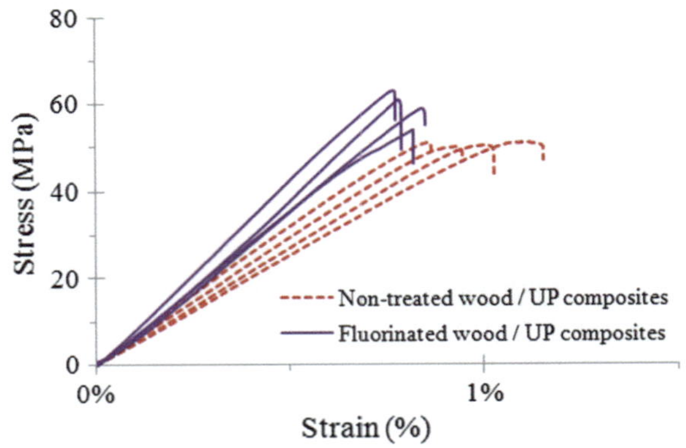

Fig. 11 Flexural stress-strain curves of composites stored 20 days under a relative humidity of 80 %

Table 5 Flexural properties of the composites stored 20 days under a relative humidity of 80 %

Composites	E (GPa)	σ_R (MPa)	A_R (%)
Non-treated wood/UP	5.8 ± 0.5	50.6 ± 0.4	1.0 ± 0.1
Fluorinated wood/UP	7.8 ± 0.6	59.2 ± 3.4	0.8 ± 0.1

stabilization after 5 h, whereas the non-treated wood flour required 24 h (Saulnier et al. 2013).

Improvement of the composite mechanical behaviour in humid conditions After a 20 day-storage under a relative humidity of 80 % and a constant temperature of 20 °C, wood polymer composites were tested to get their flexural properties. The obtained curves are shown in Fig. 11 and the corresponding values are presented in Table 5.

It can be noticed that the moduli of these samples are higher than those obtained under ambient conditions. This may be explained by the fact that the composites were not cured after processing and that humidity may have acted as an ageing agent that completed the reticulation of the polyester matrix during the storage at 80 % RH [this phenomenon has already been underlined in some studies on thermoset resins (Stamatakis et al. 2010; Odegard et al. 2011)]. Moreover, as in the case of ambient tests, fluorination increases the flexural properties of wood polymer composites. This tends to prove that, even after exposure to water, the fluorinated wood/UP composites keep good flexural properties.

Conclusions

A decrease in the hydrophilic character of wood flour was obtained by direct fluorination using molecular fluorine F_2. This can be explained by the formation of C–F bonds onto the surface of the wood flour particles and the decrease in the number of –OH groups on the wood, observed using FT-IR and ^{19}F RMN spectroscopy, and confirmed by hygroscopic characterisation. Moreover, SEM analysis showed that the treatment did not damage the physical structure of the wood. The thermal behaviour of wood flour remained stable under 100 °C after direct fluorination, which enables its use as a reinforcement for unsaturated polyester or other resins with low-temperature manufacturing characteristics.

An improvement of the tensile and flexural properties of the wood-polyester composite was obtained by the direct fluorination of the reinforcements. This can be explained by better adhesion between polyester and wood after the treatment, leading to fewer pores as shown by X-ray tomography. The composite surface energy and its capacity to absorb water were also reduced after fluorination of the reinforcements. Large scale production of low-cost fluorinated wood flour may be proceeded at low temperature (i.e. with low energy cost) and with high reaction yield because of the intrinsic reactivity of the raw materials. Additional experiments are currently being carried out to decrease even further the hydrophilicity of the composites by applying fluorination to fluorinated wood flour composites.

References

Guyonnet, R., inventor; Wood curing method. US patent 5901463, 1999 May 11.

Podgorski, L.; Chevet, B.; Onic, L.; Merlin, A.; Modification of wood wettability by plasma and corona treatments. Int J Adhes Adhes, 2000, 20, 103-111.

Ichazo, M.N.; Albano, C.; González, J.; Perera, R.; Polypropylene/wood flour composites: treatments and properties. Compos Struct, 2001, 54, 207-214.

Lu, J.Z.; Wu, Q.; Negulescu, I.I.; Wood-fiber/high-density-polyethylene composites: Coupling agent performance. J Appl Polym Sci, 2005, 96, 93-102.

Tserki, V.; Zafeiropoulos, N.E.; Simon, F.; Panayiotou, C.; A study of the effect of acetylation and propionylation surface treatments on natural fibres. Comp Part A, 2005, 36, 1110-1118.

Karmarkar, A.; Chauhan, S.S.; Modak, J.M.; Chanda, M.; Mechanical properties of wood-fiber reinforced polypropylene composites: Effect of a novel compatibilizer with isocyanate functional group. Comp Part A, 2007, 38, 227-233.

Nachtigall, S.M.B.; Cerveira, G.S.; Rosa, S.M.L.; New polymeric-coupling agent for polypropylene/wood-flour composites. Polym Test, 2007, 26, 619-628.

Dominkovics, Z.; Dányádi, L.; Pukánszky, B.; Surface modification of wood flour and its effect on the properties of PP/wood composites. Comp Part A, 2007, 38, 1893-1901.

Dányádi, L.; Móczó, J.; Pukánszky, B.; Effect of various surface modifications of wood flour on the properties of PP/wood composites. Comp Part A, 2010, 41, 199-206.

Ayrilmis, N.; Jarusombuti, S.; Fueangvivat, V.; Bauchongkol, P.; Effect of thermal-treatment of wood fibres on properties of flat-pressed wood plastic composites. Polym Degrad Stabil, 2011, 96, 818-822.

Acda, M.N.; Devera, E.E.; Cabangon, R.J.; Ramos, H.J.; Effects of plasma modification on adhesion properties of wood. Int J Adhes Adhes, 2012, 32, 70-75.

Islam, M.S.; Hamdan, S.; Jusoh, I.; Rahman, M.R. *et al.*; The effect of alkali pretreatment on mechanical and morphological properties of tropical wood polymer composites. Mater Design, 2012, 33, 419-424.

Zhang, H.; Effect of a novel coupling agent, alkyl ketene dimer, on the mechanical properties of wood–plastic composites. Mater Design, 2014, 59, 130-134.
Kharitonov, A.P.; Taege, R.; Ferrier, G.; Teplyakov, V.V. *et al.*; Direct fluorination–useful tool to enhance commercial properties of polymer articles. J Fluor Chem, 2005, 126, 251–263.
Kharitonov, A.P.; Direct Fluorination of Polymers. New York: Nova Publishers, 2008.
Maity, J.; Jacob, C.; Das, C.K.; Kharitonov, A.P. *et al.*; Fluorinated aramid fiber reinforced polypropylene composites and their characterization. Polym Composite, 2007, 28, 462-469.
Bismarck, A.; Tahhan, R.; Springer, J.; Schulz, A. *et al.*; Influence of fluorination on the properties of carbon fibres. J Fluor Chem, 1997, 84, 127-134.
Dubois, M.; Guérin, K.; Giraudet, J.; Pilichowski, J.F. *et al.*; Direct fluorination of poly (p-phenylene). Polymer, 2005, 46, 6736-6745.
Ho, K.K.C.; Beamson, G.; Shia, G.; Polyakova, N.V. *et al.*; Surface and bulk properties of severely fluorinated carbon fibres. J Fluor Chem, 2007, 128, 1359-1368.
Guérin, K.; Dubois, M.; Houdayer, A.; Hamwi, A.; Applicative performances of fluorinated carbons through fluorination routes: a review. J Fluor Chem, 2012, 134, 11-17.
Sapieha, S.; Verreault, M.; Klemberg-Sapieha, J.E.; Sacher, E. *et al.*; XRay photoelectron study of the plasma fluorination of lignocellulose. Appl Surf Sci, 1990, 44, 165-169.
Sahin, H.T.; Manolache, S.; Young, R.A.; Denes, F.; Surface fluorination of paper in CF4-RF plasma environments. Cellulose, 2002, 9, 171-181.
Müller, U.; Rätzsch, M.; Schwanninger, M.; Steiner, M. *et al.*; Yellowing and IRchanges of spruce wood as result of UV-irradiation. J Photoch Photobio B, 2003, 69, 97-105.
Popescu, M.C.; Froidevaux, J.; Navi, P.; Popescu, C.M.; Structural modifications of Tilia cordata wood during heat treatment investigated by FT-IR and 2D IR correlation spectroscopy. J Mol Struct, 2013, 1033, 176-186.
[25] Schwanninger, M.; Rodrigues, J.C.; Pereira, H.; Hinterstoisser, B.; Effects of shorttime vibratory ball milling on the shape of FT-IR spectra of wood and cellulose. Vib Spectrosc, 2004, 36, 23-40.
Isbester, P.K.; Kestner, T.A.; Munson, E.J.; High-resolution variable-temperature MAS ^{19}F NMR spectroscopy of fluorocarbon polymers. Macromolecules, 1997, 30, 2800-2801.
Aimi, K.; Ando, S.; Conformation analysis and molecular mobility of ethylene and tetrafluoroethylene copolymer using solid-state ^{19}F MAS and 1H → ^{19}F CP/MAS NMR spectroscopy. Magn Reson Chem, 2004, 42, 577–588.
Zhang, W.; Dubois, M.; Guérin, K.; Bonnet, P. *et al.;* Effect of curvature on C-F bonding in fluorinated carbons: from fullerene and derivatives to graphite. Phys Chem Chem Phys, 2010, 12, 1388-1398.
Stamm, A.J.; Wood and cellulose science. New York: Ronald press, 1964.
Jeske, H.; Schirp, A.; Cornelius, F.; Development of a thermogravimetric analysis (TGA) method for quantitative analysis of wood flour and polypropylene in wood plastic composites (WPC). Thermochim Acta, 2012, 543, 165-171.
Poletto, M.; Zattera, A.J.; Forte, M.M.C.; Santana, R.M.C.; Thermal decomposition of wood: influence of wood components and cellulose crystallite size. Bioresource Technol, 2012, 109, 148-153.
Garcia, R.A.; Amélioration de la stabilité dimensionnelle des panneaux de fibres de bois MDF par traitements physico-chimiques. PhD thesis. Université de Laval, Québec, 2005.
Li, X.; Lei, B.; Lin, Z.; Huang, L. *et al.*; The utilization of bamboo charcoal enhances wood plastic composites with excellent mechanical and thermal properties. Mater Design, 2014, 53, 419-424.
Saulnier, F.; Influence de traitements physico-chimiques des renforts sur le comportement mécanique des composites à base de co-produits de bois. PhD thesis. Université Blaise Pascal, France, 2013.
Saulnier, F.; Dubois, M.; Charlet, K.; Frezet, L. *et al.*; Direct fluorination applied to wood flour used as reinforcement for polymers. Carbohyd Polym, 2013, 94, 642-646.

Stamatakis, G.; Knuutinen, U.; Laitinen, K.; Spyros, A.; Analysis and ageing of unsatured polyester resins in contemporary art installations by NMR spectroscopy. Anal Bioanal Chem, 2010, 398, 3203-3214.

Odegard, G.M.; Bandyopadhyay, A.; Physical aging of epoxy polymers and their composites. J Pol Sci Pol Phys, 2011, 49, 1695-1716.

DSC Analysis of In Situ Polymerized Poly(Butylene Terephthalate) Flax Fiber Reinforced Composites Produced by RTM

C.M. Romão, C.M. Pereira and J.L. Esteves

Abstract The present work addresses Differential Scanning Calorimetry (DSC) analysis of neat and woven flax fiber fabric reinforced in situ polymerized polybutylene terephthalate (pCBT), produced by the isothermal RTM molding technique. A brief description of the developed RTM set-up is provided and the composite manufacturing details are presented. Thermal analysis reveals that incorporation of flax fibers leads to change in pCBT, namely leads to decrease in melting temperature, heat of fusion and heat of crystallization. The degree of crystallinity did not change significantly.

Keywords Flax fiber · CBT resin · pCBT · RTM · Composites · DSC analysis

Introduction

Thermoplastic composites offer some interesting advantages over thermoset. Improved toughness and impact resistance, greater design freedom, possibility of shorter production cycles and improved recyclability are some factors that promote industrial awareness in sectors such as aerospace, automotive, marine and sporting goods (Máirtín et al. 2001). Vegetable reinforcement incorporation further increases the range of advantages of thermoplastic matrix composites. Environmental constraints and high specific properties of some vegetal fibers (e.g. hemp and flax) are

C.M. Romão (✉)
Department of Mechanical Engineering and Industrial Management (DEMGI),
School of Technology and Management, Polytechnic Institute of Viseu, Viseu, Portugal
e-mail: romao@estv.ipv.pt

C.M. Pereira · J.L. Esteves
Institute of Mechanical Engineering and Industrial Management (INEGI),
University of Porto, Porto, Portugal

J.L. Esteves
Department of Mechanical Engineering (DEMec), University of Porto, Porto, Portugal

R. Fangueiro and S. Rana (eds.), *Natural Fibres: Advances in Science and Technology Towards Industrial Applications*, RILEM Bookseries 12,
DOI 10.1007/978-94-017-7515-1_17

pointed out as major reasons for glass fibers replacement by vegetal ones (Romão 2003, 2013). Nonetheless, the selection of the binomial thermoplastic matrix-vegetal fiber must take into account the relatively low thermal resistance of this kind of reinforcement and the processing technique that will be used. In this context, it should be noted that, on the one hand, vegetal fibers undergo thermal degradation when exposed to temperatures around 200 °C (Romão 2003, 2013; Rowell et al. 1997; Baillie 2004) and, on the other hand, thermoplastic matrices present high melt viscosity (typically 10^2–10^4 Pa/s (Máirtín et al. 2001)). These limitations prevents the manufacturing of long fiber reinforced thermoplastic composites and the combination of vegetal fibers with high performance or engineering thermoplastics (which have melt temperatures typically above 220 °C) by the traditional molding techniques. Reactive processing of thermoplastics is a recent technique, which makes use of low molecular weight mono or oligomeric precursors that impregnate the reinforcing fibers and polymerize in situ to form the desired matrix, after heated and mixed with an activator system. Due to the mentioned low molecular weight, the precursors have extremely low melt viscosities (of magnitude order of mPa/s), allowing a quick and suitable impregnation of continuous reinforcements at lower processing pressures and moderate temperatures (below 200 °C for some reactive systems). The liquid molding techniques, specific of thermoset composites, can be used. It is necessary, however, to make adjustments in the equipment as a result of the existing differences in the processing of these two types of polymeric materials (Parton 2006; Rijswijk and Bersee 2007). In these circumstances, and taking into account that some of the precursors of these polymers have melting temperatures below the thermal degradation of vegetal fibers, the use of reactive processing becomes a viable alternative for the manufacture of lighter, tougher and environmentally friendly composites with moderate mechanical properties (Romão 2013).

In the present study, a DSC analysis of pure and flax fiber reinforced in situ polymerized poly(butylene terephthalate) produced by the RTM process is made, starting from the prepolymer Cyclic Butylene Terephthalate (CBT resin). The authors selected the RTM isothermal processing since it leads to shorter production cycles and allows for the combination of CBT resin with long vegetable reinforcements.

The main goal was to investigate the possible compatibility between these two kind of materials, both very sensitive to moisture.

Materials and Methods

Materials

The polymeric precursor used in this study is the one-component CBT 160 resin, a Cyclic Butylene Terephthalate oligomer mixture that already contains the catalyst (Fig. 1), supplied by *Cyclics Corporation* (Schwarzheide, Germany) in the granulated form. It is a mixture of low molecular weight oligomers (circa dimer 40 wt%, trimer 35 wt%, tetramer 15 wt%, and pentamer and higher 10 wt% (Chen et al. 2012))

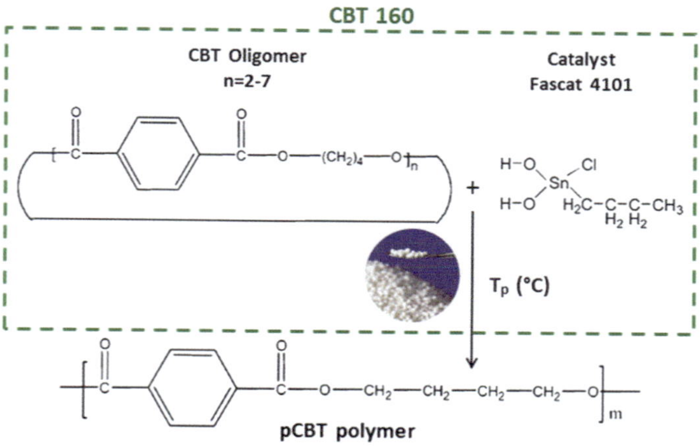

Fig. 1 CBT 160 resin and its polymerization to pCBT (Romão et al. 2013)

with two to seven butyl groups, resulting in a melting range from 120 to 160 °C (Parton 2006; Baets 2008). Before processing, the oligomers were dried for about 2 h at 80 °C in order to remove residual moisture, which could interfere with the polymerization reaction (Parton 2006). The reinforcement is a non-treated flax woven fabric, supplied by Composites Evolution (Chesterfield, United Kingdom). The woven fabric type is 4 × 4 Hopsack, Fig. 2, with a surface weight of 510 g/m^2, suitable for processing by traditional liquid molding techniques such as hand lay-up, vacuum infusion and RTM. Before processing, the fabric was dried 24 h at 80 °C.

RTM Set-up and Composite Manufacturing Details

The developed RTM set-up consists mainly of a heated mold, a system for melting the resin under nitrogen atmosphere (fusion system), a heated system to inject the

Fig. 2 Biotex flax 4 × 4 Hopsack

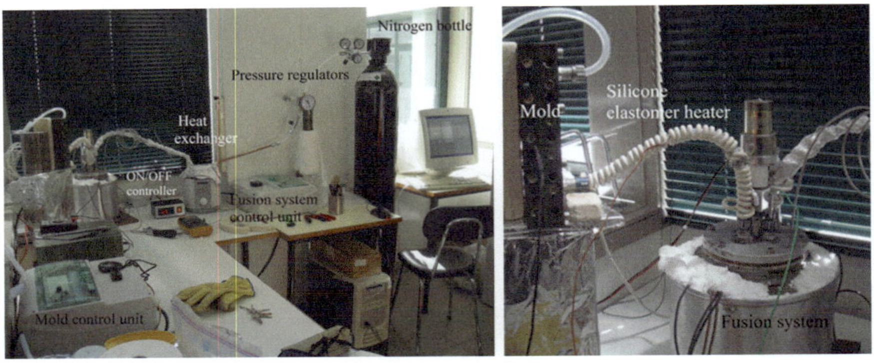

Fig. 3 The developed RTM set-up (Romão et al. 2013)

resin into the fiber bed (injection system) and three different temperature control units connected to each of these thermal systems (Fig. 3). The CBT 160 resin was heated under a nitrogen atmosphere to approximately 190 °C and then injected, under nitrogen pressure (2–50 kPa), into the closed mold at the same temperature. Once the mold was completely filled, the temperature was maintained for 30 min, in order to complete the polymerization reaction and to allow for cold crystallization (Romão 2013; Parton 2006). Composites were produced using a 32 % volume fraction of 4 × 4 Hopsack woven flax fabric. The fibers exposition time to temperature values higher than 180 °C was, in all RTM cycles, of approximately 60 min.

DSC Analysis

Different materials were investigated by Differential Scanning Calorimetry (DSC): 4 × 4 Hopsack fabric, pCBT produced by DSC and neat and reinforced pCBT produced by RTM. These tests were carried out to achieve the following objectives: (i) certifying that RTM thermal cycle does not trigger the thermal degradation of flax fibers; (ii) investigating the thermal behavior of pCBT produced by DSC, in conditions similar to the pCBT produced by RTM; (iii) investigating the thermal behavior of pCBT produced by RTM, in order to study the influence of the manufacturing process; (iv) studying the influence of flax fibers in the polymerization of CBT into pCBT and in its subsequent crystallization. These investigations were performed on a DSC Q20 V24.2 device (TA Instruments). Experiments were carried out with samples in a nitrogen environment purged at 50 ml min^{-1}, to prevent moisture and oxidative degradation. The samples cooling was accomplished with compressed air, since the device does not have a controlled cryogenic cooling system. The flax fibers and the CBT 160 samples were dried at 80 °C for 24 and 2 h, respectively, and hermetically sealed in aluminium pans before scanned.

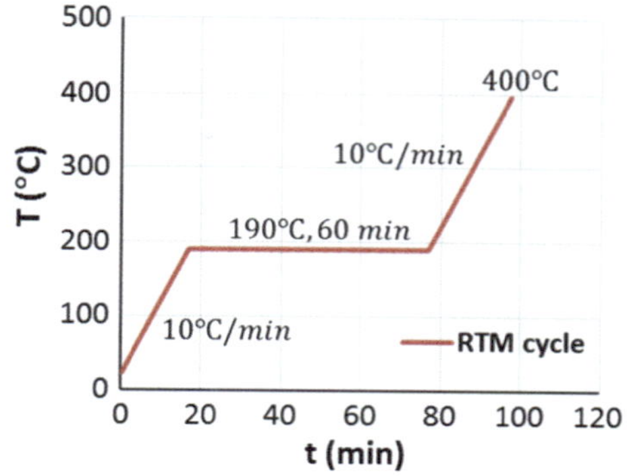

Fig. 4 DSC thermal cycle: flax fibers

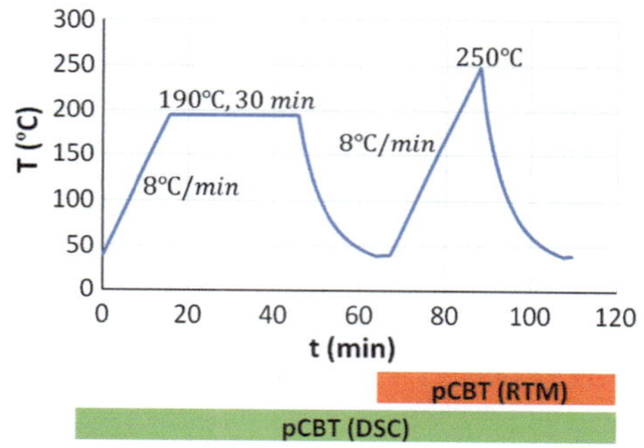

Fig. 5 DSC thermal cycles: neat/reinforced pCBT

Flax fibers were subjected to the following thermal cycle: heating to 190 °C, isothermal at 190 °C for 60 min (similar to RTM), heating ramp to 400 °C and cooling to room temperature (Fig. 4). In order to produce a sample of pCBT by DSC (with a thermal cycle similar to the pCBT produced by RTM), a CBT 160 sample was heated to 190 °C, maintained at 190 °C for 30 min and then cooled to room temperature (first heating/cooling scan). The produced pCBT was then reheated to 250 °C and cooled to room temperature (second heating/cooling scan), Fig. 5. Samples of pure and reinforced pCBT produced by RTM were only submitted to the second heating/cooling scan described previously. The data were analyzed by the Universal Analysis software package of TA Instruments, version 4.5 A.

Results and Discussion

Production

Flat plates (100 × 100 × 4 mm^3) of neat and reinforced pCBT were successfully produced using the RTM set-up and the production method described previously

Fig. 6 Pure pCBT (*left*) and flax fiber reinforced pCBT (*right*) plates produced by RTM

(Fig. 6). The neat pCBT plates exhibit a higher mold shrinkage, during which fissuration also took place. The incorporation of the reinforcement reduced the matrix shrinkage; however, the presence of microcracks randomly dispersed on the surface of the produced plates was observed. The presence of microcracks is also reported by others researchers for glass fiber reinforcements (Prabhakaran et al. 2010).

DSD Analysis

Flax Fibers

The DSD thermogram of flax fabric is provided in Fig. 7. A summary of the analysis on this thermogram is provided in Table 1. Degradation temperatures of the main components of flax fibers (lignin, hemicellulose and cellulose) were considered to be the maximum of endothermic/exothermic peaks. The broad endotherm peak at 75.71 °C (T_1) is due to the presence of residual moisture (Shinoj et al. 2010). The sharp exothermic peak at 189.99 °C is an electronic overshoot, caused by transitions in the thermal cycle stages (in this study, when the heating ramp to 190 °C ends up and the isothermal at this temperature starts and when the isothermal ends and the sample is reheated to 400 °C). The endothermic peak at 249.28 °C (T_2), indicates the propanoid side chain of lignin decomposition (Wittkowski et al. 1992). One possible reason for the absence of this peak in most DSC thermograms, published by other authors, is its overlap with the characteristic exothermic peak of hemicellulose degradation. The second exothermic peak at 266.55 °C (T_3) corresponds to hemicellulose decomposition and the third endothermic peak at 364.25 °C (T_4) to α-celullose decomposition (Ray et al. 2002). These results are similar to those obtained by other investigators (Shinoj et al. 2010; Manfredi et al. 2006; Tripathy 2009) and suggest that RTM processing does not induce thermal degradation of flax fibers, since the first thermal degradation temperature occurs at 249.28 °C.

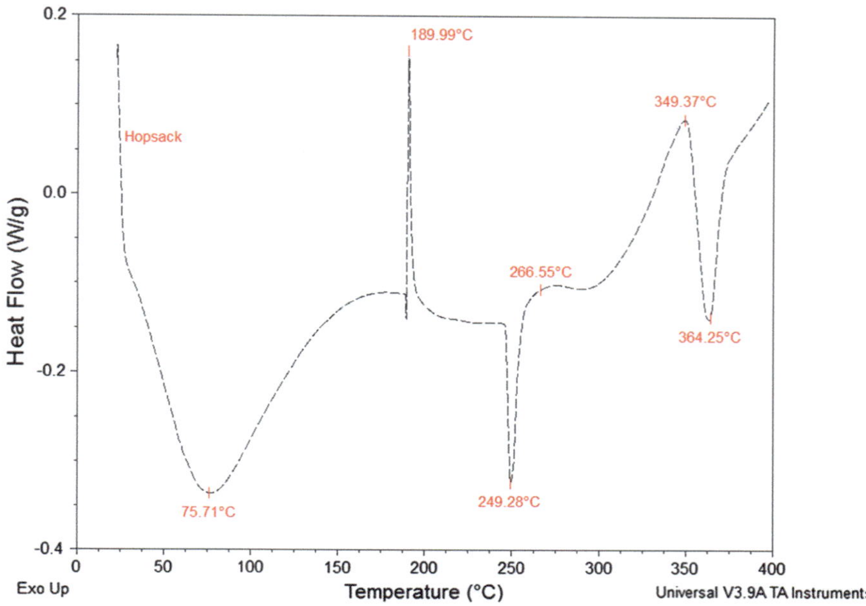

Fig. 7 DSC thermogram of flax fibers fabric

Table 1 DSC results of woven flax fabric: degradation temperatures of the main components

Sample	Heating scan	Peak temperatures (°C)			
		T_1	T_2	T_3	T_4
4 × 4 Hopsack	Similar to RTM	75.71	249.28	266.55	364.25

CBT 160 Resin

The typical DSC thermogram of CBT 160, submitted to the first thermal cycle described above, is displayed in Fig. 8. A summary of the analysis based on this thermogram is provided in Table 2. The heating curve shows the characteristic behavior of CBT, consisting of two small melting peaks at 117.20 °C (T_1) and 156.61 °C (T_3) and a broader peak at 141.42 °C (T_2). These peaks of CBT are ascribed to the presence of oligomers of different size and are in agreement with earlier observations reported by other researchers (Parton 2006; Chen et al. 2012; Ishak et al. 2007; Brunelle 2008; Lanciano et al. 2009). When the temperature is maintained at 190 °C for 30 min the CBT resin polymerize into pCBT that, at the same time, crystallizes. Because isothermal polymerization of CBT is supposed to be athermic, DSC cannot be used to investigate the reaction dynamics (Chen et al. 2012).

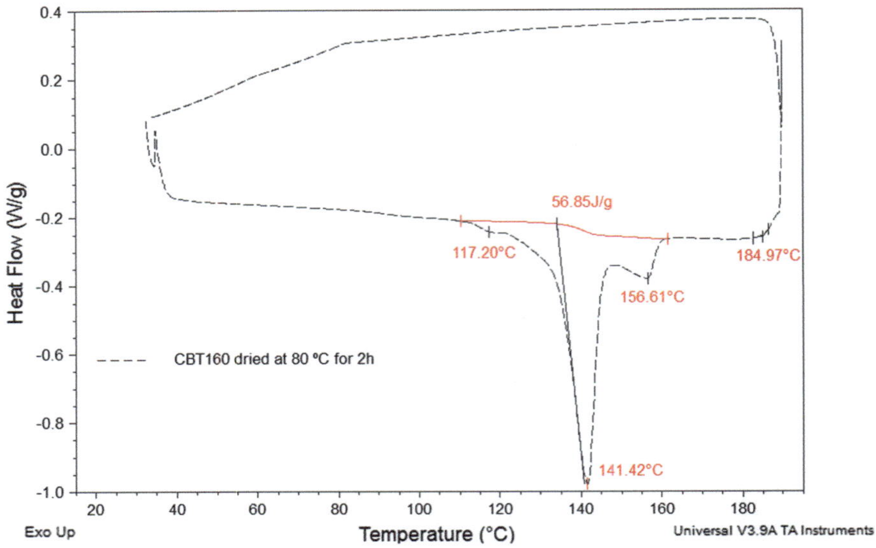

Fig. 8 DSC thermogram of CBT 160 (1st heating and cooling scan)

Table 2 DSC results of CBT160: prepolymer melting range

Sample	Heating scan	Peak temperatures (°C)		
		T_1	T_2	T_3
CBT160	Heating to 190 °C Isothermal at 190 °C for 30 min	117.20	141.42	156.61

pCBT Polymer

The DSD thermograms of pure and reinforced pCBT are provided in Fig. 9. The results of its analysis are summarized in Table 3. The melting temperature of pCBT was considered to be the maximum of the endothermic melting peak from the heating scans, whereas crystallization temperature was taken as the maximum of the exothermic peak of crystallization from cooling scans. The heat of fusion and the heat of crystallization were determined from the areas under the melting and crystallization peaks, respectively. The degree of crystallinity of pCBT was calculated as follows (Baets et al. 2008):

$$\chi_c(\%) = \frac{\left[\frac{\Delta H_f}{\Delta H_f^{\circ}}\right] \times \mathbf{100}}{\mathbf{wt\%pCBT}} \times \mathbf{100} \tag{1}$$

where ΔH_f is the heat of fusion of the pCBT and ΔH_f^0 is the heat of fusion for a 100 % crystalline PBT, which was found in literature to be 142 J/g (Baets et al. 2008);

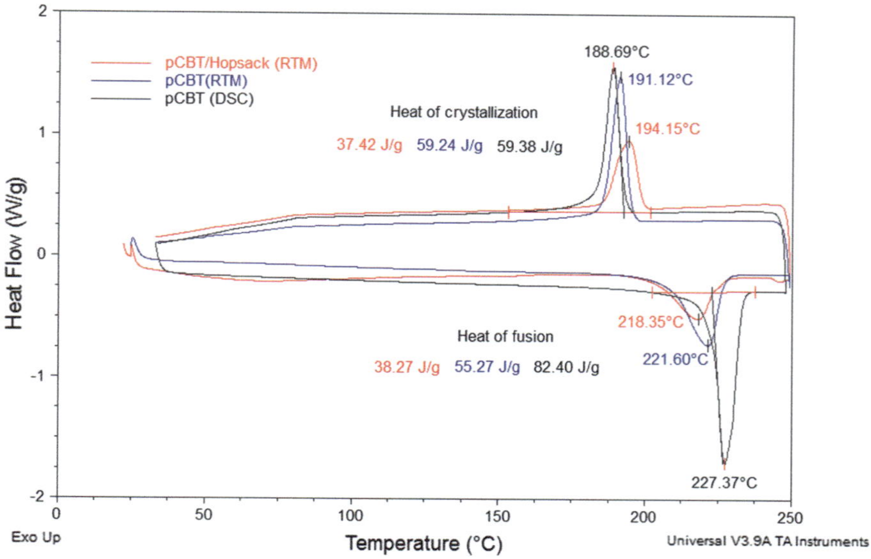

Fig. 9 DSC Thermograms (*Q-T*) of pure pCBT (DSC and RTM) and reinforced with Hopsack fabric (RTM)

Table 3 DSC results of pCBT: physical properties

Samples	Heating			Cooling	
	T_m (°C)	ΔH_f (J/g)	χ_c (%)	T_c (°C)	ΔH_c (J/g)
pCBT (DSC)	227.37	82.40	58.03	188.69	59.38
pCBT (RTM)	221.60	55.27	38.92	191.12	59.24
pCBT/hopsack (RTM)	218.35	38.27	37.96[a]	194.15	37.42

[a]Considering wt% = 71 %

the denominator is the correction parameter for a lower amount of polymer in the composites samples.

Analyzing the thermogram of pCBT produced by DSC it is clear the absence of endothermic peaks in the melting range of CBT 160, suggesting the complete conversion of CBT 160 into pCBT during the first heating cycle. The isothermal processing at 190 °C for 30 min resulted in a pCBT with a single melting temperature (T_m) of about 227 °C, which is indicative of the presence of just one type of crystal formed during the first thermal cycle (Ishak et al. 2006; Kocsis et al. 2007), and a degree of crystallinity (χ_c) nearly 58 %. While the melting temperature falls within the values published by other authors, the degree of crystallinity is expressively higher (Parton 2006; Baets et al. 2008a, b). This can be, eventually, a feature of the reactive system CBT 160 (one-part system), once the values found in literature correspond to a pCBT produced from two-part systems (where the catalyst is mixed with the CBT resin previously to the injection). The pCBT produced by

RTM exhibits lower values of melting temperature ($T_m \approx 222$ °C) and of heat of fusion ($\Delta H_m = 55.27$ J/g). The lower melting temperature may be an indicator of a lower molecular weight pCBT, a smaller size of the crystals formed during the cold crystallization and/or the presence of impurities in the polymer. Furthermore, the lower heat of fusion is a sign of the presence of smaller crystals and, consequently, of a lower degree of crystallinity of pCBT ($\chi_c \approx 39$ %). This last result suggests a lower brittleness for the pCBT produced by RTM. However, its lower melting temperature may be an indicator of a lower molecular weight, a situation that contributes to the increase of pCBT brittleness and to the decrease of the mechanical properties (Parton et al. 2005). These results can be justified by the lower precision control of the RTM processing parameters, comparing to DSC, which are known to influence the polymerization and cold crystallization reactions kinetics and, consequently, the properties of the polymer. The obtained values are also lower than those obtained by Parton (2006) and Baets et al. (2008a, b). This difference can be justified by the fact that the authors used an RTM equipment that allowed, as DSC, better-controlled conditions processing. The incorporation of flax fibers led to a more pronounced decrease in the melting temperature ($T_m = 218.35$ °C) and in the heats of fusion ($\Delta H_f = 38.27$ J/g) and crystallization ($\Delta H_c = 37.42$ J/g) of pCBT. These findings have already been reported by other researchers, who stated that the addition of fillers can influence the polymerization and crystallization reactions (Parton 2006; Lanciano et al. 2009; Kocsis et al. 2007; Baets et al. 2010; Abt et al. 2011). The change on the degree of crystallinity of pCBT was negligible ($\chi_c \approx 38$ %).

Conclusions

The developed RTM set-up enabled the successful production of materials with the desired geometry. Notwithstanding, the neat polymer showed physical properties lower than expected. The pCBT produced by RTM exhibited lower values of melting temperature and degree of crystallinity compared to that produced by DSC. The analysis of the process and of the results allowed to conclude that the polymer physical properties can be further improved by optimizing the control of the RTM processing parameters (pressure, resin heating rate to 190 °C and temperature). This control is crucial to obtain a pCBT with higher molecular weight and with an adequate crystalline morphology. The reinforcement of pCBT with Hopsack 4 × 4 flax fabric leads to a more pronounced decrease in its melting temperature, heat of fusion and heat of crystallization. These findings have been already reported by other researchers, who stated that the addition of fillers can influence the polymerization and crystallization reactions. The degree of crystallinity of reinforced pCBT did not change significantly.

Acknowledgments The authors gratefully acknowledge the funding by Ministério da Ciência, Tecnologia e Ensino Superior, FCT, Portugal, under the grant SFRH/BD/40522/2007.

References

P. Ó. Máirtín, P. McDonnell, M. Connor, R. Eder and C. Ó. Brádaigh, "Process investigation of a liquid PA-12/carbon fibre moulding system," *Composites: Part A,* vol. 32, no. 7, pp. 915-923, 2001.

C. Romão, "Estudo do comportamento mecânico de materiais compósitos de matriz polimérica reforçados com fibras naturais," University of Porto - Faculty of Engineering, Portugal, 2003.

C. Romão, "Compósitos de Matriz Termoplástica de Baixa Viscosidade Reforçados com Fibras Naturais de Origem Vegetal," University of Porto - Faculty of Engineering, 2013.

R. M. Rowell, A. R. Sanadi, D. F. Caulfield and R. E. Jacobson, "Utilization of natural fibers in plastic composites: problems and opportunities," *Lignocellulosic-plastic composites,* pp. 23-51, 1997.

C. Baillie, Ed., Green composites: Polymer composites and the environment, WoodHead publishing Limited, 2004, p. 320.

H. Parton, "Characterisation of the In-Situ Polymerisation Production Process for Continuous Fibre Reinforced Termoplastics," Katholieke Universiteit Leuven, Leuven, 2006.

K. V. Rijswijk and H. Bersee, "Reactive processing of textile fiber-reinforced thermoplastic composites – An overview," *Composites: Part A,* vol. 38, no. 3, p. 666–681, 2007.

H. Chen, W. Yu and C. Zhou, "Entropically-Driven Ring-Opening Polymerization of Cyclic Butylene Terephthalate: Rheology and Kinetics," *Polymer Engineering & Science,* vol. 52, no. 1, pp. 91-101, January 2012.

J. Baets, "Toughening of in-situ polimeryzed of cyclic butylele terephthalate for use in continuous fiber reinforced plastic composites," Katholieke Universiteit Leuven, Leuven, 2008.

C. Romão, C. M. C. Pereira and J. L. Esteves, "A Mechanical Analysis of In Situ Polymerized Poly(butylene terephthalate) Flax Fiber Reinforced Composites Produced by RTM," vol. 2013, no. Article ID 750802, p. 5, 2013.

R. T. D. Prabhakaran, T. L. Andersen and A. Lystrup, "Glass/CBT laminate processing and quality aspects," in *The 10th International Conference on Flow Processes in Composite Materials (FPCM10),* Ascona, 2010.

S. Shinoj, R. Visvanathan and S. Panigrahi, "Towards industrial utilization of oil palm fibre: Physical and dielectric characterization of linear low density polyethylene composites and comparison with other fibre sources," *Biosystems Engineering,* vol. 106, pp. 378-388, 2010.

R. Wittkowski, J. Ruther, H. Drinda and F. Rafiei-Taghanaki, "Formation of smoke flavor compounds by thermal lignin degradation," in *ACS Symposium Series 490,* Washington DC, 1992.

D. Ray, B. K. Sarkar, R. K. Basak and A. K. Rana, "Study of the thermal behavior of alkali-treated jute fibers," *Journal of Applied Polymer Science,* vol. 85, pp. 2594-2599, 2002.

L. B. Manfredi, E. S. Rodríguez, M. Wladyka-Przybylak and A. Vázquez, "Thermal degradation and fire resistance of unsaturated polyester, modified acrylic resins and their composites with natural fibres," *Polymer Degradation and Stability,* vol. 91, pp. 255-261, 2006.

A. C. Tripathy, "Characterization of flax fibres and the effect of different drying methods for making biocomposites," University of Saskatchewan, 2009.

Z. M. Ishak, Y. Leong, M. Steeg and J. Karger-Kocsis, "Mechanical properties of woven glass fabric reinforced in situ polymerized poly(butylene terephthalate) composites," *Composites Science and Technology,* vol. 67, p. 390–398, 2007.

D. Brunelle, "Cyclic Oligomer Chemistry," *Journal of Polymer Science Part A: Polymer Chemistry,* vol. 46, p. 1151–1164, 2008.

G. Lanciano, A. Greco, A. Maffezzoli and L. Mascia, "Effects of thermal history in the ring opening polymerization of CBT and its mixtures with montmorillonite on the crystallization of the resulting poly(butylene terephthalate)," *Thermochimica Acta,* vol. 493, pp. 61-67, 2009.

J. Baets, D. M. Dutoit and I. Verpoest, "Toughening of glass fiber reinforced composites with a cyclic butylene terephthalate matrix by addition of polycaprolactone," *Composites Part A,* vol. 39, no. 1, p. 13–18, 2008a.

Z. A. M. Ishak, P. P. Shang and J. Karger-Kocsis, "A modulated DSC study on the in situ polymerization of cyclic butylene terephthalate oligomers," *Journal of Thermal Analysis and Calorimetry,* vol. 84, p. 637–641, 2006.

J. Kocsis, P. Shang, Z. Ishak and M. Rösch, "Melting and crystallization of in-situ polymerized cyclic butylene terephthalates with and without organoclay:a modulated DSC study," *eXPRESS Polymer Letters,* vol. 1, no. 2, p. 60–68, 2007.

J. Baets, A. Godara, J. Devaux and I. Verpoest, "Toughening of polymerized cyclic butylene terephthalate with carbon nanotubes for use in composites," *Composites: Part A,* vol. 39, p. 1756–1761, 2008b.

H. Parton, J. Baets, P. Lipnik, B. Goderis, J. Devaux and I. Verpoest, "Properties of poly(butylene terephthatlate) polymerized from cyclic oligomers and its composites," *Polymer,* vol. 46, p. 9871–9880, 2005.

J. Baets, A. Godara and I. Verpoest, "Toughening of isothermally polymerized cyclic butylene terephthalate for use in composites," *Polymer Degradation and Stability,* vol. 95, pp. 346-352, 2010.

T. Abt, M. Sánchez-Soto, S. Illescas and J. Aurrekoetxea, "Toughening of in situ polymerized cyclic butylene terephthalate by addition of tetrahydrofuran," *Polymer International,* vol. 60, no. 4, pp. 549-556, 2011.

Parametric Study on the Manufacturing of Biocomposite Materials

A. Rubio-López, A. Olmedo, A. Díaz-Álvarez and C. Santiuste

Abstract This paper presents an analysis of the fabrication parameters that influence on the mechanical properties of biodegradable composites. Specimens were manufactured using film stacking and compression moulding process. The main analysed parameters were: constituent materials (fibres and matrix), heating temperature and pressure, thickness, fibre orientation and woven architecture. Two different polylactic acid (PLA) matrices, and four different woven fibres of Jute, Flax and Cotton were combined. Because of the parametric study, the influence of each parameter was analysed leading to an optimized manufacturing method. The results were analysed in terms of tensile strength and failure strain.

Keywords Fabrics/textiles · PLA · Mechanical properties · Biocomposites

Introduction

The high specific strength and stiffness of composite materials make them suitable for a wide range of applications in industry (Santiuste et al. 2001, 2010; Olmedo et al. 2014). The introduction of natural fibres as reinforcements and natural plastics as matrix in the last years has led to fully biodegradable composite materials (Mantia and Morreale 2011). The use of biocomposites can lead to reduction of environment impact. Their lower price and the prevision of a higher production establish a present alternative to traditional composites materials. Their low resistance to high temperatures is a limitation for some applications (Jang et al. 2012),

---

A. Rubio-López · A. Olmedo · C. Santiuste (✉)
Department of Continuum Mechanic and Structural Analysis, University Carlos III of Madrid, Avda de La Universidad 30, 28911 Leganés, Madrid, Spain
e-mail: csantius@ing.uc3m.es

A. Díaz-Álvarez
Department of Mechanical Engineering, University Carlos III of Madrid, Avda de La Universidad 30, 28911 Leganés, Madrid, Spain

R. Fangueiro and S. Rana (eds.), *Natural Fibres: Advances in Science and Technology Towards Industrial Applications*, RILEM Bookseries 12,
DOI 10.1007/978-94-017-7515-1_18

but the flammability can be reduced (Bocz et al. 2013). Another problem present in biocomposites is the poor bonding between the natural fibres and the biopolymer matrix (Bledzki et al. 2009). Thus, numerous researchers are trying to improve manufacturing processes with two objectives: first, to increase the mechanical properties of biocomposites, and second, to reduce production costs.

There is a wide range of applications for biocomposites, from the automotive sector (Müssig et al. 2006) to boats (Brouwer 2001) or electronics (Serizawa et al. 2006). Their mechanical properties call biocomposites to replace most of non-renewable materials as traditional glass-fibre composites (Jang et al. 2012; Bledzki et al. 2009; John et al. 2008; Huber et al. 2012; Summerscales et al. 2010; Voorn et al. 2001; Ochi 2006, 2008; Song et al. 2012; Porras and Maranon 2012; Graupner et al. 2009; Bodros et al. 2007). For example, Bledzki et al. (2006) and Müssig et al. (2006) have proved the viability to use natural fibres in car panels. Present industrial applications generally combine natural fibres with petrochemical plastics as PP (polypropylene) (John et al. 2008). There are also works showing that biocomposites are recyclable after it use on all this applications (Duigou et al. 2008). Recently, some authors have presented predictive models that can be used in the design of biocomposite structures (Rubio-López et al. 2015).

Some of the conventional composites fabrication techniques can be applied to manufacture biocomposites. Thermoplastic injection moulding uses a screw-type plunger to force the plastic material into a mould (Bledzki et al. 2009). Also new techniques have been developed, for example, All Cellulose-Composites (ACC) are obtained by the solution of the superficial cellulose of the fibres, and this cellulose acts as matrix of the fibres when regenerated. ACC are characterised by a total cohesion between matrix and reinforcement (Huber et al. 2012). Other fabrication procedures are heating under vacuum (Summerscales et al. 2010), RTM (resin transfer moulding) (Oksman 2001; Faruk et al. 2000), pultrusion (Faruk et al. 2000) or electrospinning for bio-nanocomposites (John et al. 2008). However, compression moulding after a film stacking of fibre and matrix plies is one of the most popular processes to manufacture biocomposites (Jang et al. 2012; Ochi 2008; Song et al. 2012; Porras and Maranon 2012; Graupner et al. 2009).

Compression moulding is characterised by its high reproducibility and low cycle time (Faruk et al. 2000). Satyanarayana et al. (2009) found compression moulding as the better option to produce high-strength biocomposites. This method also uses small amounts of energy, so the green concept is reinforced. The main parameters that influence on the mechanical properties of biocomposites manufactured using compression moulding are fibre material, matrix type, fibre quantity, heating temperature, pressure and time.

Numerous vegetal fibres have been used as reinforcement; jute, flax, and kenaf can be considered the most popular fibres. Different fully biodegradable bioplastics derived from the biomass and processed to a final plastic as PLA (poly-lactic acid) (Ochi 2006), PHA (Polyhydroxyanoates) (Satyanarayana et al. 2009) or chitosan (Julkapli and Akil 2010) can be used as matrix. The maximum processing temperature ranges between 180 and 200 °C because of the natural fibres damaging, while the melting point of natural matrix usually is around 130–230 °C (Dittenber and GangaRao 2010). Consequently, temperatures between 130 and 200 °C are compatible.

Therefore, many researchers have used different values of these parameters to manufacture biocomposites using compression moulding method. For example, Jang et al. (2012) used a 10 % coconut fibres and PLA to produce a biocomposite with a tensile strength of 65 MPa, the heating temperature was 200 °C. Ochi (2008) manufactured a PLA based biocomposite reinforced with a 70 % of kenaf fibres by compression moulding with a previous prepreg of the fibres. Temperature was 160 °C with a 5 min preheating and a hot pressing during 10 min with 10 MPa pressure, obtaining 223 MPa tensile strength. Song et al. (2012) kept at 170 °C during 5 min and applied 2.45 MPa pressure to obtain a 65 MPa tensile strength biocomposite made of 20 % hemp, 80 % PLA. Porras and Maranon (2012) manufactured a 51 % bamboo and 49 % PLA composite by compression moulding after an extrusion of the matrix with a temperature of 160 °C and a pressure of 556.7 kPa resulting 77.55 MPa tensile strength. Graupner et al. (2009) tested mixtures of different fibres and PLA, obtaining 81.8 MPa tensile strength with a 60 % Lyocell and 40 % PLA biocomposite, applying 4.2 MPa of pressure during 20 min at 180 °C. Bodros et al. (2007) followed a compression moulding after film stacking, using flax, glass fibre, hemp and sisal, applying different conditions depending on the fibre and the matrix.

However, very few studies have analysed the influence of the main manufacture parameters on biocomposites strength. Ochi (2008) realized a parametric analysis of kenaf fibre strength, with heating times from 15 to 60 min, varying heating temperatures from 25 to 200 °C, determining that at 180 °C with times of 15 and 30 min, the strength does not decrease, but it does with times of 60 min and with temperatures of 200 °C. Ochi also studied the influence of kenaf reinforcement quantity, determining optimum strength of biocomposites with around a 70 % of reinforcement. Jang et al. (2012) have studied the reinforcement quantity parameter applying different treatments to the fibres. Song et al. (2012) tested different woven architectures of hemp, one twill and the other plain, and determined that the best one is the twill. Bodros et al. (2007) compared biocomposites reinforced with flax fibres and manufactured with different matrixes: PHB (Polyhydroxybutyrate), PBS (Polybutylene succinate), PLA, PLLA (poly-L-lactide), and PP. PLA and PLLA provided the better mechanical properties in terms of stiffness and strength. In addition, they showed that flax/PLLA biocomposites could be competitive with the glass-fibre/polyester composite.

The main objective of this work is the development of a parametric study to analyse the influence of the main parameters of the compression moulding manufacture process on the mechanical properties of biocomposites. Two different PLA matrixes were reinforced with jute, flax, and cotton fibres to analyse the influence of temperature, number of plies, and pressure.

Materials

Two kinds of PLA were used as matrix: 3260HP and 10361D, booth provided by Natureworks LLC. PLA is a thermoplastic resin that was acquired in pellets form. 3260HP PLA is aimed to an extrusion process and 10361D PLA is specifically produced as a binder of natural fibres. PLA properties were obtained through tensile testing as shown in section "Experimental Tests". The PLA density is 1.24 g/cm^3 and the melt temperature is 145–170 °C. Mechanical properties of PLA are shown in Table 1.

Four woven fibres were used as reinforcement: plain weave jute (J); 2 × 1 basket weave cotton (C), 2 × 1 basket weave flax (BF); and plain weave flax (PF). No chemical pre-treatment was applied to woven fibres. Figure 1 shows plain and basket weave configurations. Table 1 shows the mechanical properties of the fibres. Tensile strength ranges from 116.5 MPa of cotton fibres to 271.62 MPa of basket weave flax.

Table 1 Mechanical properties for natural fibres woven plies and PLA matrix

Material	Tensile strength (MPa)	Young modulus (GPa)
BF fibres	271.62 ± 56.3	7.84 ± 1.64
PF fibres	225.39 ± 7.00	9.20 ± 1.70
J fibres	136.18 ± 25.7	7.39 ± 1.07
C fibres	116.50 ± 18.2	12.34 ± 1.14
PLA	47.40 ± 1.15	2.89 ± 0.15

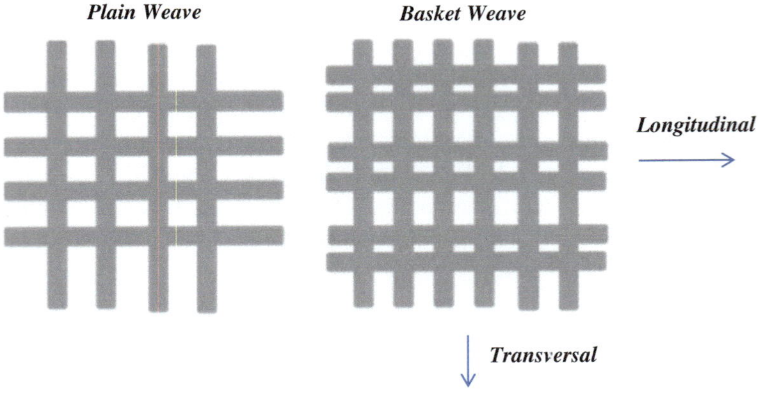

Fig. 1 Plain weave and basket weave used in this work

Matrix, fibres and biocomposites are stored and tested under atmosphere conditions 23 °C, 50 % RH.

Manufacturing

First, the PLA pellets are were placed between two thermoheated plates to obtain uniform films. Then, the matrix films were stacked alternatively with woven plies. The stacked plies were placed between the thermoheated plates. After a pre-heating time, pressure was applied using a universal test machine Servosis ME-404/100 + PCD-1065 with a limit load of 920 kN. Finally, the biocomposite panels were dried at room temperature. The values of pressure and temperature are shown in Table 2. The quantity of reinforcement was stated in a weight ratio of 65 % according to the recommendation of Ochi for compression moulding manufacturing (Ochi 2008).

As commented before, natural fibres can be degraded at temperatures higher than 180 °C, but if this temperature is applied during a short time period, fibres do not present damage. In this work, preheating time is set in 2 min, heating under pressure time in 3 min. These are the minimum times to assure PLA melting, its homogenous distribution in the laminate, and a slowly application of pressure to avoid fibre misalignment.

Experimental Tests

Tensile tests were performed in a universal test machine Instron 8516 shown in Fig. 2. The cross-head speed was set in 0.5 mm/min. Samples were cut in rectangular specimens of 120 × 30 mm. The thicknesses of the samples depend on the number of plies and the type of woven fibre, the thickness of each specimen is shown in Table 2. Tensile strength was determined with the peak force of the force-displacement curve.

Results and Discussion

Table 2 shows the values of the parameters used in each tests and the results of experimental tests in terms of tensile strength and failure strain.

Figure 3 shows the stress-strain curves obtained with specimens produced with different woven fibres at the same manufacturing conditions (tests 1, 3, 5, and 7 in Table 2). The higher tensile strength was obtained with the 2 × 1 basket weave flax (BF), following by plain weave flax (PF), jute (J), and cotton (C). These results are in agreement with the fibre strengths shown in Table 1. Thus, the biocomposite tensile strength is directly related to the fibre strength. This result is not obvious

Table 2 Tensile strength of biocomposite manufactured with different fibres, matrix, temperature, pressure, and number of plies

Plies	Thickness (mm)	Desv. (mm)	Fibre	PLA	Temperature (°C)	Pressure (Mpa)	Strength (Mpa)	Desv. (Mpa)	Max. Strain (%)	Desv. (%)	Number
2	1.35	0.00	BF	10361D	185	16	105.95	4.54	3.99	0.21	1
4	2.46	0.08	BF	10361D	185	16	102.00	4.55	2.67	0.17	2
2	0.91	0.05	PF	10361D	185	16	98.93	1.58	5.22	0.76	3
4	1.85	0.12	PF	10361D	185	16	94.83	4.48	4.76	0.20	4
2	1.20	0.00	J	10361D	185	16	73.30	3.86	2.40	0.23	5
4	2.38	0.06	J	10361D	185	16	59.93	4.84	1.71	0.13	6
2	1.40	0.00	C	10361D	185	16	58.28	0.29	6.43	0.66	7
4	2.28	0.03	C	10361D	185	16	64.43	2.44	1.64	0.07	8
2	1.15	0.04	BF	3260HP	185	32	116.33	2.42	2.77	0.08	9
2	1.24	0.03	BF	10361D	185	32	104.70	4.44	2.71	0.13	10
2	1.20	0.00	BF	10361D	175	32	102.76	0.88	2.27	0.18	11
2	1.23	0.04	BF	10361D	180	32	110.92	6.32	2.58	0.02	12
2	1.36	0.06	BF	10361D	190	32	95.32	4.69	2.48	0.13	13
2	1.33	0.13	BF	10361D	195	32	96.80	4.52	2.35	0.07	14
2	1.29	0.14	BF	10361D	200	32	99.13	7.32	3.94	0.12	15
2	1.71	0.05	BF	10361D	185	0.8	83.75	1.91	3.85	0.16	16
2	1.50	0.00	BF	10361D	185	4	95.73	5.31	2.46	0.15	17
2	1.45	0.04	BF	10361D	185	8	106.94	4.47	2.72	0.10	18
2	1.25	0.04	BF	10361D	185	24	103.55	4.51	3.26	0.35	19
2	1.23	0.10	BF	10361D	185	40	90.42	3.76	2.16	0.05	20

Fig. 2 Universal test machine Instron 8516 used to carry out the tensile tests

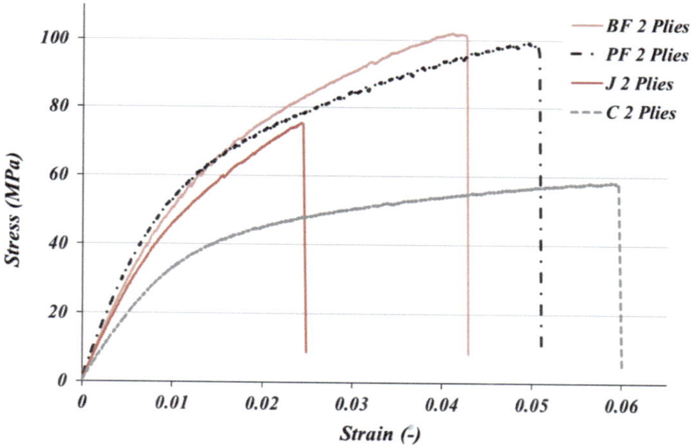

Fig. 3 Stress versus strain curves. Comparison of woven fibre

because the composite strength depends not only on the fibre strength but also on the adhesion between matrix and fibres. This means that the manufacture method is suitable for the three natural fibres used in this work.

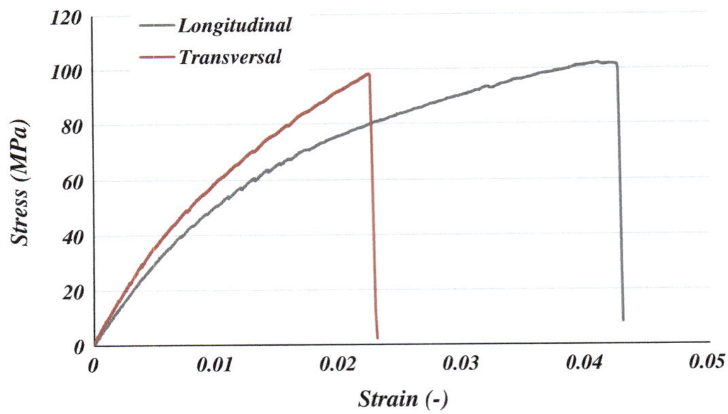

Fig. 4 Comparison between longitudinal and transversal orientation during a tensile test PLA/two layers 2 × 1 basket weave flax composites

The results according to failure strain are different. The highest failure strain was found in cotton, followed by flax, and jute. The tests showed that the plain weave lead to higher failure strain but lower tensile strength.

In the case of plain weave, such as jute and plain flax, there is no influence of the fibre orientation as the weave is equilibrated in both directions. However, in the 2 × 1 basket weave a difference is observed as a function of the orientation. Figure 4 shows the stress-strain curves obtained in a two plies PLA/flax specimen tested in different direction.

The values of tensile strength are similar, around 100 MPa. However, there is a great difference in terms of failure strain. When the specimen is tested in transverse direction, see Fig. 1, the failure strain is 2.3 %, but if the specimen is tested in longitudinal direction the failure strain is 4.2 %. The difference in failure strain at different orientations is 82 %.

The influence of woven fibre and thickness is shown in Fig. 5, where results of tests no. 1–8 are shown. The analysis of the influence of number of plies showed that there is more cohesion in bilayer biocomposites than the four layers ones, with

Fig. 5 Tensile strength obtained for different number of plies and fibres: basket weave flax (BF), plain weave flax, plain weave jute (J), and basket weave cotton (C)

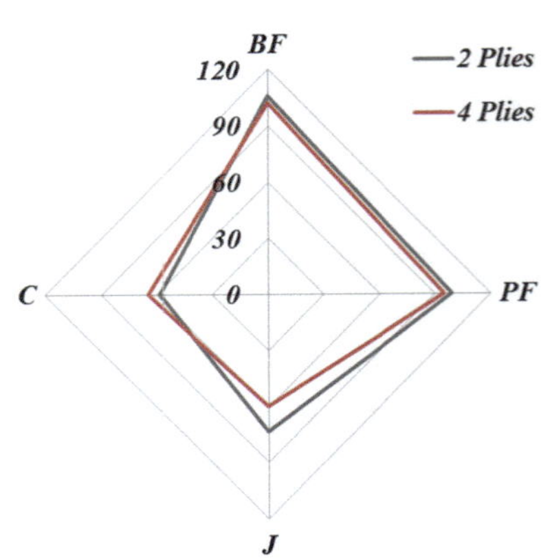

the exception of cotton. This is consistent with the cohesion problems in biocomposite materials. However, the difference between two and four layers is smaller than standard variation for flax woven biocomposites.

The two PLA matrix (3260HP and 10361D) were compared in tests no. 9 and 10. Results showed that 3260HP PLA matrix lead to higher tensile strength, 116.33 MPa, than PLA 10361D matrix, 104.70 MPa. These results are in contradiction with the supplier recommendations because 10361D is specifically designed for natural fibres binder, while 3260HP is make for extrusion processes. However, the higher properties obtained with 3260HP matrix were only valid for two layers laminates, it was not possible to spread into the inner layers when it was applied to a four layer biocomposite. Thus, 10361D PLA was selected as the best option to be used as matrix in compression moulding process.

The influence of heating temperature, from 175 to 200 °C, was analysed through comparison of tests no. 10–15, shown in Fig. 6. Heating temperatures over 200 °C produced fibre damage due to overheating. On the other hand, the matrix was not completely melted for heating temperatures below 175 °C. Thus, the temperature must be high enough for matrix melting, but no so high to produce fibres damage. As Fig. 6 shows, specimens manufactured with heating temperatures between 180 and 185 °C presented the maximum tensile strength. However, some of the specimens produced at 175 and 180 °C were discarded because of the presence of delaminations. The tensile strength corresponding to 180 °C heating temperature shown in Fig. 6, 110.92 MPa, was obtained as the average of intact specimens, however many other specimens were manufactured but not tested due to delaminations. Consequently, 185 °C was chosen as the optimum heating temperature.

The effect of pressure was studied comparing the tests no. 1, 10, 16–20, see Fig. 7. Tensile strength increases with manufacture pressure until 8 MPa, then there is plateau until a manufacture pressure of 32 MPa, finally tensile strength decreases with pressure. Pressure over 32 MPa produced fibre breakage, while pressure under

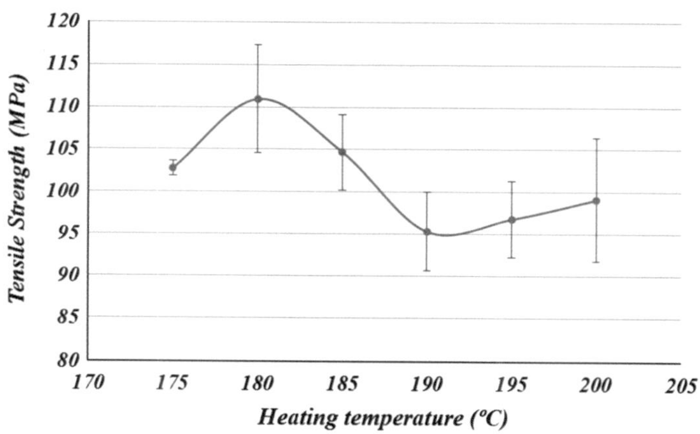

Fig. 6 Tensile strength as a function of heating temperature

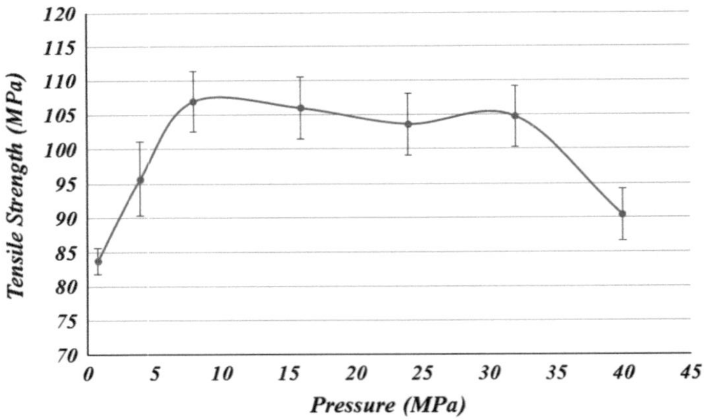

Fig. 7 Tensile strength as a function of the pressure applied in compression moulding method

8 MPa mean lack of cohesion in the biocomposite. Thus, there is a wide range of pressure, between 8 and 32 MPa, that can be used to produce biocomposites with similar mechanical properties (Fig. 7).

Conclusions

The influence of the main manufacturing parameters on the mechanical properties of biocomposites manufactured with compression moulding method was analysed leading to the following conclusions:

- Biocomposites manufactured from basket weaves show an anisotropic behaviour. Differences of 82 % were found comparing failure strain in different directions for basket weave configurations. However, the values of tensile strength were similar.
- Biocomposite strength was directly related to fibre strength. This conclusion is not obvious because this implies similar adhesion between the matrix and different fibres.
- Comparing two-layers biocomposites, higher strength was found using 3260HP PLA matrix. However, four-layers biocomposites were manufactured only with 10361D PLA matrix because 3260HP presented spreading problems.
- A heating temperature of 185 °C was found as the optimum value. Higher heating temperatures produce less-strength biocomposites because fibres are damaged due to overheating. On the other hand, matrix melting is not complete at lower heating temperatures.
- A wide range of pressure during compression moulding, from 8 to 32 MPa, led to similar values of tensile strength. Lower pressure implies lack of cohesion in the biocomposite, and higher pressure produces fibre breakage.

The present method can be used to optimize the compression moulding process to manufacture biocomposites. These results can be used to reduce the production time leading to reduction of manufacturing cost of biocomposites and increase the mechanical properties.

Acknowledgements Authors gratefully acknowledges the support of Spanish Ministry of Economy under the project DPI2014-43994-R.

References

Santiuste C, Soldani X, Miguélez MH. Machining FEM model of long fiber composites for aeronautical components. Compos Struct 2010;92(3):691-8.
Santiuste C, Barbero E, Miguélez MH. Computational analysis of temperature effect in composite bolted joints for aeronautical applications. J Reinf Plast Compos 2001;30(1):3-11.
Olmedo A, Santiuste C, Barbero E. An analytical model for predicting the stiffness and strength of pinned-joint composite laminates. Compos Sci Tech 2014;90:67-73.
Mantia FPL, Morreale M. Green composites: A brief review. Compos Part A: Appl S 2011;42 (6):579-88.
Jang JY, Jeong TK, Oh HJ, Youn JR, Song YS. Thermal stability and flammability of coconut fiber reinforced poly (lactic acid) composites. Compos Part B: Eng 2012;43(5):2434-8.
Bocz K, Szolnoki B, Marosi A, Tábi T, Wladyka-Przybylak M, Marosi G. Flax fibre reinforced PLA/TPS biocomposites flame retarded with multifunctional additive system. Polym Degrad Stab 2013;106:63-73.
Bledzki AK, Jaszkiewicz A, Scherzer D. Mechanical properties of PLA composites with man-made cellulose and abaca fibres. Compos Part A: Appl S 2009;40(4):404-12.
Müssig J, Schmehl M, von Buttlar HB, Schönfeld U, Arndt K. Exterior components based on renewable resources produced with SMC technology—Considering a bus component as example. Ind crop Prod 2006;24(2):132-45.
Brouwer WD. Natural Fiber Composites, Saving Weight and Cost with Renewable Materials Thirteenth International Conference on Composite Materials. 2001.
Serizawa S, Inoue K, Iji M. Kenaf fiber reinforced poly (lactic acid) used for electronic products. J Appl Polym Sci 2006;100(1):618-24.
John MJ, Thomas S. Biofibres and biocomposites. Carbohydr polym 2008;71(3):343-64.
Huber T, Pang S, Staiger MP. All-cellulose composite laminates. Compos Part A: Appl S 2012;43 (10):1738-45.
Summerscales J, Dissanayake N, Virk A, Hall W. A review of bast fibres and their composites. Part 2–Composites. Compos Part A: Appl S 2010;41(10):1336-44.
Voorn BV, Smit HHG, Sinke RJ, Klerk BD. Natural fibre reinforced sheet moulding compound. Compos Part A: Appl S 2001;32(9):1271-9.
Ochi S. Mechanical properties of kenaf fibers and kenaf/PLA composites. Mech mater 2008;40 (4):446-52.
Song YS, Lee JT, Ji DS, Kim MW, Lee SH, Youn JR. Viscoelastic and thermal behavior of woven hemp fiber reinforced poly(lactic acid) Composites. Compos Part B: Eng 2012;43:856–60.
Porras A, Maranon A. Development and characterization of a laminate composite material from polylactic acid (PLA) and woven bamboo fabric. Compos Part B: Eng 2012;43(7):2782-8.
Graupner N, Herrmann AS, Müssig J. Natural and man-made cellulose fibre-reinforced poly (lactic acid)(PLA) composites: An overview about mechanical characteristics and application areas. Compos Part A: Appl S 2009;40(6):810-21.
Bodros E, Pillin I, Montrelay N, Baley C. Could biopolymers reinforced by randomly scattered flax fibre be used in structural applications?. Compos Sci Technol 2007;67(3):462-70.

Ochi S. Development of high strength biodegradable composites using Manila hemp fiber and starch-based biodegradable resin. Compos Part A: Appl S 2006;37(11):1879-83.

Bledzki AK, Faruk O, Sperber VE. Cars from BioFibres. Macromol Mater Eng 2006;291 (5):449-57.

Duigou AL, Pillin I, Bourmaud A, Davies P, Baley C. Effect of recycling on mechanical behaviour of biocompostable flax/poly (l-lactide) composites. Compos Part A: Appl S 2008;39(9):1471-8.

Rubio-López A, Olmedo A, Santiuste C. Modelling impact behaviour of all-cellulose composite plates. Compos Struct 2015;122:139-43.

Oksman K. High quality flax fibre composites manufactured by the resin transfer moulding process. J Reinf Plast Compos 2001;20(7):621-7.

Faruk O, Bledzki AK, Fink HP, Sain M. Biocomposites reinforced with natural fibers: 2000–2010. Prog Polym Sci 2012;37(11):1552–96.

Satyanarayana KG, Arizaga GG, Wypych F. Biodegradable composites based on lignocellulosic fibers—an overview. Prog Polym Sci 2009;34(9):982-1021.

Julkapli NM, Akil HM. Thermal properties of kenaf-filled chitosan biocomposites. Polym-Plast Tech Eng 2010;49(2):147-53.

Dittenber DB, GangaRao HV. Critical review of recent publications on use of natural composites in infrastructure. Compos Part A: Appl S 2010;43(8):1419-29.

The Mechanical Properties of Flax Fibre Reinforced Composites

Sevhan Muge Yukseloglu and Hurol Yoney

Abstract Generally, glass and carbon fibres are most widely used materials in composites. In this study it has been aimed to use flax fibres in composite materials and to study the mechanical properties of the produced samples. The research was carried out with 100 % unbleached flax fibres at different reinforcement ratios to produce hand lay-up composite materials. The produced composite samples were then characterized using three point bending and Izod impact tests. The fracture surfaces of the reinforced composite samples were also inspected with the help of SEM. Both the mechanical tests results (impact and three point bending) and SEM micrographs indicated flax fibres as an alternative natural fibre source for developing reinforced composites for various industries. Good mechanical properties i.e. 109–165 MPa were obtained as compared to the pure resin (≈40 MPa). The flax fibre reinforced composites offered an impact strength of about 21–38 kJ/m^2. Finally, fibre/matrix-interface bonding in the composites was improved by the de-waxing process of flax fibres prior to composite fabrication.

Keywords Flax fibers · Composites · Mechanical properties · Fracture surface

Introduction

Attention to natural fibres is not new. For almost a decade there has been increasing awareness within the studies of these fibres because of their ecological concern, recyclability and environmental friendly origin. One of the earliest studies was made by Pavithran et al. (1987) where they examined the impact properties of

S.M. Yukseloglu (✉)
Department of Textile Engineering, Faculty of Technology, Marmara University, Istanbul, Turkey
e-mail: myukseloglu@marmara.edu.tr

H. Yoney
YNY Textile, Civil and Food Engineering Ltd., Kadikoy, Istanbul, Turkey

R. Fangueiro and S. Rana (eds.), *Natural Fibres: Advances in Science and Technology Towards Industrial Applications*, RILEM Bookseries 12,
DOI 10.1007/978-94-017-7515-1_19

natural fibre composites. They used sisal, pineapple, banana and coir fibres for their research and they correlated fibre properties with their impact behaviour. In recent years, natural fibres have gained further popularity in composites for engineering applications due to growing environmental concern and requirements for developing sustainable materials where they replace glass fibres as reinforcement in composite structures (Mohanty et al. 2005; Yan et al. 2012).

Some researchers forecasted (Caruls and Scholz 2011) that about 830,000 tonnes of bio-fibres will be consumed by 2020 and the share of the total reinforcement materials will go up to 28 %. Both the United States (US) Department of Agriculture and the US Department of Energy have set goals having at least 50 % of all basic chemical building blocks to be created from renewable and plant based sources by 2050 (Mohanty et al. 2005). This growth indicates that bio-composites, which are cost effective with low density, will find various places for many engineering applications in the future. Yet, some researchers (Okuba et al. 2004; Yukseloglu and Yoney 2009) studied the regenerated cellulose fibre reinforced composites i.e. bamboo fibre fillers as bio-composites. From these studies, it was observed that there is potential ability to work as a reinforced matrix in the composite structures for automotive industry. Before revealing experimental work of this current study, here is a brief summary on flax fibre.

Flax is a plant fibre which belongs to *linaceae family* and is one of the widely utilised bio-fibres. Records from Babylonia and Anatolia from 3000 BC show that flax was cultivated in Ancient Egypt for its seeds and for linseed oil; it is also one of the oldest to be extracted, spun and woven into textiles which were found in graves in Egypt dating back to 5000 BC (Lucas and Harris 1962). Flax fibres are produced in the stems of flax bast plant and are a cellulose polymer like cotton fibres. However, its structure is more crystalline which makes it stronger than cotton and it is stiffer to handle and is more easily wrinkled. Length of a flax plant extends to 100 cm which has strong fibres along its stem with average fibre diameter 10–25 μm (Bos et al. 2002). The micro-structure of a flax fibre is complex due to the hierarchical organisation at different length scale and different materials present in variable proportions (Fig. 1). From the Fig. 1, it can be seen that the thickest cell wall is S2. This wall contains numerous crystalline cellulose micro-fibrils and amorphous hemicellulose which are oriented at 10° with the fibre axis that gives the high tensile strength to the flax (Baley 2002).

Flax fibre consists of cellulose, hemicellulose, wax, lignin and pectin in different quantities were reported by many authors (Bastra 1998; Troger et al. 1998; Lilholt et al. 1999). This variation of proportions in the constituents of flax fibres is due to the plant variety, agriculture variables i.e. soil quantity, weathering conditions, level of plant maturity, quality of retting process (Charlet et al. 2009). However it is well known that flax is rich in cellulose for about 70 % of the total chemical composition and this makes it possible to be used as reinforcement in composites. As for the natural fibres such as flax, the main disadvantage is their hydrophilic nature which lowers the compatibility with hydrophobic polymeric matrices during the composite production. On the other hand, natural fibres have low mechanical properties; flax

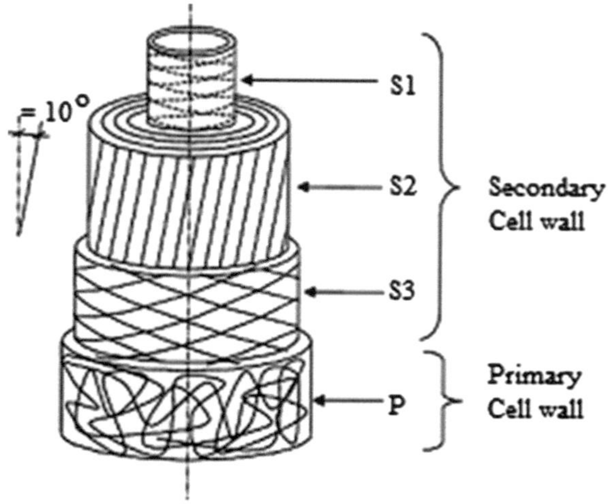

Fig. 1 The micro-structure of a flax fibre cell (Baley 2002), reproduced with permission from Elsevier

fibres do show better mechanical properties than the most natural ones. Flax looks much agnate to cotton fibre except in pigment intensity (Srinivasan et al. 2014).

It is well known that flax fibre has a higher strength than cotton fibre and retting is done to improve its mechanical properties. In the use of composite construction, flax fibres have some pros and cons which can be seen below (Jolly 2009):

- Pros:
  - Great tensile strength/low elongation
  - Vibration absorbing properties
  - Water retention and conductivity
  - Cost/kg
  - Ecologically friendly
- Cons:
  - Bad adhesion due to all substances, i.e. pectin, present in the compound
  - Degradation of the fibres above 200 °C
  - Limitation of external applications due to humidity retention
  - Regularity of the base material characteristics

Even if, natural fibres present poor dimensional stability (Brouwer 2000; Mohanty et al. 2000), when considering the cost, mechanical performance and yield into account, among various natural fibres, flax, hemp and jute are the three most promising candidates among the various natural fibres. Therefore, these three fibres can be used to replace glass fibres in composites.

Although knowing some of the above obstacles to work with natural fibres in the fabrication of composite samples; we aimed to study the flax fibres first without bleaching process and with various reinforcement ratios (total of 5 ratios) to observe their composite structure and their mechanical properties. And second, we have conducted dewaxing process earlier to composite fabrication which can be seen at the SEM images of this study.

Materials and Method

In this work, composite samples were produced with the use of flax fibres which were first combed and later were cut 80 mm in length (Fig. 2). Total of 30 single flax fibre tensile tests were measured according to the ASTM D 3822; their mean breaking strength was 1.0089 g/cm and mean breaking elongation was 6.55 %.

Table 1 compares the properties of a flax fibre, which were used as a reinforced material here, with jute, sisal and hemp. It can be noted that flax fibre has a reasonable higher tensile strength than the other three plant fibres and higher modulus among the four plant fibres shown here. A total of 5 reinforced composite samples were produced by the hand lay-up method using vinyl ester resin (Table 2) for both impact and three point bending tests.

Fig. 2 Flax fibres prepared for the composite fabrication

Table 1 Flax, jute, sisal and hemp fibre properties

Fibre	Density (g/cm^3)	Tensile strength (MPa)	Specific tensile strength (($MPa/(g/cm^3)$))	Young's modulus (GPa)	Specific Young's modulus ($GPa/(g/cm^3)$)
Flax	1.40–1.50	343–2000	245–1334	15–80	11–53
Jute	1.30–1.49	350–770	269–517	12–26	9–17
Sisal	1.33–1.50	380–635	286–423	9.4–22	7–15
Hemp	1.40–1.50	270–900	193–600	8.7–48	6–32

(Adapted from Prasad and Rao 2011; Hodzic and Shanks 2014; Placet et al. 2012)

Table 2 Mechanical properties of the vinyl ester resin

Elastic modulus (MPa)	Max-load (kgf)	Strain strength (MPa)	Elongation on Max. load (%)	Breaking strength (kgf)	Breaking elongation (%)	Width (mm)	Thickness (mm)	Flexural strength (MPa)
4425.37	8.8604	39.8249	1.07934	9.2284	1.07945	9.7900	5.3	160

Table 3 Reinforcement ratios used for the flax composite samples

Tests	Reinforcement ratio of the composite samples (%)			
Izod impact (kJ/m^2)	12		24	
Three point bending (MPa)	15	20		25

The hand lay-up is the oldest method and is also the simplest fabrication process for the composite materials. Basically, dry fibre is manually placed in a mould and subsequently resin matrix is applied. The wet composite is then rolled with hand rollers to obtain a uniform resin distribution to have better interaction between the reinforcement and matrix for a required thickness. The layered structure is then cured. The reinforcement ratios for this study are given in Table 3. 0.2 % cobalt nephthalate (as accelerator) and 2 % hardening are also mixed during the composite sample production where the fibres are laid on the mould surfaces.

All the uni-directional (UD) produced samples were then post-cured in 3 h consecutively at the temperatures of 50, 80 and 110 °C. Later, the mechanical tests (Izod and three point bending) were carried out according to the standards at room temperature (≈25 °C).

The Izod impact test and three point bending test results are given at the below section.

Results

In this research, the aim was to produce flax fibre reinforced composite materials to study their impact behaviour and three point bending strengths. Therefore, the study was carried out with 100 % unbleached flax fibres at different reinforcement ratios to produce composite materials by hand lay-up method. The results on mechanical tests and fracture surfaces of the impact materials are provided below.

Izod Impact

Izod impact test samples (dimensions: 90 × 10 × 5 mm) were prepared (Fig. 3) according to the TS EN ISO 180 (2006); the samples were tested under the 5.4 J of pendulum and with the angle of 124°. The loss of energy during impact is the

Fig. 3 Flax fibres on the mould

Table 4 Izod impact results of the flax fibre reinforced composites

Reinforcement ratio of the flax fibre composite samples (%)	Impact penetration energy (J)	(%)	Izod impact strength (kJ/m^2)
12	1.183	21.90	20.8
24	2.221	41.12	38.2

energy absorbed by the specimen during the impact. The impact results of these composite samples are presented at the Table 4.

From the Izod impact results, it can be seen that as the ratio of the samples (reinforced fibre/resin) increase the impact penetration energy of the flax fibre composite samples get high. This may also indicate that, if a reinforced flax fibre is doubled in weight at the composite material, it may improve the Izod impact strength for about 84 %. The reason for its higher impact strength may be due to more fibre arrangement in the composites. Thus, it is difficult for cracks to spread when fibre pile is much higher within the proportion of reinforced fibre/resin.

Three Point Bending

Three point bending test samples were prepared (80 × 10 × 4 mm in dimensions) according to the TS 985 EN ISO 178 (2006). The samples were tested under the 1 mm/min head speed. The results of the samples can be seen at the Table 5 and the graphs are given in Figs. 4, 5 and 6.

Table 5 Three point bending test results of the flax fibre reinforced composites

Reinforcement ratio of the flax fibre composite samples (%)	Elastic modulus (MPa)	Max. load (kgf)	Strain strength (MPa)	Elongation on Max. load (%)	Breaking strength (kgf)	Breaking elongation (%)	Width (mm)	Thickness (mm)
15	4301.00	27.4909	108.710	2.58440	27.2487	2.62783	9.8800	4.63750
20	5481.95	34.4281	142.040	2.83598	34.4218	2.83667	9.8850	4.90000
25	6671.88	38.4730	165.081	3.05564	37.8144	3.06858	9.87333	5.23667

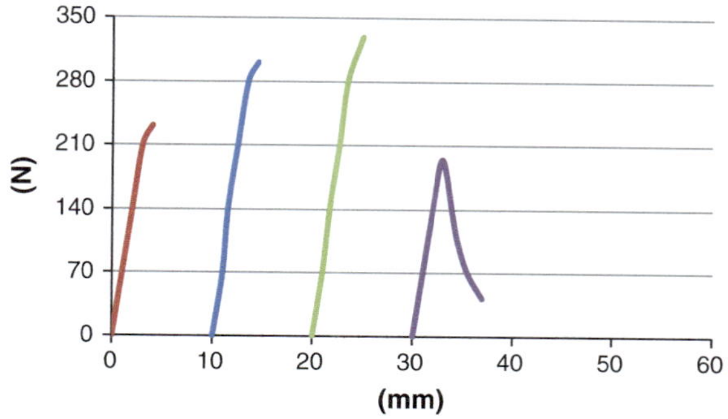

Fig. 4 Force-elongation diagram of the three point bending test of 15 % flax fibre reinforced composites

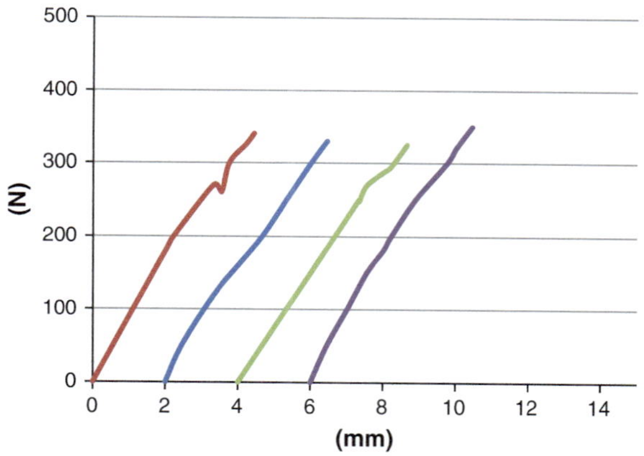

Fig. 5 Force-elongation diagram of the three point bending test of 20 % flax fibre reinforced composites

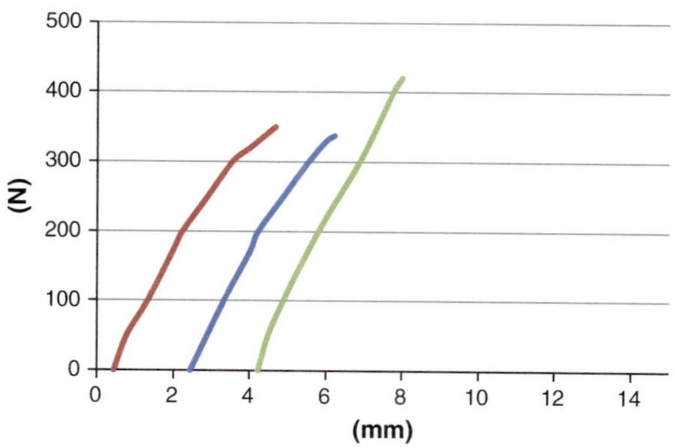

Fig. 6 Force-elongation diagram of the three point bending test of 25 % flax fibre reinforced composites

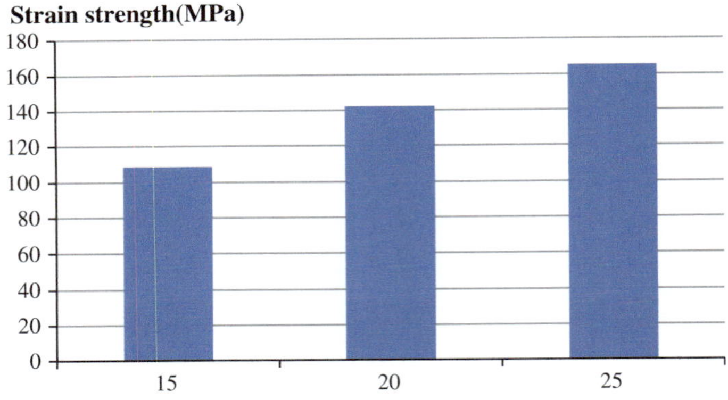

Fig. 7 Comparison of the strain strength results of the flax fibre reinforced composites

From the Fig. 7 it was observed that as the reinforcement ratio increases the strain strength of the three point bending test of the specimens' increases.

Electron Microscopy Studies

The fracture surfaces of the impact specimens were studied with the help of the scanning electron microscope JEOL JSM-5910 LV. Specimens were sputter coated with gold prior to examination. The micrographs of these produced flax fibre reinforced composite samples are given in Figs. 8 and 9.

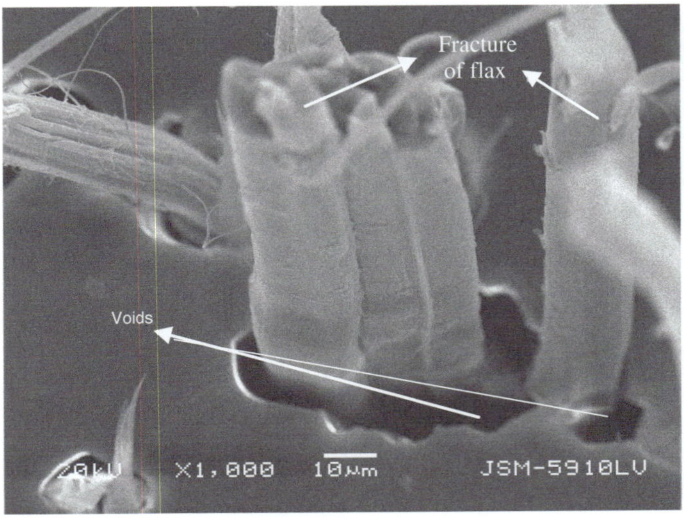

Fig. 8 SEM micrograph of the Izod impact sample of 12 % untreated flax fibre reinforcement (X1000)

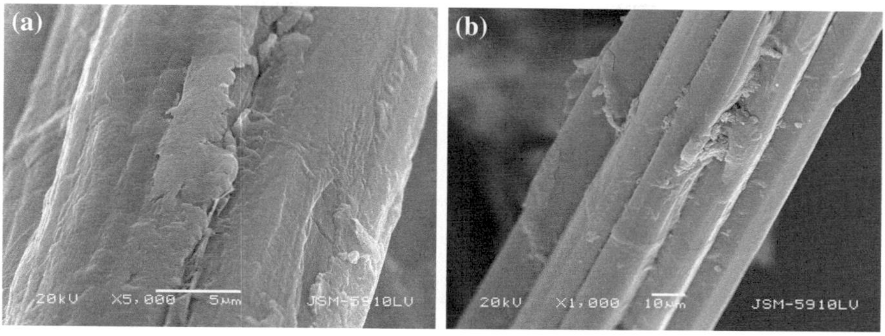

Fig. 9 Microstructure of flax fibres. **a** Untreated flax fibre with its wax and pectin content (X5000). **b** Dewaxed flax fibre structure (X1000)

Although, the untreated flax fibre reinforced composite specimens gave good mechanical properties (see Table 4 for Izod impact strength about 21–38 kJ/m^2 and see Table 5 for strain strength between 109 and 165 MPa) their matrix-inter surface cohesion revealed to be weak. This is the main problem for natural fibre/polymer composites where they present incompatibility between the hydrophilic fibres and the hydrophobic matrices. In the case of flax fibre, this leads to a poor fibre/matrix adhesion due to presence of pendant hydroxyl and polar groups in the components. Hence, nature of flax fibre leads to high moisture uptake which can lower the tensile properties of the fibres and consequently decreases their mechanical properties of the composites. This can be seen from the Fig. 8 where some of the fibre gaps at broken surfaces indicate that there is not a good inter surface mixture and therefore the flax fibre has been drawn out from the composite structure of the specimen. As the flax fibre was untreated (not de-waxed, see Fig. 9a) prior to the composite production, we therefore studied further to see if de-waxing of flax fibre reinforced composites have a good effect on the production of the specimens or not (see Fig. 10b).

For this purpose, wax in the fibre has been extracted with ether solution and therefore a better inter surface cohesion has been achieved as seen from the Fig. 10. Moreover, Fig. 10 shows breaks, brittle fractures and holes due to fibre pull out. It is thought that brittle fractures occur from the kink bands of the fibre. From the study it can be seen that the amount of voids increases with increasing flax fibre content. Thus increasing flax fibre content may also increase the module and brittleness of the fibre during the impact. Although the fibre amount in the composite material was increased to 27 % due to de-waxing process, its Izod impact strength was increased about 50 %. This indicates that, if flax fibre content increases about 3 %, the impact behaviour of the flax fibre reinforced composite gets 50 % higher than the untreated ones.

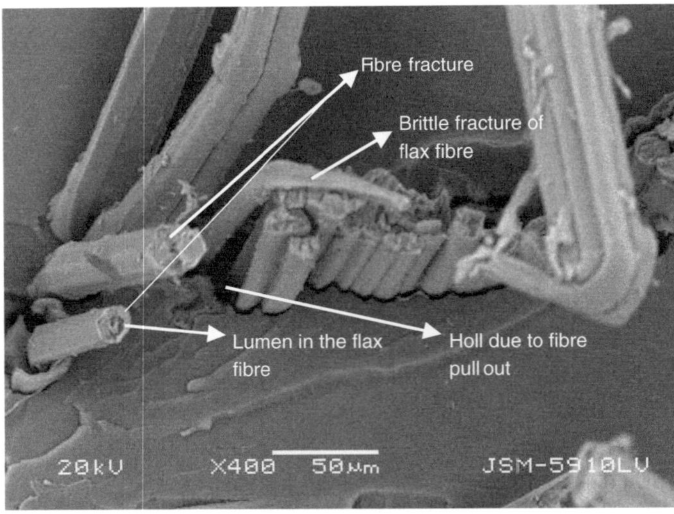

Fig. 10 SEM micrograph of the Izod impact sample of 24 % dewaxed flax fibre reinforcement (X400)

Conclusions

As a natural fibre, flax has great advantages i.e. great tensile strength/low elongation, good vibration absorbing properties, ultraviolet rays blocking properties and high water retention. Even though flax fibre has a low density (1.5 g/m^3) than glass (2.6 g/m^3) Kevlar and Carbon (1.48 g/m^3) fibres, it presents reasonably good break (between 800 and 1500 MPa) when compared to glass, Kevlar and Carbon. Keeping this in mind, in the current study, only a little amount of flax fibre (for the Izod impact: 12, 24 % and for the three point bending: 15, 20 and 25 %) was used and we have obtained good mechanical properties between 109 and 165 MPa compared to the pure resin (≈40 MPa). This indicates that flax can be a very good candidate within the natural fibre sources for the composite production. The experimental work also revealed that flax fibre can be used in fibre or sliver forms for the composite material applications.

Although the untreated flax fibre reinforced composite specimens presented good Izod impact strengths (21–38 kJ/m^2), from the SEM image the matrix-inter surface cohesion appeared to be weak. To improve this matrix-inter surface cohesion in the composites we have dewaxed the flax fibres first and later produced the composite specimens. It was observed that dewaxed flax fibre reinforced composites exhibited better impact energy absorption capability when compared to untreated flax fibre reinforced composites. As a consequence of the dewaxing process, pectin and waxy substances present in the fibre have been reduced decreasing the fibre diameter and its strength slightly. However, a suitable chemical treatment e.g. Silane may increase the tensile strength and strain of the flax fibres.

A major limitation of using flax fibres as reinforcement material might be the inferior fibre/matrix interfacial bonding than the glass fibres. This can be overcome by selecting a suitable manufacturing process and using a chemical/physical modification technique so that the mechanical properties of flax composites can get better. However, it must be noted that flax fibre becomes more sensitive to the moisture due to chemical modification. Because of this, a good drying process is important prior to production of flax fibre reinforced composites.

Finally, de-waxing is recommended prior to the composite production of flax fibres. As a consequence, this may result in a decrease in single fibre strength. But, the amount of flax fibre increases in per volume of the composites and it results in a good fibre/matrix interface bonding which improves the mechanical properties of flax fibre reinforced composites.

Because of its eco friendly production, we hope that flax fibre can find a better place for itself in the composite arena as they are a good candidate for future generation of bio-fibres, especially in automotive industry; because natural fibres possess excellent sound absorbing efficiency, are more shatter resistant and have better energy management characteristics than glass fibre reinforced composites. Therefore in future research, flax composites' durability, moisture resistance, surface modification techniques and the mechanical properties can be improved.

Acknowledgments The authors are grateful to Elsevier Publishers and authors who permitted to re-produced Fig. 1 and Table 1 from their publications. Also, the first author wishes to thank for the financial support for this research which was provided by the Scientific Research Project Unite of Marmara University project grant FEN-D-110315-0071 of 2015.

References

Pavithran C, Mukherjee P.S, Brahmakumar M, Damodaran A, Impact properties of natural fibre composites, Journal of Materials Science Letters, 1987, 6, 8 p.882-884.

Mohanty AK, Misra M, Drzal LT, Selke SE, Harte BR, Hinrichsen G. In: Mohanty AK, Misra M, Drzal LT, editors. Natural fibres, biopolymers, and biocomposites: an introduction in natural fibres, natural fibres, biopolymers, and biocomposites, Boca Raton, FL: CRC Press, Taylor & Francis Group; 2005. p. 1–36.

Yan LB, Chouw N, Yuan XW. Improving the mechanical properties of natural fibre fabric reinforced epoxy composites by alkali treatment, Journal of Reinforced Plastics and Composites, 2012, 31, 6, p.425–37.

Okuba K, Fujii T, Yamamoto Y, Development of bamboo-based polymer composites and their mechanical properties, Composite Part A: Applied Science and Manufacturing, 2004, 35, p.377-383.

Yukseloglu, SM, Yoney, H, Bamboo fibre reinforced composite structures and their mechanical properties, Tekstil ve Konfeksiyon, 2009, 19, 4, p.261-264.

Carus M, Scholz L, Targets for bio-based composites and natural fibres, Biowerkstoff report, 8th Edition, ISSN 1867-1217, March 2011 p. 24.

Lucas, A, Harris, JR, Ancient Egyptian materials and industries, Dover Publications, Mineola, New York, 4th Edition, ISBN 0-486-40446-3, 1962 p.142-146.

Bos, HL, Van Den Oever, MJA, Peters, OCJJ, Tensile and compressive properties of flax fibres for natural fibre reinforced composites, Journal of Material Science, 37, 2002, p.1683-1692.

Baley C, Analysis of the flax fibres tensile behaviour and analysis of the tensile stiffness increase, Composites Part A: Applied Science and Manufacturing, 2002, 33, 7, p.939–948.

Bastra SK. Other long vegetable fibres, In: Lewin M, Pearce EM, editors. Handbook of fibre science and technology, Fibre chemistry, New York: Marcel Dekker; 1998 p. 505–575.

Troger F, Wegener G, Seemann C. Miscanthus and flax as raw material for reinforced particle boards. Ind Crops Prod 1998, 8,2, p.113–121.

Lilholt H, Toftegaard H, Thomsen AB, Schmidt AS. Natural composites based on cellulosic fibres and polypropylene matrix. Their processing and characterization. In: Proceedings of ICCM 12, Paris; July 1999, p. 9.

Charlet K, Jernot JP, Gomina M, Bréard J, Morvan C, Baley C. Influence of an Agatha flax fibre location in a stem on its mechanical, chemical and morphological properties. Compos Sci Technol 2009, 69, 9, p.1399–1403.

Srinivasan VS, Boopathy SR, Sangeetha D, Ramnath BV, Evaluation of mechanical and thermal properties of banana-flax based natural fibre composite, Materials and Design, 2014, 60, p.620-627.

Jolly, M., http://www.norafin.com/fileadmin/user_upload/Norafin/Composites/The_Advantages_of_Flax_Fiber_in_Composites_-_Fiber_Congress_2009.pdf, accessed date: 24.12.2014.

Brouwer WD, Natural fibre composites: where can flax compete with glass?, SAMPE J,2000, 36, p.18-23.

Mohanty AK, Misra M, Hinrichsen G, Biofibres, biodegradable polymers and biocomposites: An overview, Micromolecular Materials Engineering, 2000, 276-277, 1, p.1-24.

Prasad AVR, Rao KM, Mechanical properties of natural fibre reinforced polyster composites: Jowar, sisal and bamboo, Materials and Design, 2011, 32, p.4658-4663.

Hodzic A, Shanks R, Natural fibre composites, Materials, processes and properties, Woodhead Publishing Limited, Alma Hodzic and Robert Shanks (ed), 2014, ISBN 978-0-85709-922-8 (online), p.10.

Placet V, Trivaudey F, Cisse O, Gucheret-Retel V, Boubakar ML, Diameter dependence of the apparent tensile modulus of hemp fibres: A morphological, structural or ultra structural effect?, Composites Part A, 2012, 43, p.275-287.

ASTM D 3822, Standard test method for tensile properties of single textile fibres.

TS EN ISO 180, 2006, “Plastics-Determination of Izod Impact Strength”,Turkish Standards Institute, Ankara, Turkey.

TS 985 EN ISO 178, 2006, “Plastics-Determination of Flexural Properties”,Turkish Standards Institute, Ankara, Turkey.

Eco-friendly Flax Fibre/Epoxy Resin/Composite System for Surfboard Production

Aitor Hernandez Michelena, Jasper Graham-Jones, John Summerscales and Wayne Hall

Abstract A commercially available eco-friendly epoxy resin is proposed as the matrix for a flax fibre composite surfboard as a potential "sustainable" material for component production. The study develops a method for robust production of the surfboards using flax fibre and bio-epoxy resin. The objective of the work is to achieve a balance between economical product, low weight, mechanical performance, improved impact resistance and (outside of this paper) improved environmental credentials. This study seeks to achieve a complete mechanical and thermal characterisation of the flax/eco-epoxy system as an alternative material system to a conventional S-glass/polyester system. For this reason, both systems are completely characterised and their properties compared.

Keywords Flax fiber · Bio-composite · Surf board production · Mechanical properties

Introduction

There are many references to sustainable materials used in commercial surfboard production. However, these materials have three main issues. First, the mechanical and thermal performance of sustainable materials does not match that of

---

A.H. Michelena (✉) · J. Graham-Jones · J. Summerscales
Advanced Composites Manufacturing Centre (ACMC), School of Marine Science and Engineering, Faculty of Science and Engineering, University of Plymouth, Drake Circus, Plymouth, Devon PL4 8AA, UK
e-mail: aitor.hernandezmichelena@plymouth.ac.uk
URL: http://www.plymouth.ac.uk/pages/view.asp?page=38937

W. Hall
Griffith School of Engineering, Griffith University, Gold Coast Campus, Gold Coast, QLD 4222, Australia

A.H. Michelena
5'TX", C/Bardenas Reales, n° 70, piso 4° B, 31006 Sarriguren, Spain

R. Fangueiro and S. Rana (eds.), *Natural Fibres: Advances in Science and Technology Towards Industrial Applications*, RILEM Bookseries 12,
DOI 10.1007/978-94-017-7515-1_20

petrochemical-source resins or inorganic fibres. Second, since the bio-source products are not as common the synthetics, their production costs are higher, directly affecting the final product market price. Finally, hydrophilic natural fibre (NF) compatibility with hydrophobic organic resins is not good, leading to poor interfacial properties with drastic reductions in long term performance and aging durability.

To use expanded polystyrene (EPS) foam cores, the matrix must be epoxy as EPS dissolves in the UP resin system, so conventional surfboards are produced with S-glass fibre/unsaturated polyester (UP) resin composite skins over polyurethane foam (PUF). These products are not produced from renewable materials. Once they reach their end-of-life (EOL), their recyclability is a big issue and they are mostly ground to be recycled as fillers. These factors have driven the many recent developments in Eco-surfboards.

Commercial Products include (i) Samsara Surfboards (Samsara, Australia, 2015) eco-surfboards based on flax fibre/eco-epoxy and recycled EPS foam (Waste to waves, California, 2015) (ii) Notox (Notox, France, 2015, France), who use similar technology: flax fibre from Lineo (Lineo, France, 2015), Greenepoxy55 eco-epoxy from Sicomin (Sicomin, France, 2015) and recycled EPS, and (iii) Lost surfboards (2015) (www.lostsurfboards.net/2014/lost-e-tech-surfboards). Introducing bio and recyclable materials in their surfboards production additionally there are many small surfboards workshops, e.g. 5'TX" (5TX, Euskadi, 2015) trying to introduce these eco-materials.

In this particular study, the flax fibre/bio-epoxy composite system development is focused on solving sustainability and economic issues. The point is that although there are already many finished and high quality products available in the market, they are priced as luxury items, often being two or three times the price of a conventional surfboard. Renewable materials prices are normally higher than conventional ones, but not to the extent that justifies such increment in the final product market price. For this reason, in collaboration with 5'TX"surfboards, the aim is to offer high standard product which is only 50 % more expensive that conventional surfboards.

Experimental

Materials Selection

The *FlaxPly UD150 and Woven BX150 0°/90°* flax reinforcements were sourced from Lineo (Fig. 1). The eco-friendly resin was the Super-Sap *CLR/CLS* epoxy system produced by Entropy resins Inc. (Fig. 2) and used by Samsara Surfboards. The conventional system, used as the benchmark case, was produced with Hexcel S-glass fibre (GF) woven BX125 0°/90° coded as *HexForce® 00471 800 TF970 clear* and Silmar *SIL66BQ-250A* polyester resin. Shaped foams were obtained from Artic foams (arcticfoam.us, California, 2015).

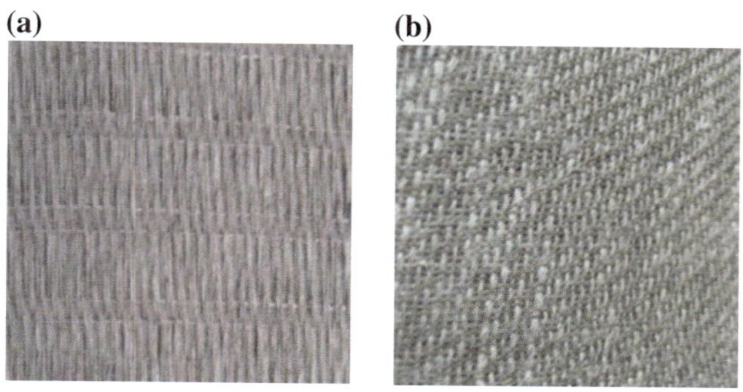

Fig. 1 **a** *FlaxPly UD150* and **b** *FlaxPly Woven BX 0/90* (100 mm × 100 mm fabrics)

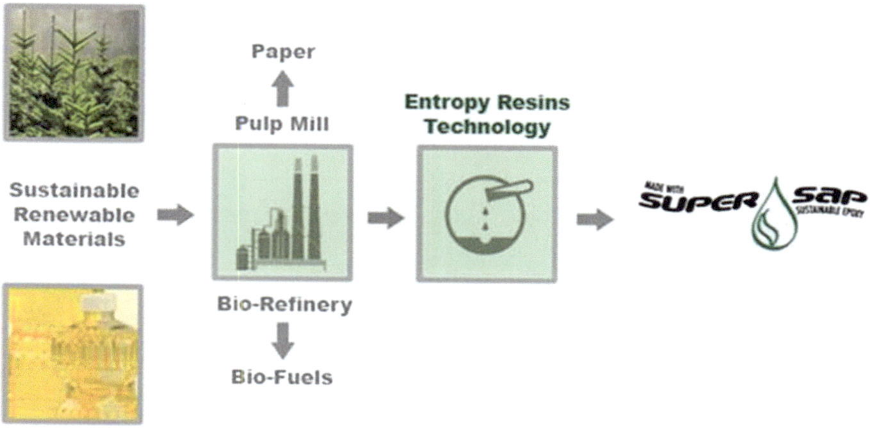

Fig. 2 SuperSap resin production process sketch (courtesy of Entropy resins Inc.)

Laboratory Process

FlaxPly/SuperSap composite produced by hand lamination and cured under vacuum bagging was considered to compensate for the lower performance of the bio-based system with more reinforcement and oven post-curing of the final product. The reference case was the Hexforce/Silmar system manufactured by hand lamination. One of the main issues is the fabric quantity increment increases the laminate/surfboard weight. However, vacuum bagging increases the fibre volume fraction (V_f) and hence reduces the resin uptake.

S-glass/UP samples were manufactured by hand-lamination and bagged with 0.9 mbar absolute vacuum applied (Table 1).

The flax/epoxy samples were manufactured by Resin Infusion under Flexible Tooling with a flow medium (RIFT II) (Summerscales 2012; Summerscales and

Table 1 Test plate manufacturing parameters

	Plate A	Plate B
Reinforcement	4 layers of fabric	8 layers of fabric
Mass of fabric (g)	200	400
Resin used (g)*	200	400
Infusion vacuum (atm gauge)	0.9	0.9
Cure cycle	2 h RT	2 h RT

* Grams of mixture, base + hardener

Searle 2005; Williams et al. 1996) (Fig. 2). The use of RIFT is necessary to improve the wetting-out and compaction of the flax fibre reinforcement stack (Fig. 3, Table 2).

Five different laminates were produced to generate the necessary samples for the complete characterization of the composite system. Plates 1–2 were produced for the longitudinal and transverse tensile test (ISO527). Plate 3 was produced for the 10° off-axis in-plane shear test (Chamis and Sinclair 1976). The thicker Plates 4/5 were produced to get longitudinal and transverse compression properties (ISO14126).

Fig. 3 Hand-lamination and infusion technique

Table 2 Test plates manufacturing parameters

	Laminate A	Laminate B
Test plates	1, 2, 5	3.4
Reinforcement	4 layers of fabric	8 layers of fabric
Mass of fabric (g)	163.35	326.70
Resin used (g)*	177	354.7
Resin preheat	40 °C for 30 min	40 °C for 30 min
Infusion vacuum (atm gauge)	0.9	0.9
Infusion time (minutes)	17	23
Cure cycle	5 min RT + 2 h @ 80 °C	2 h @ 80 °C

* Grams of mixture, base + hardener

The test samples were machined from the cured panels with a band saw then end-tabs were attached using Hexion adhesive. Tabs were produced by infusion, using two layers of 800 gsm E-glass and petrochemical epoxy resin.

Results

The mechanical characterisation involves measurement of the moduli and strengths in tension, compression and in-plane shear. The thermal characterisation includes Differential Scanning Calorimetry (DSC) and Thermo Gravimetric Analysis (TGA) (Groenewoud 2001; Haines 2002; Mulligan et al. 2000). Figure 4 shows the tested flax/epoxy longitudinal and transverse tensile specimens. Characterisation data from the FlaxPly/SuperSap CLR composite system are shown in Table 3.

For the S-glass/UP resin composite system, the properties are subject to commercial confidentiality issues. The data acquired informed the surfboard design.

Eco-surfboard Production

Based on the experimental data achieved in the testing campaign, flax fibre/bio-epoxy resin composite system surfboard lamination scheme was designed. The

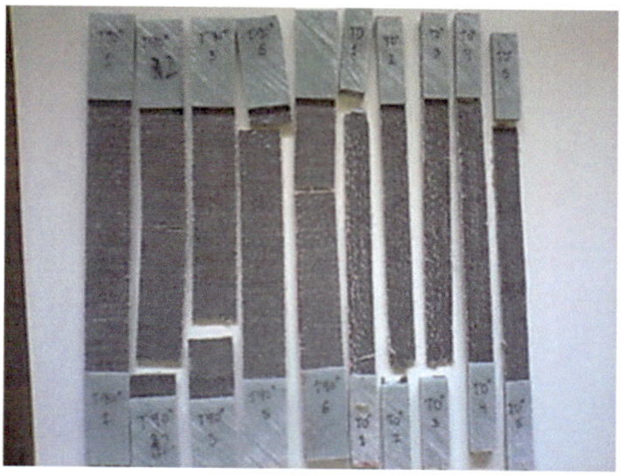

Fig. 4 Flax/epoxy composite samples after transverse tensile tests

Table 3 SuperSap CLR/FlaxPly test results for 30 % fibre volume fraction composites

Direction	Tension modulus (GPa)	Compression modulus (GPa)	Tensile strength (MPa)	Compression Strength (MPa)	Tensile strain to failure (%)	Compression strain to failure (%)
X	15.6	16.6	150	90	1.3	0.97
Y	3.3	3.37	13	55	0.11	1.55
Shear modulus		5.64 GPa		Shear strength		20.8 MPa

conventional PUF surfboards are reinforced with two layers of woven 0°/90° S-glass 125 gsm in the deck and only one layer below. See Fig. 5.

The eco-board lamination scheme is shown in Fig. 6. Basically an extra UD fabric is incorporated to each side of the PUF blank. The idea is to achieve similar rigidity introducing more reinforcing material in longitudinal direction and packing the laminate with vacuum.

For this study, the 5'TX" Mitxe commercial surfboard was been selected. Mitxe has the following characteristics; dimensions 1.88 × 0.51 × 0.06 m (6′2″ × 201/4″ × 21/4″); thruster fin system; rounded tail; 80–90 kg rider; 3 kg board weight and 0.5–1.5 m wave height (Fig. 7). Surfboard manufacturing starts with the core shape. Once the desired shape is obtained, first the surfboard bottom, then the deck is laminated with the selected composite system.

Fig. 5 S-glass lamination process

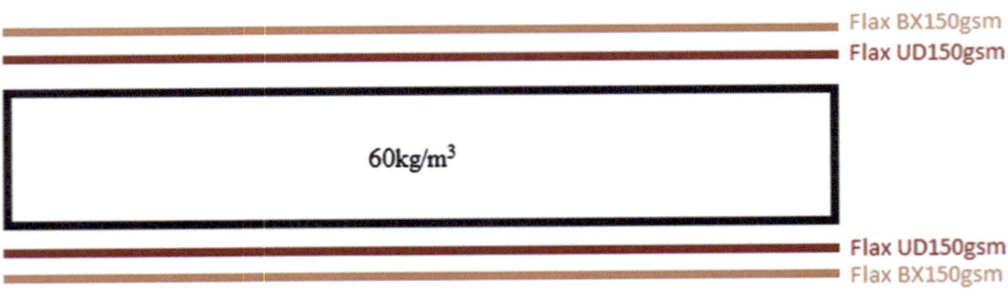

Fig. 6 Eco-board lamination scheme

Fig. 7 Glass/UP/PUF Mitxe model

In the eco-boards production, UD then BX flax fabric are laminated onto the PUF foam (Fig. 8).

The surfboard is then turned-over and the UD and BX flax deck are identically laminated. Once both sides are laminated the vacuum bag is applied (Fig. 9). It is very important in this occasion to use the peel-ply to get the necessary roughness in the composite surface for the subsequent application of the hot-coat cosmetic finish.

Hot coat application, fins/leash colocation and grinding are done in the same way as for the conventional materials.

Fig. 8 Mitxe model shape (*left*) and bottom flax lamination (*right*)

Fig. 9 Surfboard prepared for deck lamination (*left*) and finished (*right*)

Conclusion

The main objective of the current investigation is to justify the use of the bio-epoxy and flax fibre combination as a replacement for other materials with higher environmental burdens. The preliminary results suggest that the investigated sustainable system can achieve this aim.

In regards to the mechanical results, the NF materials data falls below the values achieved for the S-glass composite. However, the low values can be compensated by using more fibre and the vacuum consolidation technique. As a result, the flax fibre and bio-epoxy resin composite can achieve the same performance without an excessive increase in costs.

In the surfboard production, some manufacturing difficulties were faced. For example, the flax fibre wetting is more difficult than for GF; for this reason more resin is used in the lamination process. Due to this extra resin utilisation together with the introduction of one extra layer in the lamination scheme, it would incur a weight penalisation. This particular surfboard model has been weighed and is 1 kg (25 %) heavier than the reference model. This is a big issue since the lightness has to be respected in sport goods. Therefore, the idea is to control the resin addition and work in the correct vacuum application (Fig. 10).

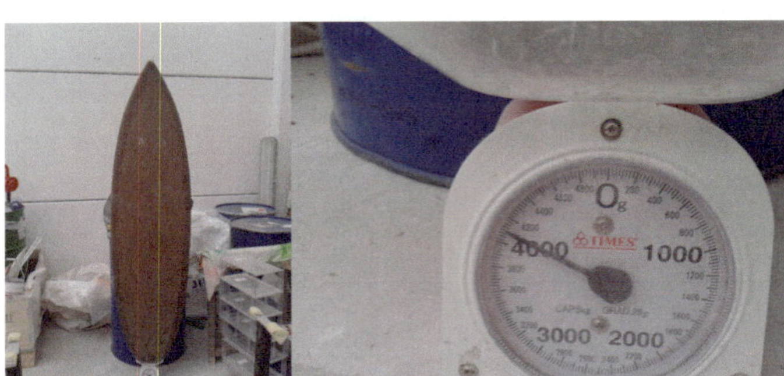

Fig. 10 Eco-surfboard weight

Additionally, since the flax fibre wetting is more difficult, the vacuum bagging process takes 25 % longer than the conventional process. This means that the prototype surfboard would be more expensive. As this is the first prototype board of the set, the company 5'TX" does not considered to be an insurmountable issue.

In general, it is anticipated that an economically competitive final product can be achieved.

Future Work

Although the static properties achieved in the experimental campaign are good, it will be necessary to ensure no interfacial properties degradation in long-term application, especially as the surfboard will be dynamically loaded in a marine environment. For this reason, the on-going Ph.D. study will consider different composite system modifications (Hernandez Michelena et al. 2015a, b, c).

(i) The flax fibre was treated with alkali solution in order to improve its compatibility with the epoxy resin . Flax fabric was immersed in 10 % NaOH solution overnight at RT, then dried for 2 days at RT before 1 h in the oven at 80 °C. Treated flax was infused with the bio epoxy resin for the production of the panel.
(ii) The bio epoxy resin formulation was in situ modified by the addition of 0.5 % BYK-LP X 22550 silane coupling agent to the resin hardener to obtain a mixture which is stable for two months according to supplier recommendations.
(iii) Both the fibre alkali treatment and resin modification were applied to the composite system.

The newly developed composites systems will be applied in the surfboards manufacturing. The idea is to use these systems and test the surfboards in practical use, basically surfing them and evaluate their performance.

The boards will be used in the sea, so it will be important to assess the durability of the composite system in the marine environment, including moisture diffusion, osmosis, blistering and environmental stress corrosion (Searle and Summerscales 1998; Davies and Rajapakse 2014), but not cavitation erosion or galvanic corrosion.

A Life Cycle Assessment (LCA) will be conducted using the environmental burdens defined in ISO/TR 14047 (ISO 2003). The Environmental Impact Classification Factors (EICF) from Azapagic et al. (2003, 2004) make these burdens more amenable to analysis (e.g. global warming potential rather than climate change). The LCA for NF surfboards will need to consider the additional vacuum systems (pump, electricity and consumables) used in the process.

Acknowledgments AHM is grateful to the **Santander Postgraduate Internationalisation Scholarship** for partial funding of this programme of work.

References

TX surf outdoor gear/sporting goods, www.facebook.com/5txsurf accessed 08 September 2015.

Artic foam, http://arcticfoam.us/, accessed 08 September 2015.

Azapagic A, Emsley A and Hamerton I. Polymers, the Environment and Sustainable Development, John Wiley & Sons, March 2003, ISBN 0-471-87741-7.

Azapagic A, Perdan S and Clift R (editors), Sustainable Development in Practice - Case Studies for Engineers and Scientists, John Wiley & Sons, May 2004. ISBN 0-470-85609-2. Second edition, 2011: ISBN 978-0-470-71872-8.

Davies P and Rajapakse YDS (editors): Durability of composites in a marine environment, Springer Solid Mechanics and its Applications series vol. 208, pp 1-13, Springer, Dordrecht, 2014. ISBN 978-94-007-7416-2. E-book ISBN 978-94-007-7417-9.

Groenewoud WM. Characterisation of polymers by thermal analysis. Elsevier, Amsterdam, 2001. ISBN_0-444-50604-7.

Haines PJ. Principles of thermal analysis and calorimetry. Royal Society of Chemistry, Cambridge, 2002. ISBN_0-85404-610-0.

Hernandez Michelena A, Graham-Jones J, Summerscales J and Hall W, Resin Transfer Moulding (RTM) production of surfboard fins with natural fibres. Part II: experimental work and testing (abstract 209), 2nd International Conference on Natural Fibers – From Nature to Market (ICNF2015), São Miguel – Azores (Portugal), 27-29 April 2015, oral presentation. CD-ROM Session 13, ISBN 978-989-98468-4-5. Book of Abstracts pages 265-266, ISBN 978-989-98468-5-2.

Hernandez Michelena A, Graham-Jones J, Summerscales J and Hall W, Eco-friendly epoxy resin/flax fibre composite system as a material for surf boards production (abstract 210), Ibid., pages 267-268,

Hernandez Michelena A, Graham-Jones J, Summerscales J and Hall W, Eco-friendly epoxy resin/flax fibre composite system interface improvement with the chemical treatments of the flax surface and resin formulation, Ibid., pages 269-270.

ISO. International Standard PD ISO/TR 14047:2003(E), Environmental Management – Life Cycle Impact Assessment – Examples of Application of ISO14042, 11 December 2003. ISBN 0-580-43112-6.

Lineo flax fibre impregnation, www.lineo.eu accessed 08 September 2015

Lost + E-tech surfboards, www.lostsurfboards.net/2014/lost-e-tech-surfboards accessed 08 September 2015

Mulligan DR, Gnanih S and Sims G. Thermal Analysis Techniques for Composites and Adhesives (Second Edition). National Physical Laboratory Good Practice Guide (GPG No. 62), Teddington, 2000.

Notox Planches de Surf, www.notox.fr accessed 08 September 2015

Samsara ecosurfboards, www.samsarasurfboards.com, accessed 08 September 2015.

Searle TJ and Summerscales J. Review of the Durability of Marine Laminates, Chapter 7 in G Pritchard (editor): "Reinforced Plastics Durability", Woodhead Publishing, Cambridge, November 1998, pp 219-266. ISBN 1-85573-320-x.

Sicomin Composires Greenpoxy 55 high biomass carbon content, http://www.matrix-composites.co.uk/prod-data-sheet/old/greenpoxy-55-ft-uk.pdf, accessed 08 September 2015.

Summerscales J. Resin Infusion Under Flexible Tooling (RIFT). In L Nicolais and A Borzacchiello (editors):Encyclopedia of Composites – second edition, John Wiley & Sons, 2012, 2648-2658.

Summerscales J, Searle TJ. Low pressure (vacuum infusion) techniques for moulding large composite structures. Proc IMechE Part L - Journal of Materials: Design and Applications, February 2005, L219(1), 45-58.

Waste to Waves, wastetowaves.org/, accessed 08 September 2015.
Williams CD, Summerscales J, Grove SM. Resin infusion under flexible tooling (RIFT): a review. Composites Part A: Applied Science and Manufacturing, July 1996, 27A(7), 517-524.
Chamis CC and Sinclair JH. 10° off-axis tensile test for intralaminar shear characterization of fiber composites. Nasa technical note. April 1976. Nasa TN D-8215.

The Use of Cellulosic Fibers Wastes to Increase the Mechanical Behaviour of Biodegradable Composites for Automotive Interior Parts

N.C. Loureiro, J.L. Esteves and J.C. Viana

Abstract The best way to obtain cellulosic fibers is to use the pulp wastes from the paper plants. The pre-processing of the fibers is a very complex and chemically heavy set of operations. At the end, a small part of the fibers will be rejected and removed from the production system. Actually these rejected fibers are being incinerated at the furnace that dries the raw-wood. This means that after a set of chemical treatments, that ends with a bleach, the fibers will be incinerated driving to a waste of money, once that all the efforts applied to the treatment are not useful, and by other side, the incineration can release some non-healthy emissions. It is possible to use these rejected fibers in composite applications where their mechanical properties can improve the matrix performance. We will present in this paper the influence of incorporation of these rejected fibers into a biodegradable PLA and PHA blended matrix to replace, in some applications, the petrol-based polymers used into the automotive interior trims. The mechanical, and thermal properties will be assessed and compared with the properties of the petrol-polymers normally used in this type of applications.

Keywords Cellulose fiber waste · Biodegradable composites · Mechanical properties · Automotive applications

N.C. Loureiro (✉) · J.L. Esteves
Institute of Mechanical Engineering and Industrial Management (INEGI), Porto, Portugal
e-mail: nuno.loureiro@fe.up.pt

N.C. Loureiro
Superior Institute of Douro and Vouga (ISVOUGA), Santa Maria da Feira, Portugal

J.L. Esteves
Department of Mechanical Engineering, Faculty of Engineering of University of Porto (FEUP), Porto, Portugal

J.C. Viana
Institute for Polymers and Composites (IPC/I3N), Polymer Engineering Department, University of Minho (UMinho), Guimarães, Portugal

R. Fangueiro and S. Rana (eds.), *Natural Fibres: Advances in Science and Technology Towards Industrial Applications*, RILEM Bookseries 12,
DOI 10.1007/978-94-017-7515-1_21

Introduction

In the last few years, biomaterials have been a target of interest at scientific, technological research and industrial level. The increasing of the environmental concern and the excessive use of petrol resources have conducted to the development of new materials, bio composites, which will achieve a great importance into the future.

It's a fact that, allied to weight reduction, the automotive industry can benefit from the inclusion of these materials in cars, not only because of the installed fear of decline and extinction of oil reserves, but also because of the strong legislation that forces an increasing incorporation of these materials in automobiles.

Since the main purpose of this work is to create a bio-composite that is suitable to replace the existing interior door trims, the work has focused on obtaining that bio-composite, taking account the raw-materials cost reduction and the maintenance of the manufacturing process of current door trims—injection molding. Allied to that, this new composite material must be biodegradable and must have a renewable-source origin.

Therefore PLA and PHA blends mechanical and morphological behavior have been studied in order to define the best matrix for use in this study (Loureiro et al. 2015). Blending PLA with PHA tries to reach the mechanical requirements. The PLA/PHA blends have the mechanical capability to replace the used petrol-base polymers but in terms of service temperature stays lower than the required value.

Subsequently it was studied the incorporation of cellulosic fiber to adjust the properties to meet the required values. The incorporation of cellulosic fibers will try to improve the service temperature and mechanical behavior.

In the north of Portugal and in the Galiza region (Spain) is possible to count 824,000 ha of Pine trees and 468,000 ha of Eucalyptus trees (Campos 2012).

To process and transform the wood in paper this euro-region counts with nine industrial plants. From these nine plants, two are thermochemical paper pulp plants. The other seven can be divided in five plants to produce MDF and two to produce fiberboard (Fig. 1).

The process to extract fibers from the wood is a chemically heavy and requires a parallel system to purify the produced wastes to decrease the environmental impact.

The process is quite similar if we are extracting pine fibers or eucalyptus fibers. The main differences are at the chemical compounds used.

In Fig. 2 it's presented a simplified diagram of a typical pulp and paper process.

At the end of this process the rejected fibers are segregated and used into the cooking stage as a boiler combustible. These rejected fibers will be used in this work.

Fig. 1 Northen Iberian Penisula wood transformation plants Geo-localization (Campos 2012)

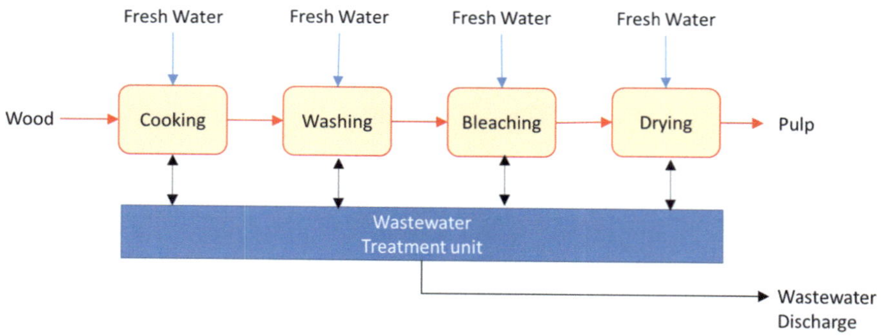

Fig. 2 Pulp process scheme (Patino and Nunez 2011)

Materials and Methods

Materials

In others works (Loureiro et al. 2015) we had presented the optimization of the ratio PLA/PHA on a blend for the composite matrix. We state that for automotive door trims the best ratio is [PLA:PHA] [70:30] weight fraction.

Table 1 Raw material manufacturer's datasheet properties values

	PLA	PHA
Melt temperature (°C)	188–210	145–155
Degradation temperature (°C)	200	200
Tensile strength at break (MPa)	48	35
Tensile elongation at break (%)	2.5	2
Tensile modulus (MPa)	2700	2950
HDT A (1.8 MPa) (°C)	60	72.5
Density	1.24	1.25
MFI (190 °C/2.16 kg) (g/600 s)	30–40	15–30

Guimarães (2009) states that for PLA/cellulosic fibers composites is not possible to incorporate more than 30 % of fiber, otherwise the PLA matrix will not be able to accommodate all the fibers.

This work presents the study of the mechanical properties of cellulosic fiber composites with a 70PLA:30PHA matrix with fiber incorporation of 10 % and 20 % wf.

The polymers used in this work were:

- PHA, under the trade name PHI002, manufactured by Natureplast, France
- PLA, under the trade name Ingeo Biopolymer 3251D, manufactured by NatureWorks LLC, USA

The material suppliers provided the data properties presented in Table 1.

The cellulosic fibers are from Portuguese eucalyptus globulus trees, and becomes from a Portuguese paper pulp plant. They have been removed of the production line after the final chemical treatment and before enter in the paper production line, which means that the fibers were bleached and disintegrated. The bulk fibers are composed essentially of cellulose (~85 %) and glucuronoxylan (~15 %). The main properties of these fibers are presented in Table 2.

Preparation of the Composites

The materials were dried at 60 °C for 24 h before processing and kept into separate Ziploc bags. The composites were compounded in Compound Injection Moulding Machine. The different ratios were adjusted by setting the feed throat of each hopper.

The mould temperature was set at 20 °C, and the injection temperature profile is presented in Fig. 3.

This temperature profile was established by combining the melting temperature of the polymers, the degradation temperature and the injection molding conditions suggested by the supplier's datasheets. A constant injection velocity of 20 mm/s (corresponding to an injection flow rate of 6.3 cm^3/s) was used.

However before injection it was necessary to extrude the bio-composite and palletize it.

Table 2 Eucalyptus globulus properties (Savastano and Warden 2000; Agopyan et al. 2005)

Average fiber diameter (μm)	10.9
Average fiber length (mm)	0.66
Tensile Strength at break (MPa)	160
Tensile elongation at break (%)	5.2
Tensile modulus (GPa)	17.4
Flexural modulus (GPa)	16
Flexural strength at break (MPa)	130
Density	1.6

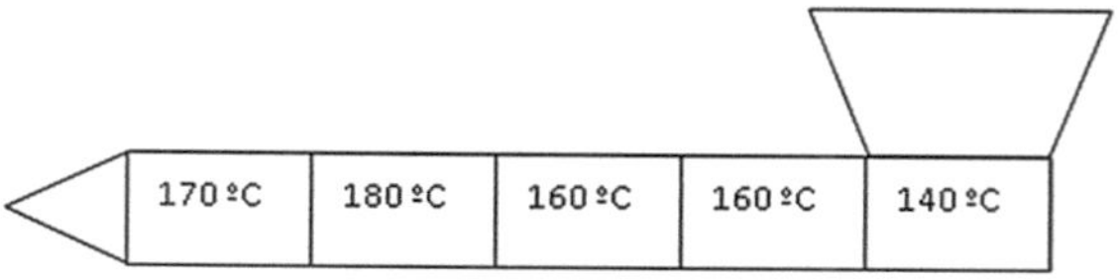

Fig. 3 Injection moulding temperature profile

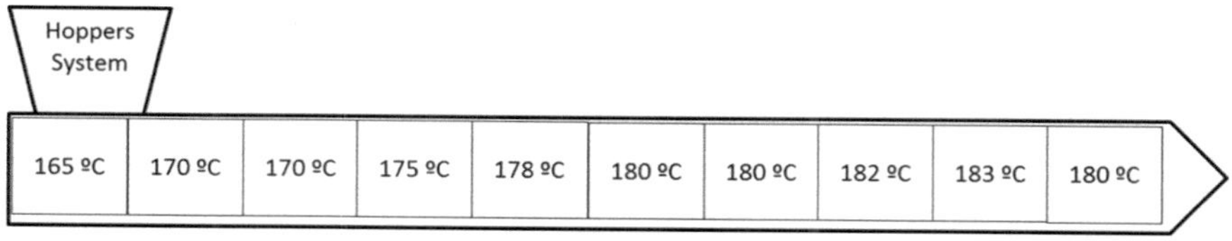

Fig. 4 Extruder temperature profile

The extruder temperature profile used is expressed into Fig. 4.

The several composite were injection molded in the form of specimens with the following geometries and dimensions (according to the respective standard):

- Flexural and HDT specimens: 12 × 150 × 6 mm parallelepiped bars
- Impact specimens: central gated discs of Ø60 × 2 mm
- Tensile specimens (type II): dog-bone geometry with a narrow section of 57 × 10 × 4 mm, and an overall dimensions of 183 × 19 × 4 mm.

Mechanical Characterization

Tensile Properties Measurements

The tensile properties were tested in a universal mechanical testing machine Shamidzu AG-X 100 kN, equipped with a 50 mm Shamidzu extensometer, according to ASTM D638 test procedures. The crosshead velocity used was of

5 mm/min, and the tests were performed in a standard laboratory atmosphere of 23 ± 2 °C and 50 ± 5 % relative humidity. A grip distance of 150 mm was used. The envisaged tensile properties were the initial modulus, the maximum/yield stress and the strain at break. At least 11 specimens were tested for each reinforced blend composition.

Flexural Properties Measurements

A Universal Tiratest 2705 5 kN Machine was used to measure the flexural properties according to ASTM D790 standard. It has been used a 3-point flexural test, with a crosshead velocity of 2.56 mm/min and a spam of 96 mm. Tests were performed in a standard laboratory atmosphere of 23 ± 2 °C and 50 ± 5 % relative humidity. The envisaged flexural properties were the initial modulus, the maximum stress and the strain at maximum stress. At least 11 specimens were tested for each reinforced blend composition.

Instrumented Impact Properties Measurements

Instrumented impact tests are performed according ISO 6603-2 standard in a CEAST Fractovis plus Impact machine (velocity of 1 m/s). All tests were carried out in standard laboratory atmosphere of 23 ± 2 °C and 50 ± 5 % relative humidity. From the force-displacement curve, the impact toughness was calculated as the area below the force-displacement graph. The impact data presented are the average of 7 measurements.

Heat Deflection Temperature (HDT) Measurements

The Heat Deflection Temperature (HDT) was measured according to ISO 75-2, in a RAY-RAN HDT apparatus. This test used the method HDT A with an applied stress state of 1.8 MPa and an increasing temperature rate of 120 °C/h. Presented HDT results are the average values of three measurements.

Experimental Results and Remarks

The mechanical properties obtained experimentally are expressed in Table 3.

Table 3 Experimental results

	Matrix [70:30] [PLA:PHA]	Biocomposite (10 % fiber)	Biocomposite (20 % fiber)
Tensile modulus (GPa)	3.4	4.8	5.6
Maximum tensile stress (MPa)	46.02	86.20	91.78
Maximum flexural stress (MPa)	62.30	68.86	76.33
Absorbed energy (J)	6.4	10.5	10.5
Heat deflection temperature (°C)	62.1	63.0	71.7

Table 4 ABS and PP general properties (Agopyan 2005)

	ABS (injection grade)	PP (injection grade)
Tensile modulus (GPa)	2.33	1.72
Maximum tensile stress (MPa)	38.4	31.3
Maximum flexural stress (MPa)	68.6	48.1
Absorbed energy (J)	22.7	10.8
Heat deflection temperature (°C)	96	63

The interior door trims are mainly injection molded into ABS or PP-copolymer, that's why is obligatory to compare the obtain results with the general ABS and PP properties.

Using online databases (www.matweb.com) it was possible to resume the main properties of ABS and PP. The results are expressed into Table 4.

Based in Tables 3 and 4 is possible to compare (Fig. 5) the properties of the raw-materials. The graphic presented in Fig. 1 will emphasize the relation of all these parameters.

Regarding the HDT, is possible to see that the ABS presents the highest value. However the value presented is an average of all ABS grades actually in the market.

It's possible to have an ABS with a HDT lower than the studied composites.

Using the figure is possible to choose the best solution since that for producing automotive parts it's going to be used the composite that presents the equal or better properties in all dimensions.

By analyzing the graphic is possible to conclude that the composite with 20 % fiber (gray shadow) is the one that fulfill the previous statement.

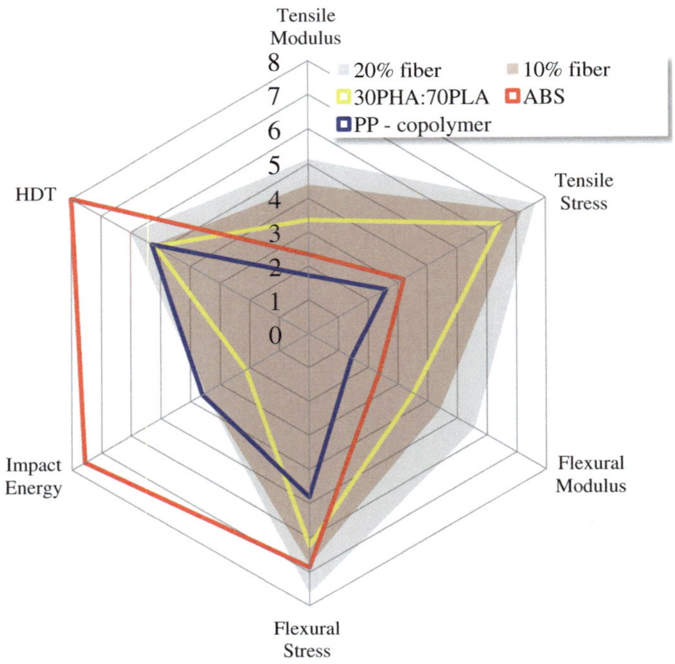

Fig. 5 Raw-materials properties comparison

Conclusion and Final Remarks

The properties of biodegradable composites can be tailored to achieve a given performance.

Composites with a [30:70] [PHA:PLA] matrix and with a fiber content of 10 and 20 % (wf) were compared with the most used petrol-based polymers for automotive interior parts (PP and ABS).

The incorporation of 20 % wf fiber leads to an eco-composite that presents the best properties among all biopolymer blends and biocomposites studied.

When compared with ABS and PP, this eco-composite, normally, presents equal or better properties, excluding the impact energy.

The feasibility of producing interior parts using the studied eco-composites has been investigated.

In conclusion it verifies that the matrix containing 70 % PLA and 30 % PHA (weight fraction) with the incorporation of 20 % of fiber improves the thermal behavior of the composite, proving that it is possible to replace the petrol-based polymers for these biocomposites without compromising the mechanical behavior of interior door trims.

It is possible to conclude that this composite can present a feasible option for replacing petrol-based polymers in some cabin interior parts applications.

Once that the composites are from renewable sources this work gives an indication of the potential use of these composites replacing the petrol-based matrix composites with synthetic reinforced fibers.

It also proved that is possible to give an added-value to rejected fibers into the pulp plants.

Acknowledgments The authors acknowledge the Portuguese Science and Technology Foundation (FCT) for the financial support provided by the project MIT-Pt/EDAM-SMS/0030/2008—Assessment and Development of integrated Systems for Electric Vehicles.

N.C. Loureiro acknowledges also the Portuguese Science and Technology Foundation (FCT) for the financial support provided by the Ph.D. grant SFRH/BD/42978/2008.

J.C. Viana also acknowledges the project PEst-C/CTM/LA0025/2011 (Strategic Project—LA 25—2011–2012).

References

N.C. Loureiro, J.L. Esteves, J.C. Viana, S. Ghosh, Mechanical Characterization of Polyhydroxyalkanoate and Poly(Lactic Acid) Blends, Journal of Thermoplastic Composite Materials, 28(2) (2015)

A.R. Campos, GreenMotion – Transfer of Ecological Materials for the Automotive Industry, 2nd Journeys of GreenMotion Project, Guimarães, Portugal (2012)

J.M. Patino, M.P. Nunez, Retrofit Approach for the Reduction of Water and Energy Consumption in Pulp and Paper Production Processes, Environmental Management in Practice, Dr. Elzbieta Broniewicz (Ed.), InTech, (2011)

H. Savastano Jr., P.G. Warden, R.S.P. Coutts, Brazilian waste fibres as reinforcement for cemente-based composites, Cement & Concrete Composites, vol.22, pp. 379-384 (2000)

V. Agopyan, H. Savastano Jr., V.M. John, M.A. Cincotto, Developments on vegetable fibre-cement based materials in São Paulo, Brazil: an overview, Cement & Concrete Composites, vol.27, pp. 527-536 (2005)

Information on: http://matweb.com (consulted at 24th February 2013)

D. Guimarães, Efeito das condições de injecção nas propriedades de PLA reforçado com fibras celulósicas, Master Thesis in Polymer Engineering, University of Minho, Portugal (2009)

Part IV
Natural Fibre Reinforced Cementitious Composites

Hemp Fibres—A Promising Reinforcement for Cementitious Materials

Ildiko Merta

Abstract Industrial hemp (*Cannabis sativa* L.) has one of the longest cultivation histories in Europe. Hemp fibres have recently been identified as viable fibre reinforcement in cementitious materials, however, there is still limited research dealing with it. In this research the potential of hemp fibres as fibre reinforcement in two types of concretes (normal- and foam concrete) was experimentally evaluated in terms of the composite compression strength, splitting tensile strength and energy absorption capacity. Primary bast hemp fibres of 5, 10, 20 and 40 mm in length with 0.5 % dosages per volume were used as reinforcement. The results show that hemp fibres could be a promising alternative to traditional fibres in cementitious composites. The addition of hemp fibres to both normal and foam concrete results in around 50 % decrease of the composites compression strength with simultaneous increases of the splitting tensile strength in foam concrete and decrease in normal concrete. Hemp fibres significantly increase the energy dissipation capacity of the composite. In the case of normal concrete up to 100 %, whereas in foam concrete the enhancement was even more significant, up to 900 %. The optimal balance between the increased energy dissipation capacity and the resulting loss in compression strength of the cementitious composite could be targeted with reinforcing it with hemp fibres of 10 mm length.

Keywords Hemp fibers · Cementitious composite · Compressive strength · Fracture behaviour

Introduction

Concrete is the most widely used man-made material in the world. Its yearly consumption is estimated at 3 tons per person on the planet. Concrete belongs to the group of so-called cementitious composite materials because its main ingredient is

I. Merta (✉)
Institute of Building Construction and Technology, Technische Universität Wien, Karlsplatz, 13/206, 1040 Vienna, Austria
e-mail: ildiko.merta@tuwien.ac.at

R. Fangueiro and S. Rana (eds.), *Natural Fibres: Advances in Science and Technology Towards Industrial Applications*, RILEM Bookseries 12,
DOI 10.1007/978-94-017-7515-1_22

Portland cement. Cementitious composites are quasi-brittle materials that are strong in compression but weak under tension load. These composites have low tensile strength, low tensile strain capacity resulting in poor crack resistance and energy absorption capacity (toughness) under tensile load.

It has long been recognized that these deficiencies of the material could be very effectively captured by reinforcing the cementitious matrix with fibre reinforcement. The concept of using horse hair in mortar and straw in the production of mud brick goes back to ancient times. The first modern fibre-reinforced product invented by Ludwig Hatschek in the late 19th century was cement roofing sheets reinforced with asbestos fibres. However, the urgent need to replace asbestos fibres because of their health risk gave rise to the development of modern alternative fibre reinforcement in the early 1950s.

The most typical fibres used in cementitious composites as fibre reinforcement are steel, synthetic or glass fibres (Balagur and Shah 1992; Bentur and Mindess 2007). These fibres are short, randomly distributed macro- or micro fibres. With the addition of fibre reinforcement the transition from brittle behaviour to quasi-ductile or even ductile behaviour with increased toughness and energy absorption capacity could be achieved. It was reported that with the incorporation of fibre reinforcement in cementitious materials the cracking due to plastic shrinkage and drying shrinkage in the material as well as its permeability could be significantly reduced, the impact and abrasion resistance were increased, and the tensile strain capacity of the material substantially improved (Bentur and Mindess 2007; Johnston 2010; Mobasher 2011). Generally fibres do not increase the strength of concrete and are not capable to replace the structural reinforcement.

However, traditional steel and synthetic fibres are highly dependent on virgin material resources and are extremely resource- and energy-intensive to manufacture. Their production contributes to high pollution and carbon dioxide emissions. The environmental awareness in the construction industry and the need for reducing raw material consumption and carbon emissions related to cementitious building materials is, however, apparent. Governmental laws are forcing both the academic and industrial sectors to develop a new generation of energy-efficient fibre reinforced cementitious building materials with enhanced environmental sustainability. Because of the vast quantities of concrete used worldwide, concrete is a large raw resource consuming material, demands high energy use in production and generates high pollution. Consequently it is responsible for a vast carbon footprint.

Extensive research is in progress for finding alternative fibres to conventional steel and synthetic fibres as well as in developing a new generation of environmentally friendly and sustainable cementitious composites. Recently natural plant fibres, such as sisal, coir, flax, etc. have been considered as possible sustainable substituent fibres for steel, synthetic or glass fibres (Swamy 1990; Savastano et al. 2000; Tolêdo Filho et al. 2003; Silva et al. 2010a, b; Torgal and Jalali 2011; Sierra Beltran 2011; Merta et al. 2011; Merta and Tschegg 2013).

Natural fibres produced from locally available renewable resources are easily available and cheap, biodegradable, recyclable and non-abrasive, and there is no concern with health and safety during their handling (Ranalli 1999; Müssig 2010). A very important aspect is that natural fibres are substantially lighter in comparison to steel fibres. Composites made with natural fibres have low energy and material consuming production and decreased manufacturing costs. The employment of natural fibres ensures the development of economically and ecologically sustainable composite materials that are considerably less dependent on non-renewable energy resources. It is a viable direction, a way to increase the greenness and sustainability of cement based materials.

However, there are still obstacles in bright use of natural fibres in cementitious materials that need deep understanding and further research. Their variable mechanical and physical properties often lead to unpredictable properties of the composite. The high moisture absorption capacity and durability problem of fibres in the alkaline environment of the cement matrix is the major concern in successful application of natural fibres in cementitious composites. The degradation of fibres in the cement matrix occurs as a consequence of the dissolving of the lignin and the hemicellulose in the middle lamellae of the fibres through the alkaline pore water. The composite may undergo a reduction in strength and toughness as a result of weakening of the fibres by a combination of alkali attack and mineralization through the migration of hydration products to lumens and spaces (Bergström and Gram 1984; Filho et al. 2000; Juárez et al. 2007; Kriker et al. 2008; Sedan et al. 2008; Troëdec et al. 2009; Merta et al. 2012; Filho et al. 2013; Hamzaoui et al. 2014; Wei and Meyer 2015).

Industrial hemp (*Cannabis sativa L.*) has one of the longest cultivation histories in Europe. Hemp fibres have recently been identified as viable fibre reinforcement in cementitious materials, however, there is still limited research published dealing with it (Li et al. 2004, 2006; Dalmay et al. 2010; Jarabo et al. 2012; Awwad et al. 2012; Merta and Tschegg 2013; Merta 2014; Hamzaoui et al. 2014). In this research the potentials and restrictions of hemp fibres as fibre reinforcement in two types of concretes were experimentally evaluated in terms of the composite compression strength, splitting tensile strength and energy absorption capacity.

Materials and Methods

Concrete and Hemp Fibres

Two different concrete matrixes were investigated in this research: normal concrete and foam concrete. The aggregate for normal concrete was river gravel with maximum particle size of the fine aggregate (sand) 4 mm and of coarse aggregate 12 mm. The cement/sand/coarse aggregate ratio was 1/1.5/2.5. Cement CEM II with a water to cement ratio of $w/c = 0.5$ was employed. The mean compressive strength of normal concrete was 43 N/mm^2. The concrete was mixed by a laboratory drum-mixer and compacted on a vibrating table. The specimens were demoulded 24 h after casting and stored in water until testing at the age of 28 days.

Foam concrete or also aerated concrete is one special type of lightweight concretes. It is a cement-bonded material made by blending extremely fluid cement paste (slurry) into which stable, pre-formed foam manufactured on site, is injected. Due to the very large air content in the closed cell structure (up to 80 % air) foam concrete is a material with very low density (200–800 kg/m^3) and relatively low compressive strength values of 2–6 MPa. The low density with high air content results in one of the most superior properties of foam concrete, the outstanding thermal- and acoustic insulating characteristics accompanied with excellent fire resistance. Thermal conductivity is reported to be 0.07–0.11 W/m C which is several times lower than normal weight concrete.

In this research foam concrete with the following mix design proportion were used: cement 250 kg/m^3, water 125 kg/m^3 (*w*/*c* ratio = 0.5) and foaming agent 0.45 l/m^3. No aggregate was used. The specific weight of the foam concrete was 300 kg/m^3 and its mean compression strength 0.28 N/mm^2. The foam concrete was produced by the company Ormet S.p.A. in Italy with a mobile mixing station. The specimens were demoulded 24 h after casting and stored in wet sand until testing at the age of 28 days.

As fibre reinforcement primary bast hemp (*Cannabis sativa* L.) fibres obtained locally in Hódmezővásárhely in Hungary from cultivated plants were used. The tensile strength of the fibres was in range of 300–800 N/mm^2 (Ranalli 1999; Müssig 2010) with a diameter of 15–50 μm. The bulk density of the fibres was 1.103 g/cm^3, which is slightly under the value reported in the literature (Ranalli 1999; Müssig 2010).

The fibres were cut in four different lengths: 5, 10, 20 and 40 mm and added to the concrete matrix in dosage of 0.5 % per volume (Fig. 1). In both normal concrete and foam concrete the hemp fibres were added to the concrete matrix in dry condition at the end of the mixing time and then mixed for an additional 2–3 min. Altogether the research is based on 4 different fibre reinforced foam concrete matrixes and 4 different fibres reinforced normal concrete matrixes. As a reference, plain concrete matrixes were cast.

Fig. 1 Primary bast hemp (*Cannabis Sativa* L.) fibres cut in form of fibre reinforcement for concrete

Experimental Program

The composites compression strength was tested in accordance with EN 12390-3 on three cubic specimens. The fracture mechanical properties of the composite were determined with the employment of the Wedge Splitting Test (WST) (Tschegg 1986) according to ÖNORM B 3592. It is a stable fracture mechanics test that is able to accurately determine the post-peak response of the material in terms of load-crack mouth opening displacement (CMOD) diagrams. Figure 2a shows the setup of the WST.

For each different concrete matrix a series of 6 identical cubic specimens with dimensions of 150 × 150 × 130 mm^3 were cast. At the top of the specimens, a 30 mm long and 3 mm wide starter notch was cut. The rectangular groove on the upper side of the cube, needed for the load transmission pieces, was achieved by gluing on two marble pieces (Fig. 2).

The specimens were positioned on a narrow linear support and the two load transmission pieces and a slender wedge are inserted into the groove. The vertical load produced by the testing machine is transferred by the load transmission pieces from the wedge into the specimen and leads to the splitting of the specimen or to Mode I (opening mode) type fracture. The CMOD is determined at the height of the load application line on both sides by linear variable differential transformers (LVDT). The tests were carried out with a mechanical testing machine from Schenck RSA with a load capacity of 100 kN and a stiffness of 125 kN/mm.

All tests were carried out under displacement control with a constant cross-head speed of 0.5 mm/min at a temperature of 21 °C and a relative humidity of around 50 %. For each specimen the load-CMOD diagrams were recorded by means of a

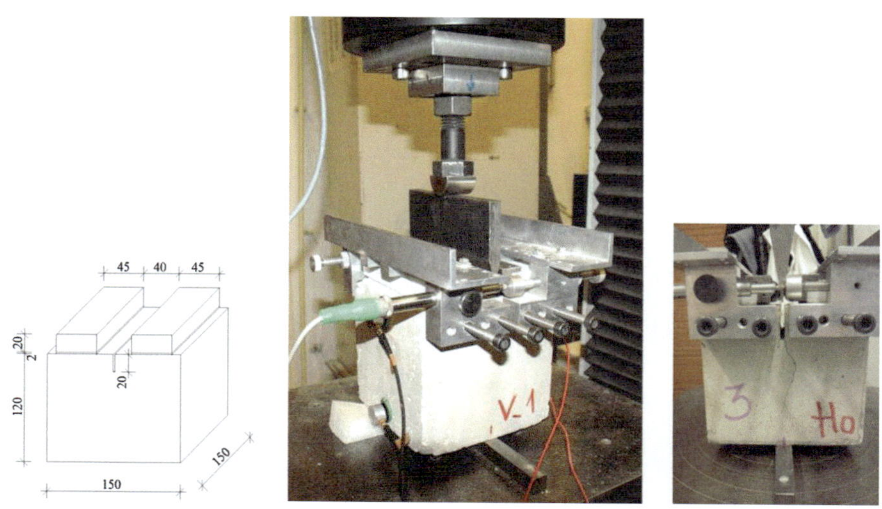

Fig. 2 Geometry of the specimens and setup of the Wedge Splitting Test according to Tschegg (1986)

state-of-the-art data acquisition system. From the load-CMOD diagrams the fracture mechanical parameters such as the stress at the peak load and the energy dissipation capacity of the material were determined.

Results and Discussion

Compression Strength

With the addition of hemp fibres to both normal and foam concrete matrixes a loss in compression strength can be seen (Fig. 3). Similar behaviour has been reported by Li et al. (2006) who found that the addition of hemp fibre in dry condition to the matrix weakens the performance of the composite.

Generally, the addition of hemp fibres results in up to 50 % loss of the composites compression strength compared to the compression strength of non-reinforced (plain) concrete specimens (Fig. 4). In case of normal concrete reinforced with hemp fibres the most distinctive decrease in compression strength is observed at a fibre length of 10 mm. If the fibre length increases (to 20 and 40 mm) the composites compression strength slightly increases (Fig. 3a). The same trend could be seen in case of foam concrete specimens reinforced with hemp fibres, however, here an even more pronounced compression strength loss is observed if the fibre's length increases beyond 10 mm (Fig. 3b). Foam concrete has a sensitive porous structure and a considerable lower (approximately 10 times lower) compression strength than normal concrete. The addition of fibres (especially the long ones) has probably more distinctive influence on the homogeneity of the matrix resulting in stronger compressive strength loss. It seems that for both normal and foam concrete specimens a good balance between the beneficial effect of fibre reinforcement and the resulting loss in compression strength of the composite could be obtained if the fibre's length is no longer than 10 mm.

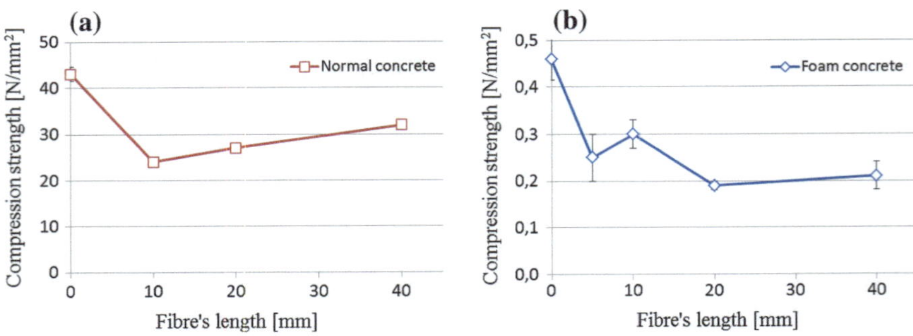

Fig. 3 Compression strength of **a** normal and **b** foam concrete specimens reinforced with hemp fibres of different lengths

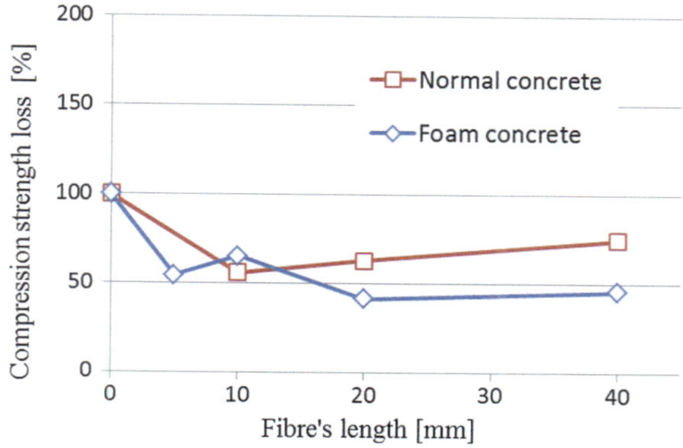

Fig. 4 Loss of compression strength of normal- and foam concrete specimens reinforced with hemp fibres of different lengths compared to plain concrete specimens

Load—Crack Mouth Opening Displacement Curve

A typical load versus crack mouth opening displacement (CMOD) response of the composite recorded during the WST is shown in Fig. 5. The load-CMOD curve indicates a stable crack propagation through the entire loading range and does not show any discontinuous parts. The first part of the diagram, until shortly before the peak load, is the elastic-linear range where both the matrix and the fibres behave linearly. The stiffness of the composite is dominated by the matrix properties. At the end of the linear part the initiation and propagation of the first cracks occur and a deviation from the linear part follows. The fibres are bridging across the crack and the stress transfer between the matrix and fibres is realized through adhesion, friction and mechanical interlock. The first mechanism that is overcome is the bond between matrix and fibres. In the post cracking zone (after the peak of the load-CMOD curve) the major stress transfer governing mechanisms are the friction

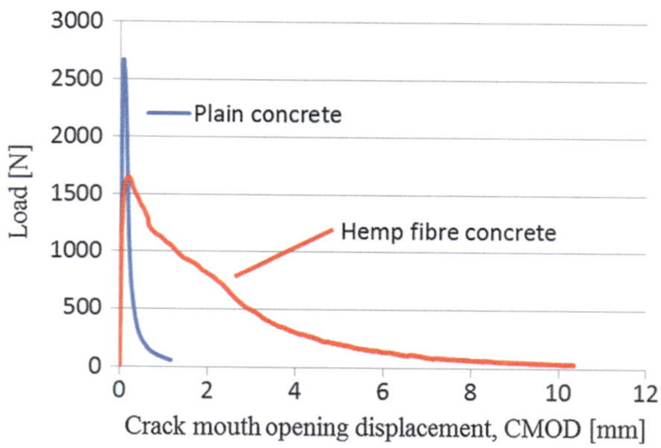

Fig. 5 Load versus crack mouth opening displacement (CMOD) diagram of plain concrete and concrete reinforced with hemp fibres recorded during the Wedge Splitting Test

and mechanical interlock during the pull-out phase of the fibres from the matrix. These mechanisms are responsible for the enhanced energy dissipation capacity of the composite. In case of hemp fibre reinforced concrete a significant increase in the toughness compared to plain concrete is noticeable in the post cracking zone which is quantified by the increased area under the load-CMOD curve. Generally with addition of fibre reinforcement to concrete its strength (peak of the load-CMOD curve) could not been enhanced.

Splitting Tensile Strength

The stress at the peak load (peak of the load-CMOD diagram in Fig. 5) is the so called splitting tensile strength and it is an indirect measure of tensile strength of the composite. The splitting tensile strength of normal- and foam concrete specimens reinforced with hemp fibres of different lengths are shown in Fig. 6.

With the addition of hemp fibres to normal concrete and foam concrete a fundamentally different influence on the splitting tensile strength was observed. While in the case of normal concrete the splitting tensile strength decreases with addition of fibres (Fig. 6a), the in case of foam concrete a distinctive increase of the splitting tensile strength was observed (Fig. 6b).

In the case of normal concrete the increase of the fibre's length (from 10 to 20 mm) negatively influences the splitting tensile strength of the composite and up to 40 % splitting tensile strength loss is observed (Fig. 7). With further increase of the fibre's length (to 40 mm) the splitting tensile strength significantly increases and reaches almost the same value as specimens without fibres (Figs. 6a and 7).

In the case of foam concrete a very different trend in splitting tensile strength was observed. The addition of hemp fibres significantly improved (up to 80 %) the composite's splitting tensile strength (Fig. 6b). The most distinctive increase was observed in the case of short fibres (5 mm) followed by a slight decrease of the composite's splitting tensile strength with increase of the fibre's length. The composites splitting tensile strength (if reinforced with long fibres of 20 and

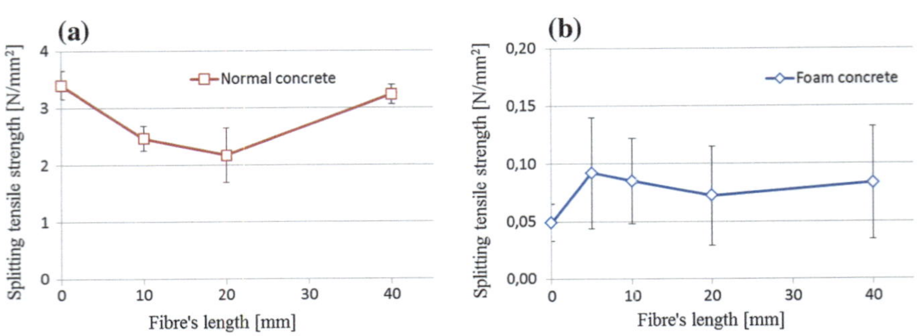

Fig. 6 Splitting tensile strength of **a** normal and **b** foam concrete specimens reinforced with hemp fibres of different lengths

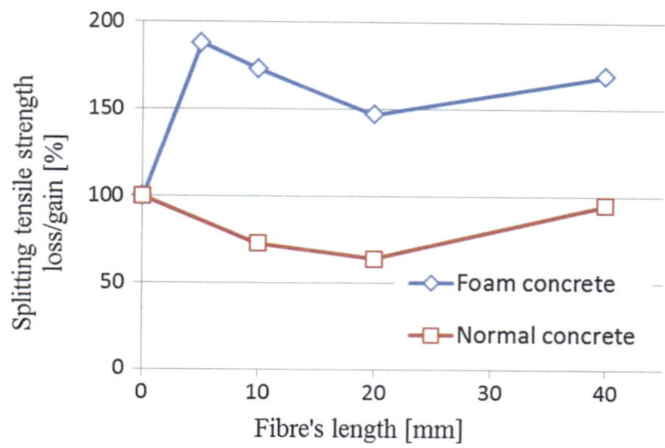

Fig. 7 Loss/gain in splitting tensile strength of normal and foam concrete specimens reinforced with hemp fibres of different lengths compared to plain concrete specimens

40 mm) is still about 50 % higher compared to the splitting tensile strength of non-reinforced foam concrete. The addition of hemp fibres to a filigree and porous matrix of foam concrete has somehow a beneficial effect on the composite and increases the first crack strength of the matrix.

However, in the case of foam concrete specimens very high dispersion in test results was observed compared to those of normal concrete specimens. The incorporation of fibres into the foam concrete could disturb the closed pore structure and homogeneity in some parts of the matrix. This could result in strong variation of the specimen's response under external load.

Energy Dissipation Capacity

The energy dissipation capacity (or toughness) of the material represents the energy dissipated per unit of newly created surface area during the material's fracture. It is calculated as the area under the load-CMOD diagram up to a defined CMOD. In this research some specimens reached relatively high CMOD values of up to 12 mm and the fracture surfaces were still not separated. For a consistent comparison of the results the dissipated energy of all specimens was related to the CMOD value of 10 mm. The specific dissipated energy is the dissipated energy divided by the area of the fracture plain.

With the addition of hemp fibres the energy dissipation capacity of both foam- and normal concrete significantly increase (Fig. 8). In the case of normal concrete, the addition of hemp fibres of 10 mm already increases the fracture energy of plain concrete approximately by 100 % (Fig. 8a). The further increase of the fibre's length (to 20 and 40 mm) does not have any influence on the energy dissipation capacity of the composite anymore and it remains constant. In the case of foam concrete, however, the addition of hemp fibres significantly improves the composites energy dissipation capacity (Fig. 8b). Already the incorporation of short fibres of 5 mm in length increases the energy dissipation capacity of plain foam

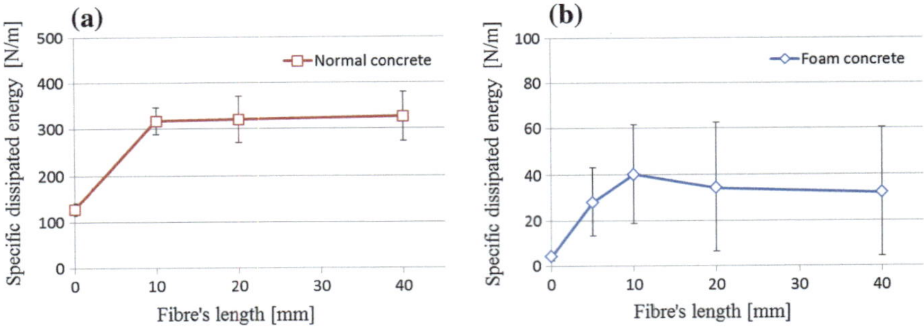

Fig. 8 Specific dissipated energy of **a** normal and **b** foam concrete specimens reinforced with hemp fibres of different lengths

Fig. 9 Enhancement of the energy dissipation capacity of normal- and foam concrete specimens reinforced with hemp fibres of different lengths compared to plain concrete specimens

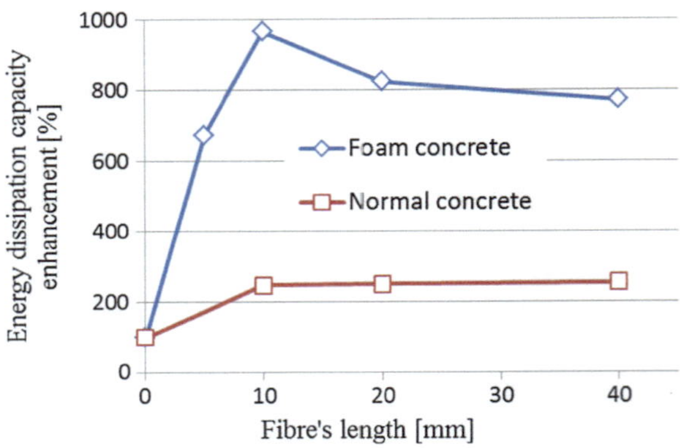

concrete by 600 % whereas longer fibres (10 mm) increase the energy dissipation capacity of the composite by additional 300 % (Fig. 9). With a further increase of the fibre lengths (to 20 mm) the energy dissipation capacity of the composite decreases by 100 % and remains the same in the case of 40 mm fibres also.

Generally, in both normal and foam concrete the sufficient fibres length is 10 mm for targeting the maximal energy dissipation capacity of the composite. Fibres longer than 10 mm do not improve the energy dissipation capacity anymore since only the part of the fibres bridging the crack in the concrete matrix are effective in stress transfer.

Conclusions

In this research the reinforcing effect of primary bast hemp (*Cannabis sativa* L.) fibres on two cementitious materials, normal and foam concrete, was experimentally evaluated in terms of the composite compression strength, splitting tensile strength and energy absorption capacity. Based on the parameters studied the following conclusions could be drawn:

1. The addition of hemp fibres to both normal and foam concrete results in around 50 % decrease of the composite's compression strength.
2. The addition of hemp fibres to normal concrete results in 40 % decrease of the composite's splitting tensile strength. In contrary, foam concrete increases its splitting tensile strength up to 90 % with the addition of hemp fibres.
3. With the addition of hemp fibres the energy dissipation capacity of normal concrete increased up to 100 %, whereas in foam concrete the enhancement was even more significant, up to 900 %.
4. For both normal and foam concrete the optimal balance between the increased energy dissipation capacity and the resulting loss in compression strength of the composite is obtained with hemp fibres of 10 mm in length.
5. Based on these results it could be concluded that hemp fibres are a promising alternative fibre reinforcement to traditional steel and synthetic fibres in cementitious composites.

Acknowledgments This research was carried out with the support of the Hertha Firnberg post-doc fellowship of the author provided by the Austrian Science Fund (FWF) under the grant nr. T 503-N13. Part of the research was enabled with the support of the Austrian Research Promotion Agency (FFG) under the Innovation Check program with grant nr. 840149.

References

Awwad E., Mabsout M., Hamad B., Farran M.T., Khatib H., (2012) Studies on fiber-reinforced concrete using industrial hemp fibers, Construction and Building Materials, Volume 35, pp 710-717, ISSN 0950-0618

Balagur PN, Shah SP. (1992) Fibre reinforced cement composites. Civil Engineering Series. McGraw-Hill.

Bentur A, Mindess S. (2007) Fibre reinforced cementitious composites. Modern Concrete Technology Series, CRC Press, Taylor & Francis Group.

Bergström S. G., Gram H. E., (1984) Durability of alkali-sensitive fibres in concrete, International Journal of Cement Composites and Lightweight Concrete, Volume 6, Issue 2, pp. 75-80

Dalmay, P., Smith, A., Chotard, T., Sahay-Turner, P., Gloaguen, V., Krausz, P., (2010) Properties of cellulosic fibre reinforced plaster: influence of hemp or flax fibres on the properties of set gypsum. J. Mater. Sci. 45, 793–803.

Filho J. A. M., Silva F. A., Toledo Filho R. D., (2013) Degradation kinetics and aging mechanisms on sisal fiber cement composite systems, Cement and Concrete Composites, Volume 40, pp 30-39, ISSN 0958-9465

Filho R. D. T., Scrivener K., England G. L., Ghavami K., (2000) Durability of alkali-sensitive sisal and coconut fibres in cement mortar composites, Cement & Concrete Comp., 22, pp. 127-143

Hamzaoui R., Guessasma S., Mecheri B., Eshtiaghi A.M., Bennabi A., (2014) Microstructure and mechanical performance of modified mortar using hemp fibres and carbon nanotubes, Materials & Design, Volume 56, pp 60-68, ISSN 0261-3069

Jarabo R., Fuente E., Monte M.C., Savastano Jr. H., Mutjé P., Negro C., (2012) Use of cellulose fibers from hemp core in fiber-cement production. Effect on flocculation, retention, drainage and product properties, Industrial Crops and Products, Volume 39, pp 89-96, ISSN 0926-6690

Johnston C.D, (2010), Fiber-Reinforced Cements and Concretes, ISBN 90–5699-694-0, Taylor and Francis

Juárez C., Durán A., Valdez P., Fajardo G., (2007) Performance of "Agave lecheguilla" natural fiber in portland cement composites exposed to severe environment conditions, Building and Environment, 42, pp. 1151-1157.

Kriker, A. Bali, G. Debicki, M. Bouziane, M. Chabannet, (2008) Durability of date palm fibres and their use as reinforcement in hot dry climates", Cement & Concrete Composites, 30, pp. 639-648.

Li, Z., Wang, L., Wang, X., (2004) Compressive and flexural properties of hemp fiber reinforced concrete. Fibers Polym. 5 (3), 187–197.

Li Z., Wang X. and Wang L. (2006) Properties of hemp fiber reinforced concrete composites. Composites: Part A, 37, 497 – 505, doi:10.1016/j.compositesa.2005.01.032

Merta I., Tschegg E., Tschegg S., Kolbitsch A., (2011) Fracture Machanics of Concrete Reinforced With Hemp, Straw and Elephant Grass Fibres, Composite Materials: Key to the Future, 18th International Conference on Composite Materials, Jeju Island, Korea

Merta I., Kopecsko K., Tschegg E. (2012), Durability of hemp fibers in the alkaline environment of cement matrix, Fibre Reinforced Concrete: challenges and opportunities ISBN: 978-2-35158-132-2, 8th RILEM INTERNATIONAL SYMPOSIUM ON FIBRE REINFORCED CONCRETE (BEFIB 2012), Guimarães, Portugal

Merta, I. Tschegg, E.K. (2013) Fracture energy of natural fibre reinforced concrete, Original Research Article Construction and Building Materials, Volume 40, 991-997

Merta I. (2014) Hemp Fibre Reinforced Foam Concrete: Discussion of the Wedge Splitting Test Method, International Journal of Research in Engineering and Technology, 0X, 1 - 5.

Mobasher B, (2011) Mechanics of Fiber and Textile Reinforced Cement Composites, ISBN 978-1-4398-0660-9, CRC Press, Taylor & Francis Group

Müssig J. (2010) Industrial applications of natural fibres: structure, properties and technical applications (Wiley series in renewable resource). West Sussex (United Kingdom): John Wiley & Sons.

ÖNORM B 3592 Determination of cut-through-tensile splitting strength and specific fracture energy of building materials, combinations of building materials and composites - Wedge splitting method

Ranalli P. (1999) Agronomical and physiological advances in hemp crops. Advances in hemp research. New York: The Haworth Press, 61–84.

Savastano, J.H., Warden, P.G., Coutts, R.S.P., (2000) Brazilian waste fibres as reinforcement for cement-based composites. Cem. Concr. Compos. 22, 379–384.

Sedan, D., Pagnoux, C., Smith, A., Chotard, T., 2008. Mechanical properties of hemp fibre reinforced cement: Influence of the fibre/matrix interaction. J. Eur. Ceram. Soc. 28, 183–192

Sierra Beltran MG. (2011) Ductile cement-based composites with wood fibres – material design and experimental approach. PhD Thesis. ISBN 978-90-9026518-6. TU Delft, Netherlands.

Silva FA, Tolêdo Filho RD, Melo Filho JA, Rego Fairbairn E.M. (2010) Physical and mechanical properties of durable sisal fiber–cement composites. Constr Build Mater; 24(5): 777–85.

Silva FA., Mobasher B, Tolêdo Filho RD. (2010) Fatigue behavior of sisal fiber reinforced cement composites. Mat Sci Eng: A; 527(21–22): 5507–13.

Swamy, R.N., (1990) Vegetable fibre reinforced cement composites – a false dream or a potential reality? In: Sobral, H.S. (Ed.), Proceedings, 2nd International Symposium on Vegetable Plants and their Fibres as Building Materials. Rilem Proceedings 7. Chapman and Hall, London, pp. 3–8.

Tolêdo Filho R D; Ghavami K; England G L; Scrivener K (2003). Development of vegetable fibre-mortar composites of improved durability. Cement & Concrete Composites, 25, 185–196.

Torgal F.P., Jalali S (2011), Fibrous and Composite Materials for Civil Engineering Applications Natural fiber reinforced concrete, University of Minho, Portugal, edited by R Fangueiro, Woodhead Publishing Limited

Troëdec,M., Peyratout, C.S., Smith,A., Chotard, T., (2009).Influence of various chemical treatments on the interactions between hemp fibres and a lime matrix. J. Eur. Ceram. Soc. 29, 1861–1868.

Tschegg, E.K., (1986) Equipment and appropriate specimen shapes for tests to measure fracture values. Austrian Patent Nr. 390328, Austrian Patent Office, 31.1.1986.

Wei J., Meyer C., (2015) Degradation mechanisms of natural fiber in the matrix of cement composites, Cement and Concrete Research, Volume 73, pp 1-16, ISSN 0008-8846

Tensile and Bond Characterization of Natural Fibers Embeeded in Inorganic Matrices

Bahman Ghiassi, Arezou Razavizadeh, Daniel V. Oliveira, Vera Marques and Paulo B. Lourenço

Abstract Innovative composite materials made of continuous fibers embedded in mortar matrices have been recently received attention for externally bonded reinforcement of masonry structures. In this regards, application of natural fibers for strengthening of the repair mortars is attractive due to their low specific weight, sustainability and recyclability. This paper presents experimental characterization of tensile and pull-out behavior of natural fibers embedded in two different mortar-based matrices. A lime-based and a geopolymeric-based mortar are used as sustainable and innovative matrices. The obtained experimental results and observations are presented and discussed.

Keywords Flax fiber · Reinforced mortar · Mechanical properties · Bond characteristics · Pull-out test

Introduction

Innovative composite materials made of continuous fibers embedded in mortar matrices have been recently received attention for externally bonded reinforcement of masonry structures. These materials are usually identified as FRM, TRM, or TRC in the literature. Fiber Reinforced Mortars (FRMs) are proposed to overcome or reduce the disadvantages related to the use of FRPs such as durability and compatibility with the masonry substrate (D'Ambrisi et al. 2013; Carozzi et al. 2014; De Felice et al. 2014). Mortar based matrices exhibit a significant heat resistance, allows vapor permeability, can be applied either at low temperatures or

B. Ghiassi (✉) · A. Razavizadeh · D.V. Oliveira · P.B. Lourenço
ISISE, University of Minho, Guimarães, Portugal
e-mail: bahmanghiassi@civil.uminho.pt

V. Marques
University of Minho, Guimarães, Portugal

R. Fangueiro and S. Rana (eds.), *Natural Fibres: Advances in Science and Technology Towards Industrial Applications*, RILEM Bookseries 12,
DOI 10.1007/978-94-017-7515-1_23

on wet surfaces, have a better compatibility with masonry substrates and allow for reversibility. The latter is of great importance in case of strengthening of historical structures.

FRMs are usually constructed with continuous fibers embedded in cementitious mortars for application to concrete structures (Mobasher 2012). Lime-based mortars are however preferable in case of masonry and historical structures as they are more sustainable and compatible with the substrate (De Felice et al. 2014). The fabrics are usually in the form of grids made of multifilament carbon, glass or steel yarns. The use of natural fibers such as flax, hemp and jute has also received recent interest for strengthening of masonry structures (Codispoti 2013). Application of natural fibers in Civil Engineering applications is attractive due to their low specific weight, sustainability and recyclability, although durability of these systems still remains open.

Despite the recent interest on the use of FRMs for strengthening purposes, information regarding their mechanical performance and long-term durability is still lacking. The performance of FRM strengthened masonry elements is strongly dependent on the mechanical properties of FRM, the bond behavior between the fibers and mortar matrix, and the bond behavior between the mortar and the masonry substrate (Sueki et al. 2007; Carozzi and Poggi 2015).

This paper presents preliminary results of an experimental campaign aimed at mechanical characterization of FRMs made of flax natural fibers. The tests include tensile characterization of FRM composite coupons and pull-out tests for characterization of the fiber/mortar bond behavior. A pozzolanic lime-based and a geopolymeric-based mortar were used as two sustainable and innovative matrices. Lime-based mortar was selected as a commercially available mortar usually used for strengthening of masonry structures. On the other hand, it has been shown that natural fibers embedded in lime-based mortars present better performance under aggressive environmental conditions (Olivito et al. 2014). Geopolymeric-based mortar was produced in the laboratory condition as an innovative mortar which has received extensive attention during the last years but their durability and long-term performance are still unknown.

Tensile tests were performed on flax single yarn, flax fabric and flax fabrics embedded in mortar. Six specimens were prepared and tested for each material model. The pull-out tests were performed on flax fabrics embedded in the mortar with 50 and 100 mm bonded lengths. Five specimens were prepared for each mortar type and bonded length resulting in total twenty pull-out tests. The results were presented and discussed critically throughout the paper.

Materials and Specimens

The materials consisted of a commercial pozzolanic lime-based and a geopolymeric-based mortar as the matrix and a commercial bidirectional flax fabric as the reinforcing material. The pozzolanic lime-based mortar was a commercially available mortar selected as a sustainable solution for strengthening of masonry and historical structures. The geopolymeric-based mortar was produced in the laboratory as an innovative and sustainable matrix based on the activation of alkaline materials rich in silica and alumina, whose mechanical properties can be considerably higher than those of conventional binders.

Flax fiber specimens were prepared for mechanical characterization of materials and characterization of bond at fiber/mortar interface. Mechanical characterization tests included compressive and flexural tests on mortar and tensile tests on non-impregnated flax yarn filament and fabric, and fiber reinforced mortar specimens. The specimens were prepared and stored in the laboratory conditions (20 °C and 60 % R.H.) for three months before performing the mechanical tests to assure that the mortar was completely cured. Cylindrical specimens with 50 mm diameter and 100 mm height and prismatic specimens with dimensions of $40 \times 40 \times 160\ mm^3$ were prepared for mortar compressive and flexural tests, see Fig. 1. The specimens were prepared according to the specifications provided by EN 1015-11 (2007).

The single yarn and fabric (with 50 mm width) specimens were cut with 250 mm length, Fig. 2. 50 mm aluminum tabs were glued at both ends of the specimens for gripping the specimens during the tests which led to reduction of the specimens' free length to 150 mm. The fiber reinforced mortar specimens were prepared according to the geometrical details shown in Fig. 3.

The geometrical details of the pull-out tests specimens are shown in Fig. 4. Fiber fabrics with 50 mm width were embedded in the cylindrical mortar specimens with different bonded lengths. Again, aluminum tabs were used at the end of the fabric free end for gripping the specimens. The free length of the fabrics was impregnated with epoxy resin to avoid unfortunate failure during the tests. Two sets of specimens with different bonded lengths of 50 and 100 mm were prepared for each mortar type. The cylinders height was chosen equal to the bonded length according to the shown details.

(a) (b)

Fig. 1 Mortar test specimens: **a** prismatic samples; **b** cylindrical samples

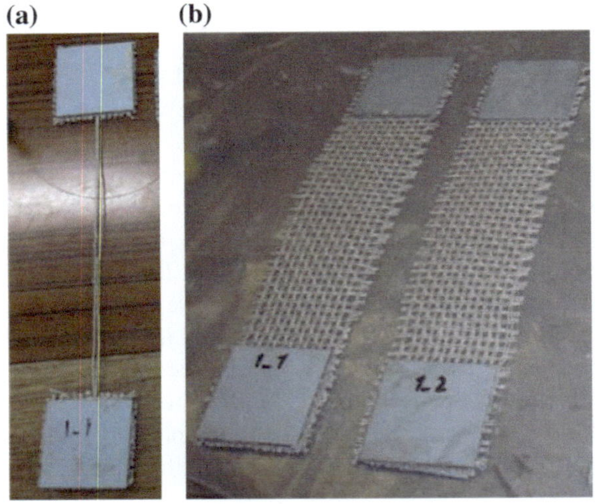

Fig. 2 Specimen details for tensile tests: **a** single yarn; **b** fabric test

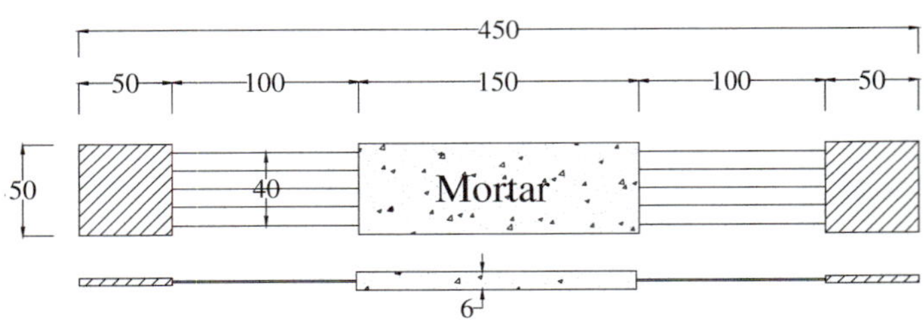

Fig. 3 Geometrical details of fiber reinforced mortar specimens

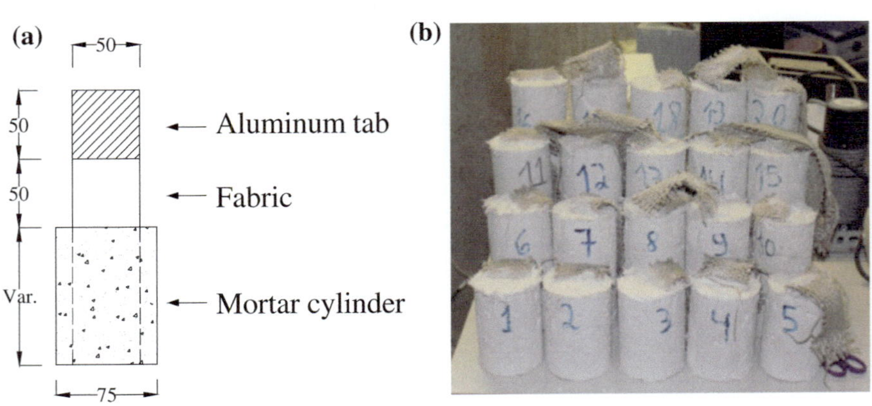

Fig. 4 Pull-out test specimens: **a** geometrical details; **b** specimens

Test Methods

The tests included compressive and flexural tests on mortar specimens, direct tensile tests on single flax yarns, flax fabric and fiber reinforced mortar specimens and fiber pull-out tests.

The mortar compressive and three-point bending tests were performed by means of a universal testing machine under force controlled conditions and according to EN 1015-11 (2007), see Fig. 5.

The tensile tests were performed on single flax yarn, flax fabric and fiber reinforced mortar specimens by means of a universal testing machine, see Fig. 6. All the tests were performed under displacement control conditions at the rate of

(a) (b)

Fig. 5 **a** Compressive tests on mortar cylinders; **b** three-point bending tests

(a) (b) (c)

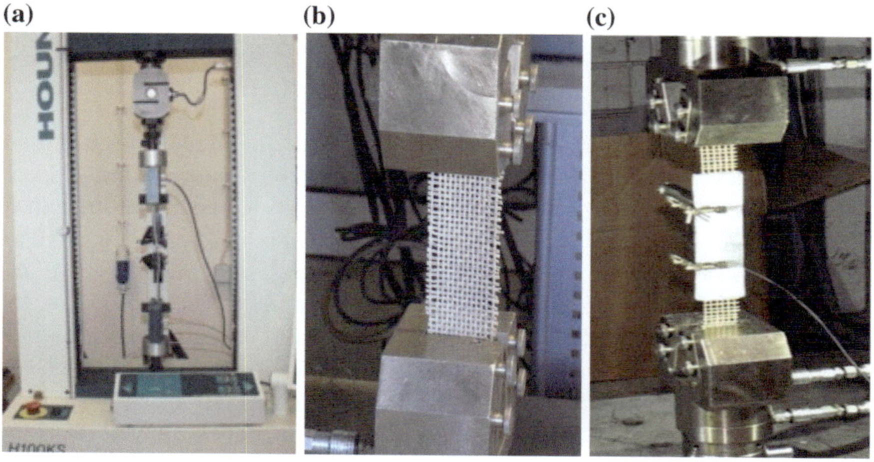

Fig. 6 Tensile tests setup: **a** single yarn tests; **b** fabric test; **c** FRM test

Fig. 7 Pull-out test setup

2 mm/min so that the results will be comparable. The applied load was recorded by a load cell integrated in the testing machine, while deformation was monitored by a clip gauge placed on the middle of the specimens.

Fiber pull-out tests were performed to characterize the bond behavior between flax fabrics and mortars. The specimens were fairly clamped to a supporting frame to avoid any movement during the tests, see Fig. 7. The fabrics were pulled from the mortar monotonically with a velocity of 0.3 mm/min. The tests were driven under displacement control conditions using a LVDT placed at the loaded end of the flax fabric. The resulting load was measured by means of a load cell.

Results and Discussion

Mortar Properties

The experimental mechanical properties of the mortars (three months curing) are listed in Tables 1 and 2. The values are presented as average of six tested specimens. It can be observed that the lime-based mortar has a compressive strength of 12.7 MPa, almost two times higher than the geopolymeric-based mortar (6.5 MPa). The elastic modulus of the mortars is 3247.5 and 7480.1 MPa for lime-based and geopolymeric-based mortars, respectively. The flexural tensile strength of the lime-based mortar, obtained from three-point bending test, is 6.1 MPa, being much larger than the geopolymeric-based mortar (1.12 MPa). The CoVs of the results are relatively small and in the acceptable range of materials testing. Although the mechanical properties of the geopolymeric-based mortar is lower than the lime-based mortar, it seems suitable for application to historical and masonry structures.

Table 1 Mortars compressive properties

Specimen	E (MPa)	CoV (%)	f_c (MPa)	CoV (%)
Lime-based mortar	3247.5	14.6	12.7	11.8
Geopolymeric-based mortar	7480.1	8.8	6.5	4.1

Table 2 Mortars tensile flexural properties

Specimen	f_t (MPa)	CoV (%)
Lime-based mortar	6.1	15.2
Geopolymeric-based mortar	1.23	16.8

Tensile Tests Results

The summary of the results obtained from the tensile tests is presented in Table 3. Higher stiffness, but lower tensile strength is observed in the fabric specimens in comparison to single yarn tests. On the other hand, the tensile strength has increased in the fiber reinforced composite specimens in comparison to the bare flax fabrics. This can be due to the increased integrity of the system due to the mortar presence in the embedded specimens. The stiffness and tensile strength of both composite specimens (lime-based and geopolymeric-based) are similar.

The stress-strain curves of the FRM specimens are shown in Fig. 8. It should be noted that these curves are obtained based on the total deformation of the specimens

Table 3 Uniaxial tension test results

Specimen type	E (MPa)	CoV (%)	f_t (MPa)	CoV (%)
Single yarn	5913.6	18.0	193.7	3.5
Fabric	8182.6	12.4	134.4	5.3
FRM (LM)	5454.4	6.2	259.8	5.1
FRM (GPM)	5636.5	13.9	237.2	18.2

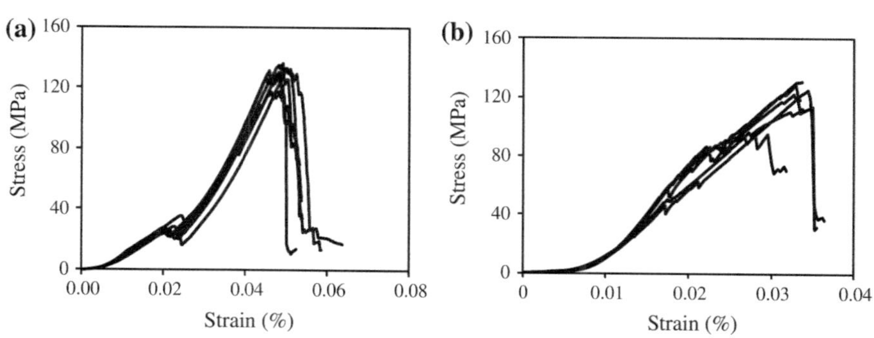

Fig. 8 Tensile stress-strain curves obtained for fiber reinforced mortar specimens: **a** lime-based composite; **b** geopolymeric-based composite

Fig. 9 Typical failure mode of the specimens in tensile tests

consisting of the reinforced and unreinforced parts, see Fig. 3. In general, it can be observed that the lime-based composites have higher stiffness after mortar cracking in comparison to geopolymeric-based composite. The typical failure mode of the specimens was tensile rupture of the fibers after cracking of the mortar layer. The cracks were uniformly distributed along the specimens, as can be seen in Fig. 9, showing a good stress transfer between the fibers and mortar in both mortar types. As aluminum tabs were used at both ends of the specimens, no fiber slipping occurred during the tensile tests.

Pull-Out Tests Results

The summary of pull-out test results is presented in Table 4. The results are presented as the average of five tested specimens. The CoV of the obtained results is in an acceptable range of pull-out tests. It can be also observed that the matrix type and bonded length significantly affect the pull-out strength of the specimens.

In the specimens with lime-based mortar (FRM-LM), slipping of the fibers is observed when the bonded length is 50 mm. However, in the specimens with 100 mm bonded length, slipping of the fibers occurred in the beginning of the tests followed by the fibers tensile rupture. The maximum pull-out force, F_{max} in Table 4, has increased from 0.39 kN in the specimens with 50 mm bonded length to 1.26 kN

Table 4 Pull-out test results

Bond length (mm)	FRM (LM)			FRM (GPM)		
	F_{max} (kN)	CoV (%)	Failure mode*	F_{max} (kN)	CoV (%)	Failure mode*
50	0.39	18.85	S	0.68	14.92	S + R
100	1.26	15.84	S + R	0.82	13.85	S + R

**S:* Fibers slipping
**R:* Fibers tensile rupture

in the specimens with 100 mm bonded length. This increase of the maximum pull-out force together with the observed change of failure mode can be the evidence of having the effective bonded length between 50 and 100 mm.

In the specimens with geopolymeric-based mortar (FRM-GPM), the tensile rupture of the specimens has occurred in both bonded lengths. Therefore, the maximum pull-out force was not expected to change significantly. The observed small difference between the observed values can be attributed to the materials and specimens variability. According to the observed experimental failure modes, the effective bonded length in the specimens with geopolymeric-based mortar is less than 50 mm.

Conclusions

Use of inorganic based matrices, especially cement-free or lime-based, are highly recommended for strengthening of masonry and historical structures. This paper presented the results obtained from mechanical characterization of mortar-based composites made of natural fibers and two innovative matrices. The tests included mechanical characterization of mortar as well as tensile and pull-out bond tests on flax fiber reinforced mortar specimens. The selected mortars were a pozzolanic lime-based mortar and a geopolymeric-based mortar. The lime-based mortar was a commercially available mortar used for strengthening of masonry structures, while the geopolymeric-based mortar was produced in the laboratory as a sustainable and innovative matrix. The tensile tests were performed on single yarn, fabrics and fiber reinforced mortar specimens. Fiber reinforced mortar specimens showed higher tensile strength in comparison to the bare fabrics due to the presence of mortar. The failure occurred in the specimens due to tensile rupture of the fibers after a uniform development of tensile cracks along the mortar layers. Characterization of the tension stiffening behavior of FRMs made of natural fibers with the aim of full field measurement techniques is of great interest for numerical modeling approaches at this stage and can be the subject of future studies. The results from the pull-out tests showed that the mortar type can significantly affect the bond performance and failure mode. The effective bond length in both sets of specimens was observed to be in less than 100 mm. A comprehensive characterization of the bond-slip behavior and development of suitable bond constitutive laws are still required.

Although the geopolymeric based mortar had lower mechanical properties in comparison to the lime-based mortar, the tensile and pull-out behavior seems satisfactory for application to historical masonry. Use of natural fibers for development of innovative composite materials seems interesting although issues such as wetability of the fibers and durability of the system remain open.

Acknowledgments The first author gratefully acknowledges the funding by Fundação para a Ciência e a Tecnologia (FCT), under the grant SFRH/BPD/92614/2013. The assistance of the technicians of the Structures Laboratory (LEST), where the experimental work took place, is also acknowledged.

References

D'Ambrisi A, Feo L, Focacci F. Experimental analysis on bond between PBO-FRCM strengthening materials and concrete. Compos Part B 2013;44:524–32.

Carozzi FG, Milani G, Poggi C. Mechanical properties and numerical modeling of Fabric Reinforced Cementitious Matrix (FRCM) systems for strengthening of masonry structures. Compos Struct 2014;107:711–25.

[3] De Felice G, De Santis S, Garmendia L, et al. Mortar-based systems for externally bonded strengthening of masonry. Mater Struct 2014;47:2021–37.

Mobasher B. Mechanics of Fiber and Textile Reinforced Cement. CRC Press, 2012.

Codispoti R. Mechanical performance of natural fiber-reinforced composites for the strengthening of ancient masonry. PhD thesis, University of Calabria, 2013.

Sueki S, Soranakom C, Mobasher B, Peled A. Pullout-slip response of fabrics embedded in a cement paste matrix. J Mater Civ Eng 2007;19:718–27.

Carozzi FG, Poggi C. Mechanical properties and debonding strength of Fabric Reinforced Cementitious Matrix (FRCM) systems for masonry strengthening. Compos Part B 2015;70:215–30.

Olivito RS, Cevallos OA, Carrozzini A. Development of durable cementitious composites using sisal and flax fabrics for reinforcement of masonry structures. Mater Des 2014;57:258–68.

EN 1015-11. Methods of test mortar for masonry. Part 11: Determination of flexural and compressive strength of hardened mortar. 2007.

Eco-Efficient Earthen Plasters: The Influence of the Addition of Natural Fibers

José Lima and Paulina Faria

Abstract Clayish earth-based mortars are being recognized, all over the world, as eco-efficient products for plastering. Apart from being a product with low embodied energy when compared to other types of plasters, their application on the interior surface of walls may give a strong contribution for the health and comfort of inhabitants. As part of an ongoing research regarding earth-based plasters this work assesses the influence of the addition of two types of natural fibers—oat straw and typha fiber-wool—on the characteristics of plastering mortars made with a clayish earth. Mechanical and physical characteristics were tested, showing that addition of these fibers contribute to decrease linear drying shrinkage and thermal conductivity, as well as promoting the adhesion strength of plaster to the substrate. The improvement of mechanical resistance reveal to be dependent on the type of fiber added while the hygroscopic capacity of the plaster is maintained regardless of the fiber additions.

Keywords Natural fibers · Earthen plasters · Mechanical properties · Hygroscopic capacity

Introduction

There is an increasing international interest regarding earth-based mortars due to its recognition as eco-efficient products, namely for indoor plastering, since they can contribute to improve some import aspects of building performance related to health and comfort of inhabitants, as well as building sustainability.

J. Lima (✉)
Faculty of Architecture, University of Lisbon, Alto da Ajuda, Portugal
e-mail: jose.lima.ferreira@gmail.com

P. Faria
CERIS-ICIST, Department of Civil Engineering, NOVA University of Lisbon, Caparica, Portugal

R. Fangueiro and S. Rana (eds.), *Natural Fibres: Advances in Science and Technology Towards Industrial Applications*, RILEM Bookseries 12,
DOI 10.1007/978-94-017-7515-1_24

Earth-based mortars can contribute to building life cycle sustainability mainly due to its low embodied energy. That derives from the use of raw clay as a natural binder, without the need of any heat treatment, and from wide availability of local resources of clayish earth all over the globe, allowing the mitigation of ecological impact of long distance transportations (Melià et al. 2014). Another contribution of earth-based mortars for building life cycle sustainability comes from the fact that earth-based mortars, produced without addition of any chemical stabilizer like cement or lime, can be easily recycled just with low consumption of mechanical energy for grinding and the addition of water. Furthermore the use of clayish earth as raw material does not directly lead to any pollutant emissions during building life cycle, therefore mitigating contamination hazards even in an eventual end of life disposal (Minke 2006).

Due to high hygroscopicity of the clay minerals (Botelho-Costa 1973; Gomes 1988), earth-based mortars present a high adsorption and desorption capacity, particularly when compared to other type of mortars for interior plastering (Minke 2006). This capacity allows earth-based plasters to act as a moisture buffer, contributing to balance the relative humidity of the indoor environment of buildings (Minke 2006; Maddison et al. 2009; Liuzzi et al. 2013).

This high humidity buffering capacity of earth-based plasters promotes the comfort and health of inhabitants. Firstly because a high relative humidity environment increase the discomfort associated with the perception of heat or cold (Moret-Rodrigues et al. 2009). Secondly because, by balancing the relative humidity of the indoor environment, earth-based plasters contribute, in one hand, to mitigate health conditions associated with high relative humidity, like infections, allergies or asthma, and in another hand, contribute also to mitigate the probability of mucosal membranes irritation and inflammation associated to exceedingly dry indoor environments (WHO 2009).

Earth-based plasters may also contribute to indoor air quality since clay can act as a passive removal material (PRM), lowering indoor ozone concentrations, and therefore lowering the probability of occurrence of indoor ozone reaction with other building materials. This may lead to lower concentrations of oxidized reactions products, which could be toxic and irritating to mucosal membranes and other tissues (Lamble et al. 2011; Darling et al. 2012).

Despite the increase of interest in earth-based plasters there is still no European standards for this type of mortars. However, in Germany, the *Deutsches Institut für Normung* (DIN) recently released DIN 18947 (DIN 2013), a national standard specifically devoted to earth-based mortars. Thenceforth that standard has been followed by several scientific studies regarding the assessment of mechanical and physical properties of this type of mortars (Delinière et al. 2014; Faria et al. 2015; Lima and Faria 2014).

As part of an ongoing research regarding earth-based plasters specifically formulated with clayish earth extracted from Algarve sedimentary basin, which revealed a high potential for interior plastering (Lima and Faria 2014), the present study also follows that DIN standard and aim to assess the addition of two natural fibers to the plastering mortar, namely oat straw and cattail seed fibers (*Typha latifolia L.*), also known as typha fiber-wool. While oat straw is a relatively common raw material in earthen architecture construction systems, such as adobe, cob or wattle and daub, the use of typha fiber-wool as a reinforcement for earth-based plasters only got the focus of international scientific community in recent years (Maddison et al. 2009; Georgiev et al. 2013).

For this assessment six mortars were formulated with the same volumetric proportions of clayish earth and siliceous sand, respectively 1:3. One was considered the reference mortar, without any fiber addition, while the other five mortars were added with different amounts of cut oat straw or typha fiber-wool. The mortars were assessed in fresh state considering water contents, wet density, flow table consistency. The plaster samples were assessed in harden state, considering linear shrinkage, flexural and compressive strength, adhesive strength, thermal conductivity and adsorption and desorption capacity.

Materials

All the six mortars assessed in this study were formulated with the same clayish earth extracted from a clay quarry located in the East sector of *Barrocal* sub-region of Algarve, in Southern Portugal. This sub-region is set in the highest area of Algarve sedimentary basin and presents a high concentration of clayey soils with a mineralogy dominated by the ilite clay mineral, mainly as a consequence of sedimentogenesis process based in a marine environment. The selected clay quarry integrates a group of clay quarries located in geologic formations from the Rhaetian/Hettangian age, in the transition from Triassic to Jurassic period, that present the very high predominance of the ilite clay mineral (Manuppella et al. 1985; Lima and Faria 2014).

The ilite clay mineral concentration was the key factor to select the clayish earth for the development of this study since it is characterized by a significant water vapor adsorption capacity, combined with low swelling when wetted. These characteristics are due to ilite alumino-silicate crystalline structure arranged in a succession of tetrahedron/octahedron/tetrahedron layers, with a interlayer space mainly occupied by potassium cations responsible for the lowered swelling (Botelho-Costa 1973; Gomes 1988). These properties are most important for interior plastering because maximize the moisture buffering capacity of the plaster and simultaneously mitigate the chance of the occurrence of shrinkage cracking during their drying.

The clayish earth was extracted and was mechanically disaggregated and sieved at 2 mm. The particle size distribution of the disaggregated clayish earth was analyzed by dry sieving according to standard EN 1015-1 and is presented in Fig. 1.

The siliceous sand selected for developing this experiment was extracted from a quarry located in *Santiago do Cacém*, in *Setubal* region, South West cost of Portugal. The particle size distribution of the selected sand was analyzed by dry sieving, as used for mortar's formulation, according to standard EN 1015-1 and is presented in Fig. 1.

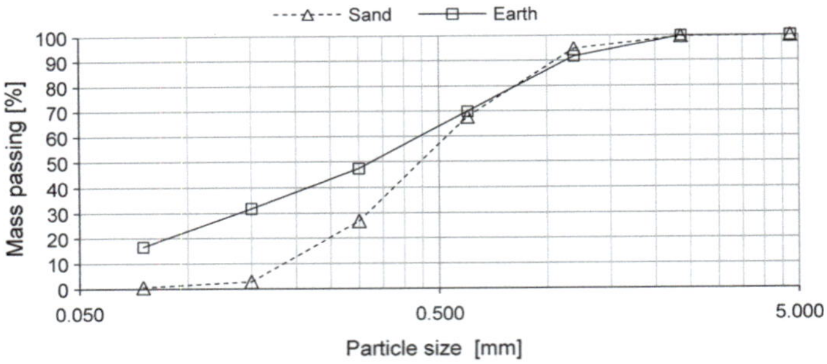

Fig. 1 Particle size distribution curves of clayish earth and sand by dry sieving

The oat straw fibers added to two of the mortars were collected in a farm straw supplier and were cut to a maximum length of 20 mm. After cutting, the oat fibers were allowed to dry in open-air, inside plastic containers protected with a fine aluminum mesh (<2 mm), during two summer weeks in Algarve hot and dry climate.

The typha spadixes that provided the typha fiber-wool for this experiment were harvested at the end of plant vegetative cycle (autumn) in a free water stream in hills Northeast sub-region of Algarve, where *Typha latifolia L.* is widely present in water stream ecosystems (Fig. 2).

After harvesting the typha spadixes were also allowed to dry in open-air, inside plastic containers protected from the wind with a fine aluminum mesh on top (<2 mm), till complete expansion of seeds (Fig. 3). The expansion process of the typha seeds is spontaneous under certain conditions of warm temperature and dry relative humidity, and provide a ready-made dry fiber that can be easily added to mortar formulations dry mixture as a reinforcement, in order to mitigate the chance of occurrence of shrinkage cracking during the drying of the plaster (Maddison et al. 2009; Georgiev et al. 2013).

Fig. 2 **a** Free water stream in hills Northeast sub-region of Algarve, **b** Typha spadixes development, **c** Dry typha spadixes at the end of the vegetative cycle, **d** Typha seed spreading from a spadix

Fig. 3 Typha spadixes drying and expansion

Mortar's Formulations and Fresh State Characterization

For the development of this study six mortars were formulated with the same volumetric proportions of clayish earth and siliceous sand, respectively 1:3. One of the mortars, referenced as REF, was formulated without any fiber addition to be taken as a reference mortar, while the other five mortars were added with different amounts of cut oat straw and typha fiber-wool. Mortars referenced as OF10 and OF20 were added with volumes of 10 and 20 % of oat straw respectively (percentage of the total volume of clayish earth and sand). Mortars referenced as TF20, TF40 and TF80 were added with volumes of 20, 40 and 80 % of typha fiber-wool respectively.

The standard DIN 18947 (DIN 2013) was followed for preparing and mixing the mortars, as well as for fresh state characterization. The mortars were prepared with the minimum amount of water needed in order to achieve the flow table consistency defined in the DIN 18947 (175 ± 5 mm) and ensuring a very good workability. Their proportions are registered in Table 1, in terms of volumetric ratios and weight ratios of clayish earth, sand, fibers and water contents. Wet density and flow table consistency were assessed.

Hardened State Characterization

The hardened state characterization of the six mortars also followed the DIN 18947 (DIN 2013). Prismatic 160 mm × 40 mm × 40 mm samples and circular 90 mm diameter and 20 mm thick samples were prepared in metallic molds, as well as samples of 20 mm mortar layer applied on hollow bricks. The mortars were assessed in terms of linear drying shrinkage, density, flexural, compressive and adhesive strength, as well as thermal conductivity and water vapor adsorption and desorption capacity. The first four tests used the prismatic samples, for adhesive strength the mortar layer on brick samples were used, while the circular samples were used for thermal conductivity.

The dynamic adsorption and desorption test was also conducted according to the DIN 18947 (DIN 2013). For each mortar formulation three plaster samples were

Table 1 Composition and fresh state mortars characterization

Mortar	Volume ratios				Weight ratios					
	Earth (%)	Sand (%)	Fiber[a] (%)	Water[a] (%)	Earth (%)	Sand (%)	Fiber (%)	Water[b] (%)	Density[c] (kg/m^3)	Consistency[d] (mm)
REF	25.0	75.0	0.0	19.6	21.6	78.4	0.0	12.8	2130.7	173.2
OF10	25.0	75.0	10.0	22.1	21.5	78.0	0.5	14.4	2007.0	169.6
OF20	25.0	75.0	20.0	25.0	21.4	77.6	1.0	16.2	1936.9	162.8
TF20	25.0	75.0	20.0	21.7	21.6	78.3	0.1	14.2	2101.9	172.4
TF40	25.0	75.0	40.0	22.5	21.6	78.3	0.1	14.7	2079.3	174.6
TF80	25.0	75.0	80.0	25.6	21.6	78.2	0.3	16.7	2054.9	171.3

[a]Percentage of volume added considering the total volume of clayish earth and sand
[b]Percentage of mass added considering the total mass of clayish earth, sand and fibers
[c]Fresh state density
[d]Flow table consistency

Fig. 4 Climatic chamber with mortar samples for adsorption and desorption test

produced, with a surface area of 1000 cm^2 (500 mm × 200 mm) and a thickness of 15 mm. Each sample was prepared in a metallic molds to guarantee that adsorption and desorption would occur only in the top exposed surface.

The samples were stabilized in a climatic chamber (Fig. 4) at 50 % relative humidity (RH) and 23 °C. After the stabilization of the samples the climatic chamber condition was set to 80 % RH for the adsorption test phase. The samples were weighted at time intervals defined on the standard, respectively: 0.5, 1.0, 3, 6 and 12 h. The adsorption test was extended till 24 h, beyond the 12 h interval defined in the standard, in order to achieve a more comprehensive understanding of the adsorption behavior of the samples. After 24 h the samples were weighted and the condition of the chamber was changed back to 50 % RH, forcing the samples to a desorption phase, that was assessed with the same time interval protocol during another period of 24 h.

Results and Discussion

The flow table consistency of all mortars complied with DIN 18947 (DIN 2013) and it can be seen that the addition of both types of fibers decreased the wet density (Table 1). The decrease was lower with the addition of the typha-wool, most probably due to the small size of those fibers and the densification imposed on the mortar matrix, when compared with the coarser oat fibers.

The hardened state characterization reveals that all the six mortars present a significantly low linear drying shrinkage (less than 1.0 %—Fig. 5), which is consistent with the low expansibility of the clayish earth used, which has a mineralogy dominated by illite clay mineral. As expected the addition of fibers to the mortars formulations contribute to decrease the samples linear shrinkage when compared

Fig. 5 Linear drying shrinkage

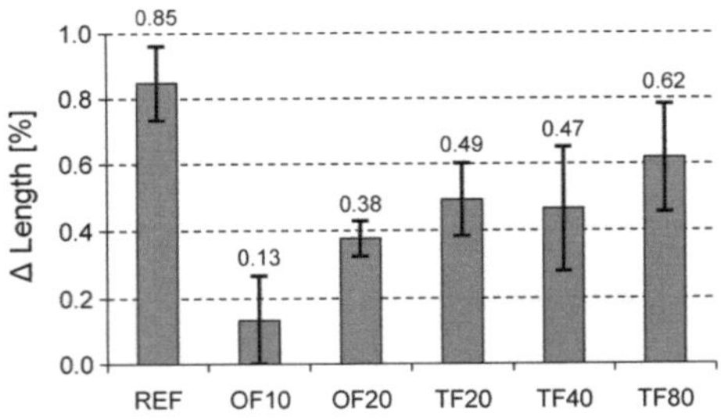

with the reference mortars. However the increase of the fiber addition reveals the opposite effect, most probably due to the correspondent necessary increase of water to the mortar formulation, in order to achieve workability and flow table consistency defined in DIN 18947 (Fig. 5).

Regarding hardened state bulk density (Fig. 6) according to DIN 18947 (DIN 2013), it can be noticed the same tendency of the wet density. The reference mortar, formulated only with clayish earth and sand, falls within density class 2.0 (from 1.81 to 2.0 kg/dm^3). As expected the two mortars with addition of 10 and 20 % of oat straw present lower bulk density than the reference mortar, belonging to density class 1.8 (from 1.61 kg/dm^3 till 1.8 kg/dm^3), while the three mortars with addition of 20, 40 and 80 % of typha fiber-wool did not present a significant bulk density decrease, falling within the same density class of the reference mortar.

The thermal conductivity test results (Fig. 7) reveals, as expected, that the increase of the added volume of oat fibers promote the decrease of the thermal conductivity of the mortars, which is consistent with the decrease of the mortars

Fig. 6 Hardened state bulk density

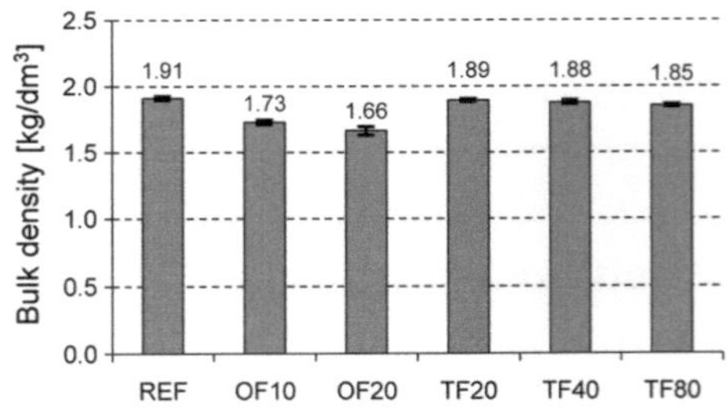

Fig. 7 Thermal conductivity

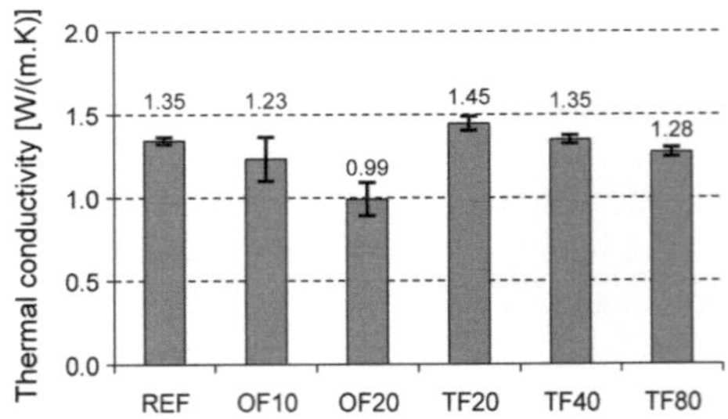

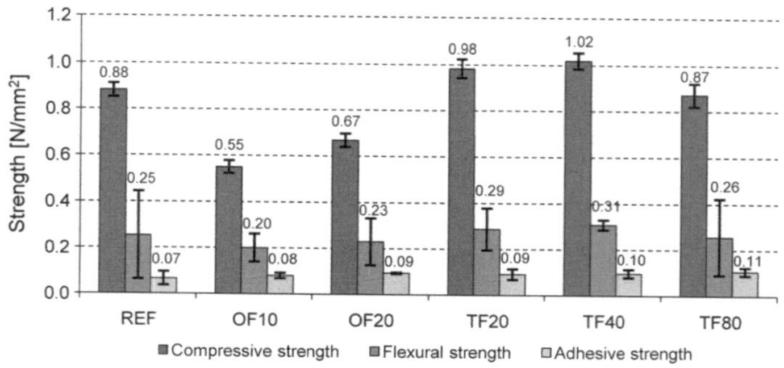

Fig. 8 Compressive, flexural and adhesive strength

bulk density. The same can be observed on mortars with the addition of typha fiber-wool; however when the results obtained by those mortars are compared with the reference mortar it can be remarked that the mortar with the lowest typha fiber-wool volumetric addition achieved a slightly higher thermal conductivity than the reference mortar. Again, that may be due to a densification of the matrix pores with the addition of a small amount of this type of fibers.

Regarding flexural and compressive strength (Fig. 8), mechanical tests showed that the reference mortar did not achieve (however by a small difference) the minimum values of mechanical resistance defined in the class S-I of the DIN 18947 (compressive strength ≥ 1.0 N/mm^2 and flexural strength ≥ 0.3 N/mm^2). Concerning the addition of fibers, these mechanical tests reveal that the two types of natural fibers assessed on this study have a significantly different effect over flexural and compressive strength of the mortars.

The addition of oat straw remarkably decreased both flexural and compressive strength (from 9 to 21 % on flexural strength and from 24 to 38 % on compressive strength). This result was not expected and it seems to be related with the lower bulk density of these mortars, which leads to mortars with reduced volumetric clay contents. However, and as expected, the increase of added volume of oat fiber, from 10 to 20 %, promotes some improvement in both flexural and compressive strength.

The mortars with additions of 20 and 40 % volume of typha fiber-wool showed an increase on both flexural and compressive strength (respectively, an increase of 13 and 22 % on flexural strength and an increase of 11–15 % on compressive strength), while the mortar with the addition of 80 % volume of the same fiber did not present significant differences. The results of the first two mortars are consistent with the expected reinforcement contribution of typha fiber-wool, and also with the bulk density presented by these mortars, similar to the density of the reference mortar. This fact allows to assume that their clay contents is also similar. The difference observed on mortar with an addition of 80 % volume of typha fiber-wool may be justifiable due to the substantial amount of added water needed to process this mortar in order to achieve workability and the range of flow table consistency defined in DIN 18947.

From among the six mortars assessed in this study only the mortar with an addition of 40 % volume of typha fiber-wool achieved the minimum values of mechanical resistance defined in class S-I of the DIN 18947 (compressive strength ≥1.0 N/mm^2 and flexural strength ≥0.3 N/mm^2). This flexural and compressive strength results suggest that the mechanical resistance of these mortars should be improved, which considering the low expansibility of the clayish earth used (and low shrinkage) could be achieved through the formulation of mortars with higher concentrations of clayish earth content.

Nevertheless, according to DIN 18947 (DIN 2013), the adhesion strength of all six mortars has complied with the minimum value of the resistance class S-I (≥0.05 N/mm^2). Furthermore the two mortars with higher addition of typha fiber-wool achieved the resistance class S-II (≥0.10 N/mm^2). These results showed also that all the mortars with the addition of natural fibers present higher adhesion strength than the reference mortar, what is extremely important for plastering mortars. And among those mortars it is also possible to observe a trend of a slight increase of adhesion strength along with the increase of fiber addition.

The results of the dynamic adsorption and desorption test (Fig. 9) indicate that all six mortars present a very high adsorption and desorption capacity. According to DIN 18947 (DIN 2013) all mortars assessed in this study achieved adsorption values within the water adsorption class WS-III, the higher adsorption class defined in the standard (≥60.0 g/m^2 of adsorbed water vapor after 12 h at temperature of 23 °C and a relative humidity of 80 %). The high adsorption and desorption capacity of the assessed mortars is consistent with previous research from the authors regarding the characterization of earth-based mortars formulated with different proportions of the same clayish earth, reveling clay content as the key factor on plaster adsorption and desorption capacity (Lima and Faria 2014).

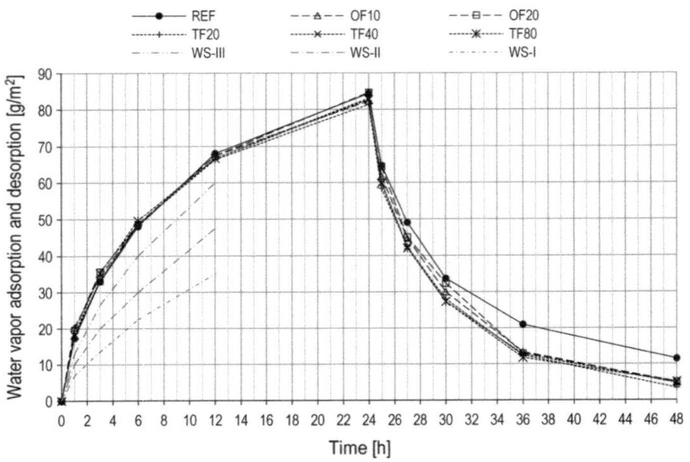

Fig. 9 Water vapor adsorption and desorption

These results also show that all the six mortars have a similar dynamic behavior of adsorption and desorption, however showing some desorption delay, or hysteresis, being this effect most pronounced on the reference mortar. This desorption delay can generate a lag effect between consecutive cycles of adsorption and desorption, which could lead, over a certain period of time, to the decrease of adsorption and desorption capacity of the plaster, or even to its full moisture saturation.

In this study, although the different additions of cut oat fiber and typha fiber-wool showed to contribute to decrease the desorption delay, it was not possible to foresee from this results a clear influence of the amount and type of fibers added to mortars on their adsorption and desorption capacity, since all mortars with fiber additions reveal similar adsorption and desorption rates over time.

Conclusions

The results from this study showed that the addition of cut oat straw or typha fiber-wool can contribute to decrease linear drying shrinkage of a earth-based plaster, even when a clayish earth with low expansibility is used. However from a certain proportion, the added fiber reveals the opposite effect, most probably due to the correspondent necessary increase of water to the mortar formulation.

As expected the increase of the added volume of oat straw or typha fiber-wool promote the decrease of the thermal conductivity of the mortars, which is consistent with the decrease of the mortars density. However more research is needed on this subject particularly regarding the influence of the addition of typha fiber-wool on mortar's microstructure.

The addition of the two types of natural fibers assessed on this study reveals to have a significantly different effect over flexural and compressive strength of mortars. The addition of oat straw remarkably decreased both flexural and compressive strength, while the addition of typha fiber-wool showed the opposite effect increasing both flexural and compressive strength. This may be due to the distinct particle size of the two types of fibers and different inclusion on the mortar's matrix. Furthermore only the mortar with an addition of 40 % volume of typha fiber-wool achieved the minimum values of mechanical resistance defined in DIN 18947 (DIN 2013). This results suggest that the mechanical resistance of these mortars should be improved, which, considering the low expansibility of the clayish earth used, could be achieved through the formulation of mortars with lower sand proportion and, therefore, higher clay content, what will be assessed in a future experimental campaign.

Nevertheless all the six mortars has complied with DIN 18947 (DIN 2013) regarding adhesion strength, with positive reference being made for the mortars with typha fiber-wool addition, which achieved a promising adhesion strength, one of the most important properties for plasters.

As expected the adsorption and desorption test showed that assessed mortars present a very high adsorption and desorption capacity. All the mortars, regardless the different added volume of the two vegetal fibers, reveal a similar dynamic behavior of adsorption and desorption; however, showing some desorption delay. In consecutive cycles of adsorption and desorption this desorption delay can lead to the decrease of adsorption capacity of the plaster, or even to their moisture saturation. More research is needed in this subject in order to assess the extent of the desorption delay effect over long periods in real conditions, and also regarding the influence of the amount of fibers added to mortars on their adsorption and desorption capacity.

Acknowledgments The authors gratefully acknowledge the help of Vitor Silva on the experimental campaign held at NOVA University of Lisbon.

References

Botelho-Costa J (1973) Caracterização e constituição do solo. 7ª edição. Lisboa: Fundação Calouste Gulbenkian. ISBN 972-31-0073-8.

CEN European Committee for Standardization (1998) Methods of test for mortar for masonry. Part 1: Determination of particle size distribution (by sieve analysis). EN 1015-1, Brussels, Belgium.

Darling EK, Cros CJ, Wargocki P et al (2012) Impacts of a clay plaster on indoor air quality assessed using chemical and sensory measurements. Build Environ., 57, p-370–376.

Delinière R, Aubert JE, Rojat F et al (2014) Physical, mineralogical and mechanical characterization of ready-mixed clay plaster. Build. Environ., 80, p. 11–17.

DIN Deutsches Institut für Normung (2013) Earth plasters – Terms and definitions, requirements, test methods. DIN 18947, Berlin.

Faria P, Santos T, Aubert J-E (2015) Experimental characterization of an earth eco-efficient plastering mortar. Journal of Materials in Civil Engineering (in press), Doi: 10.1061/(ASCE) MT.1943-5533.0001363.

Georgiev G, Theuerkorn W, Krus M, Kilian R (2013) Cattail-reinforced clay plasters in building heritage preservation and new constructions. In: Correia M, Carlos G, Rocha S, editors. Vernacular Heritage and Earthen Architecture. CRC Press, p. 415–20.

Gomes C (1988) Argilas: o que são e para que servem. 1ª edição. Lisboa: Fundação Calouste Gulbenkian.

Lamble SP, Corsi RL, Morrison GC (2011) Ozone deposition velocities, reaction probabilities and product yields for green building materials. Atmospheric Environ., 45(38), p. 6965–6972.

Lima J, Faria P (2014) Earthen plasters: The potential of the clayey soils of Barrocal Region in Algarve. In: 40th IAHS World Congress on Housing – Sustainable Housing Construction. (CD-Rom ID-217, ISBN 978-989-98949-1-4).

Liuzzi S, Hall MR, Stefanizzi P, Casey SP (2013) Hygrothermal behaviour and relative humidity buffering of unfired and hydrated lime-stabilised clay composites in a Mediterranean climate. Build. Environ., 61, p. 82–92.

Maddison M, Mauring T, Kirsimae K, Mander U (2009) The humidity buffer capacity of clay-sand plaster filled with phytomass from treatment wetlands. Build. Environ., 44, p. 1864–1868.

Manuppella G, Moreira J, Grade J, Moura A (1985) Contribuição para o conhecimento das características das argilas do Algarve. Porto: Estudos, Notas e Trabalhos do Serviço de Fomento Mineiro e Laboratório da D.G.G.M. Tomo 27, p.59–75.

Melià P, Ruggieri G, Sabbadini S, Dotelli G (2014) Environmental impacts of natural and conventional building materials: a case study on earth plasters. J. Clean. Prod., 80, p. 179–186.

Minke G (2006) Building with earth: design and technology of a sustainable architecture. Basel: Birkhauser - Publishers for Architecture. 199 p. ISBN-13: 978-3-7643-7477-8. ISBN-10: 3-7643-7477-2.

Moret-Rodrigues A, Canha-Piedade A, Braga A (2009) Térmica de edifícios. 1ª edição. Alfragide: Edições Orion. ISBN 978-972-8620-13-4.

WHO World Health Organization (2009) Guidelines for indoor air quality: dampness and mould. Copenhagen: World Health Organization - Regional Office for Europe. ISBN 978-92-890-4168-3.

Part V
Innovative Applications of Natural Fibres

Poly Lactic Acid Fibre Based Biodegradable Stents and Their Functionalization Techniques

Rita Rebelo, Nívea Vila, Sohel Rana and Raul Fangueiro

Abstract Commercial stents, especially metallic ones, present several disadvantages, and this gives rise to the necessity of producing stents with different materials, like natural polymers, in order to improve their biocompatibility and minimize the disadvantages of metallic ones. Another way to improve the biocompatibility and create a stent with less thrombogenic and inflammatory materials is through surface functionalization. This possibility allow to provide new functionalities to materials, like drug release or higher biocompatibility. One of the most used technique for stents functionalization is Layer by Layer (LBL). The principle of this technique is based on the alternating adsorption of materials with opposite charge or functional groups, in order to form integrated thin films. This review paper discusses some applications of natural-based polymers in stents, namely polylactic acid (PLA) for stent development and chitosan for biocompatible coatings of stents. Furthermore, some effective stent functionalization techniques will be discussed, namely Layer by Layer technique.

Keywords PLA fiber · Artificial stent · Chitosan · Biocompatible coating · Layer by layer technique

Introduction

Cardiovascular diseases (CVD) are the leading cause of death worldwide. Atherosclerosis of the coronary arteries is one of the most principal cardiovascular diseases and it is caused by the accumulation of cholesterol (fat produced by the body) in the arterial walls, obstructing blood flow. This accumulated cholesterol can

R. Rebelo (✉) · N. Vila · S. Rana · R. Fangueiro
Centre for Textile Science and Technology (2C2T), University of Minho, Guimarães, Portugal
e-mail: ritarebelo@det.uminho.pt

R. Fangueiro and S. Rana (eds.), *Natural Fibres: Advances in Science and Technology Towards Industrial Applications*, RILEM Bookseries 12,
DOI 10.1007/978-94-017-7515-1_25

cause a clot, leading to severe circulatory problems. Artificial stents receive great value due to these coronary artery diseases, as they are used as implants inside the blood vessels, in order to keep them open, until the blood circulation normalizes, and they are used in 70 % of angioplasties (Vila 2009).

Once stents are apply in cardiovascular area, they are subject to an extended set of requirements, in order to be safe. The main requirement include acceptable biocompatibility and hemocompatibility. Furthermore, it is extremely important to be ease of implantation, fabrication and sterilization, as well as mechanical strength, radiopacity, longitudinal flexibility, corrosion resistance and having high strength and high radial expansion ability to recover (Vila 2009; Commandeur et al. 2006; Dyet et al. 2000).

The approval of the first cardiovascular stent by the FDA (Food and Drug Administration) occurred in 1994. Since then, stents have been evolved, in terms of implantation mechanism, materials, methods of manufacture and coatings.

Stents can be divided into conventional or pharmacologic. The first ones are endovascular fasteners, whose function is to maintain dilated artery. Pharmacologic stents belong to the second generation of stents and are incorporated with drugs. These stents are available commercially since 2002 (Rebelo et al. 2014).

In most of the cases, stents can be found in stainless steel and others metal alloys. However, metal stents present some disadvantages like corrosion, restenosis and bleeding complications (França and Pereira 2008). Furthermore, after stent implementation, proteins from blood or tissue adsorb onto stent's surface and can lead to formation of a biofilm and cause a bacterial infection. Usually, this infection results in prolonged hospitalization and, therefore, a higher medical costs, or even in the death of the patient (França and Pereira 2008).

Despite the high initial success rate of bare metal stents, in order to combat the problems arising from conventional stents. Thus, the trying of prevent restenosis resulted in several biocompatible or therapeutic stent coatings, like heparin, endothelial cells, bare polymers, and drug-eluting polymers (Commandeur et al. 2006).

Biodegradable polymeric stents obtained from natural raw materials such as PLA (Poly (lactic acid)) and chitosan begun to be considered.

PLA is a biodegradable aliphatic polyester synthesized from lactic acid which is obtained from renewable sources such as milk, corn, sugarcane, tapioca and sugar beet, among other (Calabia and Tokiwa 2007). Due to its biocompatible and biodegradable nature, PLA has been widely used in biomedical applications in area as implants, sutures and drug delivery systems (Lili 2007). Lactic acid is one of the simplest chiral molecules and exists as two stereoisomers: D - and L-lactic acid. The L-form (levorotatory) differs from Form D (dextrorotatory) for their effect on polarized light, as observed in Fig. 1.

Chitosan (Fig. 2) is a cationic polysaccharide produced by deacetylation of chitin, a polysaccharide found in the exoskeleton of crustaceans, obtained from an alkalizing process at high temperatures (Dias et al. 2008).

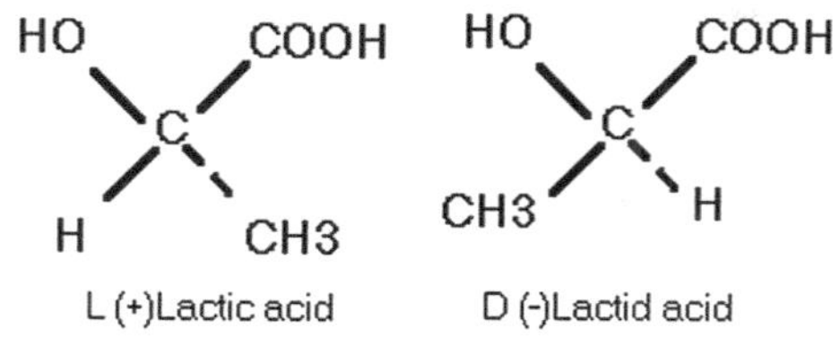

Fig. 1 Molecular structures of PLLA and PDLA

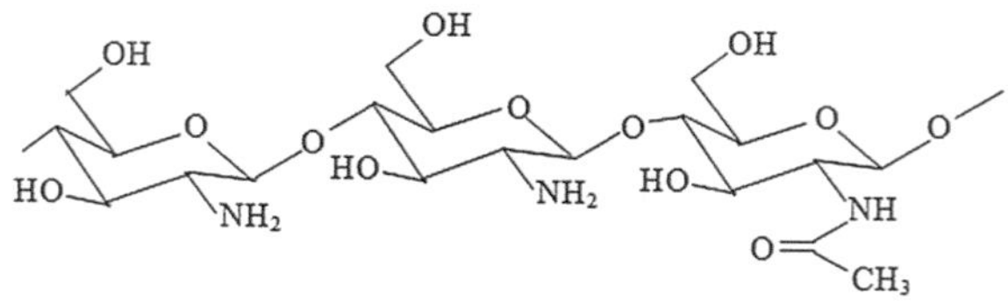

Fig. 2 Molecular structure of chitosan

Chitosan is a biodegradable and biocompatible natural polymer that has been widely used in different forms (gels, films, particles, membranes or scaffolds) in a large number of applications, ranging from biomedical to industrial areas (Balan and Verestiuc 2014).

Furthermore, due to its characteristics, chitosan has an important role in cell attachment and growth, and it is vastly used as matrix in tissue engineering, as well as in stent applications (Meng et al. 2007; Zhu et al. 2004; Reis et al. 2008).

The stent surface characteristics, such as, hydrophobicity, roughness, porosity and chemical composition, have a direct influence in proteins adsorption (Hoiby et al. 2010; Pavithra et al. 2008). Thus, it is, extremely, important to control and be able to modify the surface characteristics.

In order to minimize the reaction of the human body and fight the adhesion of microorganisms to stent surface. Stent coating alters the surface's chemistry and can be used per se or as a platform for local drug delivery (Commandeur et al. 2006).

The development of stent coatings with new materials, like natural polymers, is taking place (Knetsch and Koole 2003).

Natural-Based Stents

PLLA Stents

There is an explicit risk on the use of stents with metallic materials, once structure and composition of these materials provokes the risk of restenosis and can damage the blood vessels (Grag and Serruys 2010).

Thus, the use of stents made of biodegradable material has become an advantageous alternative. One of the most commonly used biodegradable polymers, due to their properties, is PLLA. The polymeric stent structures has a lower radial force when compared with metallic and should be compensated for thicker structures.

Some of natural based stents are already marketable (Fig. 3). Tamai et al. (2000) used PLLA stents in human cases (Igaki-Tamai stent). This stent has a thickness of 0.17 mm and has a zigzag helical coil pattern and were implanted in 15 patients. During 30 days no thrombosis or major cardiac events were observed, and follow-ups with angiography and intravascular ultrasounds were realized 1 day, 3 months and 6 months after stent implementation. No major cardiac event, except for repeat angioplasty, developed within 6 months (Tamai et al. 2000).

Nishio et al. (2012) evaluated the long-term safety of the Igaki-Tamai stent over 10 years, where 50 patients with 63 lesions were treated electively with 84 Igaki-Tamai stents.

Generally, clinical follow-up for cardiac events and rates of scaffold thrombosis was analyzed with the results of angiography and intravascular ultrasound.

At the end of the study, authors conclude that the rate of major cardiac events were similar to those of metallic stents, and no stent recoil was observed, suggesting a long-term safety of Igaki-Tamai stent (Nishio et al. 2012).

Absorb® (VHL) is a fully bioresorbable stent made of PLLA, coated with PDLLA that uses the everolimus drug, and it was approved in 2011 for use in Europe. The second generation of this stent (BVS 1.1) uses PLLA, as well, however, has some modifications in polymer processing and scaffold design that leads to a higher radial strength and mechanical integrity (Grag and Serruys 2010; Ormiston et al. 2008).

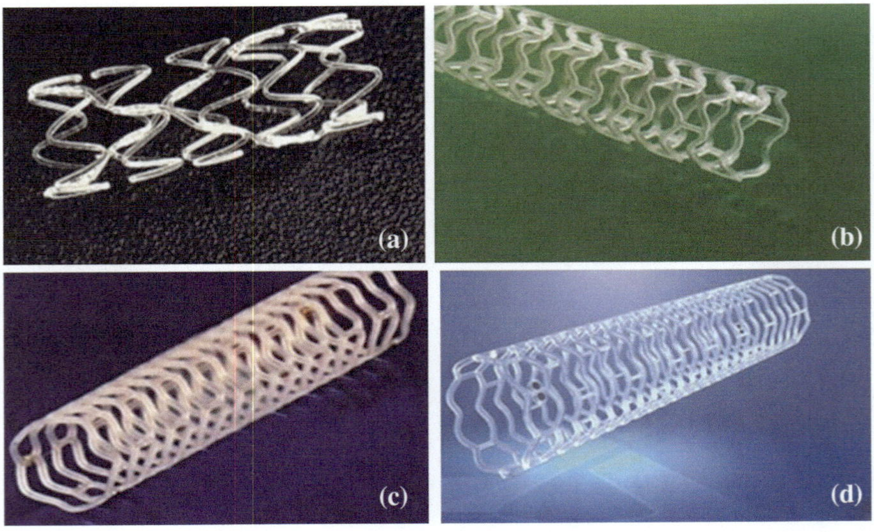

Fig. 3 Commercial stents: **a** Igaki-Tamai stent® (Tamai et al. 2000), **b** Absorb stent® (Gogas et al. 2012), **c** Amaranth stent® (Muramatsu et al. 2013), **d** Stent DESolve® (Muramatsu et al. 2013)

Table 1 Overview of PLLA stents (Muramatsu et al. 2013)

Stent	Material	Coating material	Drug	Radiopacity	Strut thickness, μm	Stent-to-artery coverage, %	Duration of radial support	Resorption time, months	Current status
Igaki-Tamai	PLLA	None	None	Gold markers	170	24	6 months	24	CE approved
Absorb BVS 1.1	PLLA	PLLA	Everolimus	Platinum markers	157	26	6 months	24	CE approved
Amaranth	PLLA	None	None	None	150–200	N/A	3–6 months	N/A	FIM initiated
DESolve	PLLA	PLLA	Myolimus	Metallic markers	150	N/A	N/A	12–24	FIM completed

CE Conformité Européenne; *FIM* First-in-man

Bioresorption of Absorb® (BVS) was evaluated, by optical coherence tomography (OCT), through the implantation of the stent in coronary arteries of pigs. The follow-up occurred immediately after the implantation, at 28 days, 2 years, 3 years and 4 years.

At 28 days, OCT showed 82 % of the struts as preserved box, and 18 % as an open box appearance. At 2 years, histological analysis demonstrated that the polymeric strut voids to be replaced by proteoglycan-rich matrix, with polylactide residues at low levels.

After 4 years, it was possible to observe by OCT that 51.2 % of the struts were classified as dissolved bright box and 48.8 % as dissolved black box. The remnant struts were hardly detectable by histology, appearing as foci of low-cellular-density connective tissue (Gogas et al. 2012).

Furthermore, there is another natural based polymer stents as Amaranth® PLLA and DESolve®. The first one is a cardiac stent made of PLLA. It is being tested so far and guarantees a radial structure for 3–6 months and a total resorption at 12 month.

DESolve® is another device manufactured with PLLA and coated with the same polymer. It also use as a DES (a drug-eluting stents) and deliveries myolimus as drug. This resorption time varying from 12 to 24 months.

The main characteristics of PLLA stents are presented in Table 1.

Chitosan Stents

Lauto et al. (2001) developed a coil and self-expandable stent produced from chitosan thin films, which were inserted into the rat prostate. Nevertheless, this model was not considered viable for application in blood vessels (Lauto et al. 2001).

A genin-crosslinked chitosan polymer stent has been developed (also known as genipin stent) and the mechanical device's properties were determined. The results indicate to the cyclic crosslinking structures formed inside the stent matrix contributed to the improvement of its mechanical properties. Moreover, the acceptable radial compression loading of the genipin stent was higher when compared to metallic stent (Fig. 4).

Genipin stent were inserted in rabbit infrarenal abdominal aortas and after 3 months no thrombosis was observed and a nearly intact layer of endothelial cells was observed on vessel wall (Chen et al. 2009).

In order to evaluate its possibility as a drug delivery vehicle, sirolimus (an anti-proliferative drug) was loaded. It was observed that the release of sirolimus possesses its original activity in inhibiting smooth muscle cell proliferation. These findings suggest that the genipin stent with enhanced mechanical strength can be used as a drug delivery stent (Chen et al. 2009).

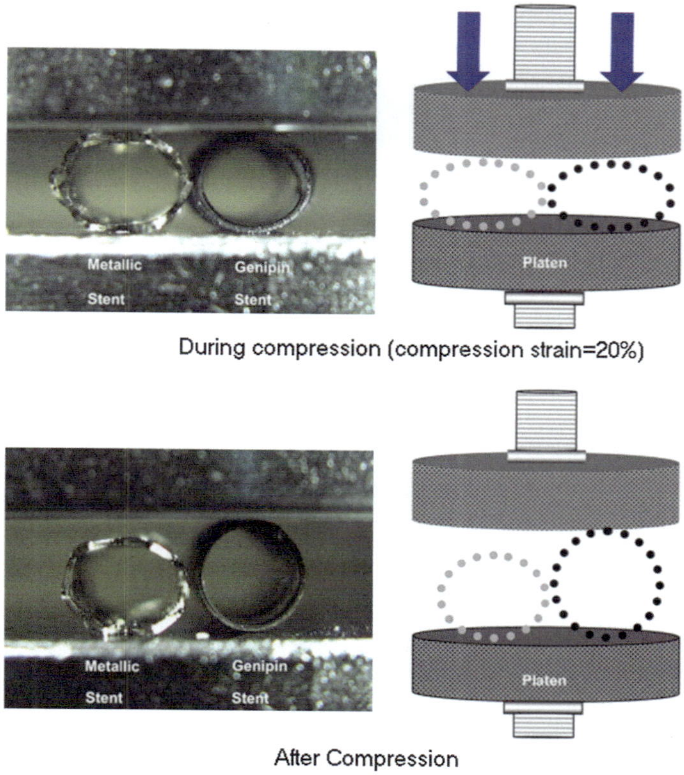

Fig. 4 Elastic deformation of the metallic stent and the genipin stent under a compression strain of 20 % (Chen et al. 2009)

Stents Functionalization

The interest in surfaces' functionalization is not new and it have been studied since the appearance of technology in the antiquity. The possibility of functionalize a material and provide it with additional functionalities, like drug release or major biocompatibility, assumes particular importance in biomaterials. Regarding stents applications, with surface modification it is intended to improve the stent's biocompatibility, by modifying of their surfaces with less thrombogenic and inflammatory materials, and reduce the infection and the occurrence of thrombosis and restenosis, by coating the stent with therapeutic agents, which are released, over time, after stent implantation (Thierry et al. 2003).

Therefore, there are three different kinds of stent coatings: biocompatible coatings, drug-delivery coatings and polymer-free coatings/surfaces. Biocompatible coatings comprehend, mostly, inorganic materials, like silver, gold or carbon, and have as principal function increase the stent biocompatibility, serving as a barrier to ion release by metallic stents and, usually, these coatings have no capacity to carry drugs.

To avoid this disadvantage, were developed biocompatible polymer coatings, which can vary and release therapeutic agents. These coatings can be non-biodegradable polymers or biodegradable polymers like: poly(lactic acid) (PLA), poly(glycolic acid) and their copolymer, poly(lactic-co-glycolic acid) (PLGA).

In order to understand and control, more adequately, the surface chemistry leaded researchers to deposit single molecular layers at surfaces. Thus, in the first half of 20th century, the functionalization research was restricted to the deposition of monolayers. The need to obtain better and alternative deposition strategies leaded to alternated adsorption of opposite charged polymers, which quickly escalate to one of the most used functionalization technique used, Layer by layer (LBL) (Michel et al. 2012).

Layer by Layer

Layer by layer (LBL) technique was introduced in 1991 and since then it became one of the most popular techniques for preparation of nanoscale films.

Films are formed by the depositing of alternating adsorption of materials containing complementary charged or functional groups to form integrated thin films.

Usually, a species with opposite charge of the substrate is the first layer to be adsorption and, this way, the charge of the substrate is reverse. This process is repeated forming several layers, until the desired thickness is achieved (Fig. 5) (Wang et al. 2008).

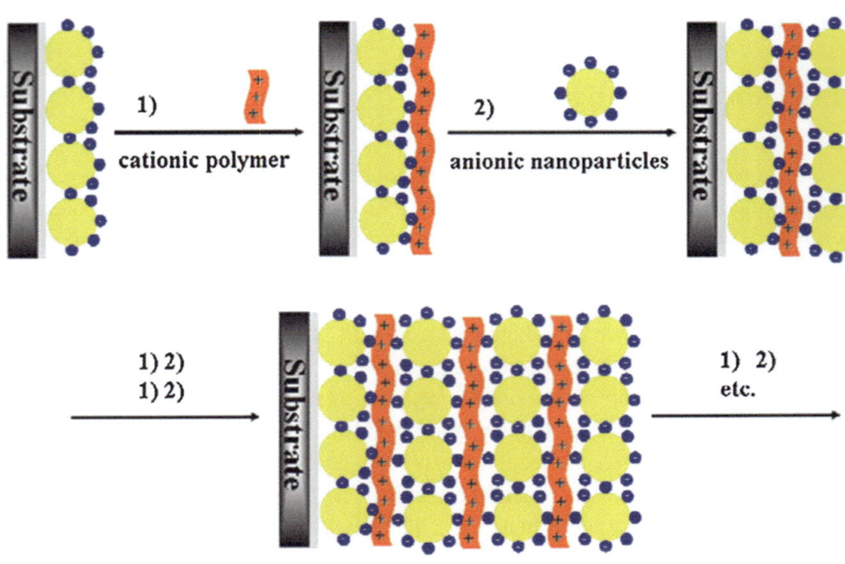

Fig. 5 Layer by layer technique (Rebelo et al. 2014)

The major advantage of this technique relies on its simplicity and versatility as it can be easily automated, through the control of several parameters: type of species adsorbed, the number, and, consequently, the thickness of layers and the conditions employed during the assembly process.

The flexibility of this technique allows the use of a variety of materials, such as polymers, nanoparticles, proteins, among others; and not only on planar substrates but also on nanoparticles, colloids and pores of those materials. The attachment between the substrate and the adsorption species can be through electrostatic interactions but also through hydrogen bonding, hydrophobic interactions, covalent bonding, and complementary base pairing (Cheng and McCarthy 1997).

This technique has been studied and employed on stents, by deposition of a thin film and one of the most used natural polymers in coatings is chitosan.

Chitosan Coatings

Meng et al. (2009) coated a 316 L stainless steel stent surface, using, layer by layer technique, with chitosan and heparin to promote the acceleration of re-endothelization and healing process after coronary stent implementation in a porcine iliac artery (Meng et al. 2009).

Chitosan and heparin were chosen because of their oppositely charged polymer electrolyte characteristic and, this way, they can be used in layer by layer method. Furthermore, the stent surface layer, coated with chitosan and heparin, could be efficient in the re-endothelization process, once this coated can achieve a balance between endothelial cell affinity and thrombus resistance, after stent implementation.

The hemocompatibility and endothelial cell compatibility was evaluated in vitro and in vivo, through a porcine coronary injury model.

The results showed that the coating was found to be safe and promote endothelial cell compatibility and hemocompatibility to the stent surface.

More recently, Shen et al. (2012) developed an asymmetrical coating system with PLLA and chitosan and heparin (C/H LBL-SES). The abluminal surface of the stent was coated with PLLA carrying sirolimus, while the luminal surface was coated with chitosan and heparin, through layer by layer technique, as observed in Fig. 6.

The developed stents, as well as bare metal stents (BMS) and sirolimus-eluting stents (SES), were implanted into porcine coronary arteries. The evaluation and follow up of stents were made at 7, 14 and 28 days (Fig. 7).

After 14 days of implementation, the re-endothelization of C/H LBL-SES was almost complete, despite the re-endothelization of SES was, approximately, 50 %.

Regarding the diameter stenosis, after 28 days of implementation, the behavior of C/H LBL-SES was comparable to SES and superior to BMS.

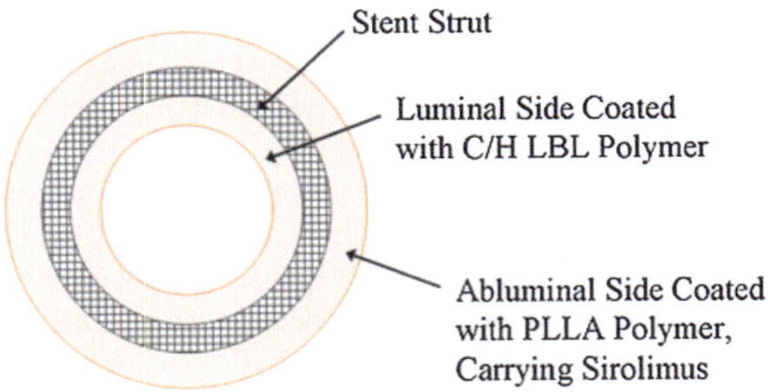

Fig. 6 Asymmetrical design for the developed stent (Shen et al. 2012)

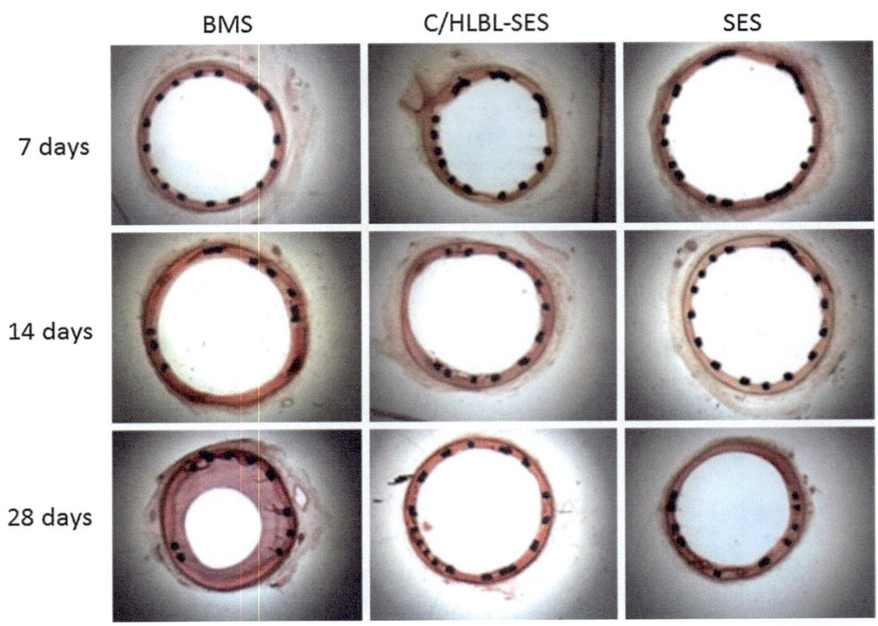

Fig. 7 Histopathology and histomorphometry analysis at 7, 14 and 28 days follow-up (Shen et al. 2012)

Even though in the follow up days C/H LBL-SES suffered a greater vasoconstriction than bare metal stents, it presented a better behavior than SES.

After the study, authors conclude that the designed stent successfully inhibited neointima hyperplasia during the 28 days follow-up and no thrombotic events occurred. Furthermore, the developed C/H LBL-SES showed a good behavior and may be a promising stent, with superior performance than traditional DES (Shen et al. 2012).

Conclusions

On account of natural polymers' characteristics, like biocompatibility, non-toxicity and biodegradability, their application in biomedical areas including stent applications, is rapidly expanding with huge potentialities.

Natural polymers based stents or stents coated with natural polymers present good results and improve biocompatibility, minimizing the disadvantages of metallic stents.

The surface modification and inclusion of natural coatings in stents is a major innovative advancement, since, in this way, problems like restenosis, infection or review surgeries can be avoided.

Acknowledgments The authors gratefully acknowledge the funding by Ministério da Ciência, Tecnologia e Ensino Superior, FCT, Portugal, under grant SFRH/BD/90321/2012.

References

N. Vila, "Braided Hybrid Stents Design," in *Master Thesis on Design and Marketing*, University of Minho, 2009.

S. Commandeur, H. M. Van Beusekom and W. J. Van Der giessen, "Polymers, Drug Release, and Drug-Eluting Stents," *Journal of Interventional Cardiology,* vol. 19, no. 6, pp. 500-506, 2006.

J. Dyet, W. Watts, D. Ettles and A. Nicholson, "Mechanical Properties of Metallic Stents: How do These Properties Influence the Choice of Stent for Specific Lesions?," *Cardiovascular and Interventional Radiology,* vol. 23, no. 47-54, 2000.

R. Rebelo, R. Fangueiro, S. Carvalho, M. Henriques and S. Rana, "Methods of incorporation antimicrobial agents in stents," *International Journal of Engineering Science and Innovative Technology (IJESIT),* vol. 3, no. 2, pp. 409-422, 2014.

L. França and A. Pereira, "Update on vascular endoprostheses (stents):from experimental to clinical pratice.," *Jornal Vascular Brasileiro, 7,* 2008.

B. Calabia and Y. Tokiwa, "Production of D-lactic acid from sugarcane molasses, sugarcane juice and sugar beet juice by Lactobacillus delbrueckii," *Biotechnology letters,* vol. 29, pp. 1329-1332, 2007.

Z. Lili, "Síntese e caracterização do copolímero triblocoanfifílico biodegradável Poli (L, L- Lactídeo-stat e carpolactona)-bloco- poli (óxido de etileno)- bloco-Poli (L, L- Lactídeo- stat e carpolactona)," *Master thesis on Engineering, University of São Paulo,* 2007.

F. Dias, D. Queiroz, R. Nascimento and M. Lima, "Simple system for preparation of chitosan microspheres," *Química Nova,* vol. 31, no. 1, 2008.

V. Balan and L. Verestiuc, "Strategies to improve chitosan hemocompatibility: A review.," *European Polymer Journal,* pp. 171-188, 2014.

S. Meng, Z. Liu, L. Shen, Z. Guo, L. Choud, W. Zhong, Q. Du and J. Ge, "The effect of a layer-by-layer chitosan–heparin coating on the endothelialization and coagulation properties of a coronary stent system," *Biomaterials,* vol. 30, p. 2276–2283, 2009.

Y. Zhu, C. Gao, X. Liu, T. He and J. Shen, "Immobilization of biomacromolecules onto aminolyzed poly(L-lactic acid) toward acceleration of endothelium regeneration," *Tissue Engineering,* vol. 10, pp. 53-61, 2004.

L. Reis, N. Neves, J. Mano, M. Gomes, A. Marques and H. Azevedo, Natural-based polymers for biomedical applications, Woodhead Publishing Limited, 2008.

N. Hoiby, T. Bjarnsholt, M. Givskov, S. Molin and O. Ciofy, "Antibiotic resistance of bacterial biofilms," *International Journal Antimicrobial Agents,* vol. 35, pp. 322-332, 2010.

D. Pavithra and M. Doble, "Biomedical Materials," *Biofilm formation, bacterial adhesion and host response on polymeric implants—issues and prevention,* vol. 3, pp. 1-13, 2008.

M. Knetsch and L. Koole, "New strategies in the development of antimicrobial coatings: the example of increasing usage of silver and silver nanoparticles," *Polymers,* vol. 2, pp. 340-366, 2003.

S. Grag and P. Serruys, "Coronary Stents Current Status," *Journal of the American College of Cardiology,* vol. 56, pp. S1-S42, 2010.

H. Tamai, K. Igaki, E. Kyo, K. Kosuga, A. Kawashima, S. Matsui, H. Komori, T. Tsuji, S. Motohara and H. Uehata, "Initial and 6-month results of biodegradable poly-l-lactic acid coronary stents in humans," *Circulation,* vol. 102, no. 4, pp. 399-404, 2000.

S. Nishio, K. Kosuga, K. Igaki, M. Okada, E. Kyo, T. Tsuji, E. Takeuchi, Y. Inuzuka, S. Takeda, T. Hata, Y. Takeuchi, Y. Kawada, T. Harita, J. Seki, S. Akamatsu, S. Hasegawa, N. Bruining, S. Brugaletta, S. de Winter, T. Muramatsu, Y. Onuma, P. Serruys and S. Ikeguchi, "Long-Term (> 10 Years) Clinical Outcomes of First-in-Human Biodegradable Poly-l-Lactic Acid Coronary Stents," *Circulation,* vol. 125, pp. 2343-2353, 2012.

J. Ormiston, P. Serruys, E. Regar, D. Dudek, L. Thuesen, M. Webster, Y. Onuma, H. Garcia-Garcia, R. McGreevy and S. Veldhof, "A bioabsorbable everolimus-eluting coronary stent system for patients with single de-novo coronary artery lesions (ABSORB): a prospective open-label trial," *The Lancet,* vol. 371, pp. 899-907, 2008.

B. Gogas, V. Farooq, Y. Onuma and P. Serruys, "The ABSORB Bioresorbable Vascular Scaffold: An Evolution or Revolution in Interventional Cardiology?," *Hellenic Journal of Cardiology,* vol. 53, pp. 301-309, 2012.

T. Muramatsu, Y. Onuma, Y. Zhang, C. Bourantas, A. Kharlamov, R. Diletti, V. Farooq, B. Gogas, S. Garg, H. García-García, Y. Ozaki and P. Serruys, "Progress in Treatment by Percutaneous Coronary Intervention:The Stent of the Future," *Revista Española de Cardiología,* vol. 66, no. 6, p. 483–496, 2013.

A. Lauto, M. Ohebshalom, M. Esposito, J. Mingin, P. Li, D. Felsen, M. Goldstein and D. Poppas, "Self-expandable chitosan stent: design and preparation," *Biomaterials,* pp. 1869-1874, 2001.

M. Chen, C. Liu, H. Tsai, W. Lai, Y. Chang and H. Sung, "Mechanical properties, drug eluting characteristics and in vivo performance of a genipin-crosslinked chitosan polymeric stent," *Biomaterials,* vol. 30, pp. 5560-5571, 2009.

B. Thierry, F. Winnik, Y. Merhi, J. Silver and M. Tabrizian, "Bioactive Coatings of Endovascular Stents Based on Polyelectrolyte Multilayers," *Biomacromolecules,* vol. 4, pp. 1564-1571, 2003.

M. Michel, V. Toniazzo, D. Ruch and V. Ball, "Deposition Mechanisms in Layer-by-Layer or Step-by-Step Deposition Methods: From Elastic and Impermeable Films to Soft Membranes with Ion Exchange Properties," *ISRN Materials Science,* vol. 2012, pp. 1-13, 2012.

Y. Wang, A. Angelatos and F. Caruso, "Template Synthesis of Nanostructured Materials via Layer-by-Layer Assembly," *Chemistry of Materials,* vol. 20, no. 3, 2008.

W. Cheng and J. McCarthy, "Layer-by-Layer deposition: a tool for polymer surface modification," *Macromolecules,* vol. 30, p. 1997, 78-86.

L. Shen, Y. Wu, F. Zhang, L. Wu, C. Dong, Y. Gao, A. Sun, Y. Zou, J. Qian, J. Sun, W. Zhong and J. Ge, "Assessment of an asymmetrical coating stent with sirolimus released from ablumial matrix in porcine model," *Clinical Research in Cardiology,* vol. 101, pp. 917-927, 2012.

S. Meng, Z. Liu, W. Zhong, Q. Wang and Q. Du, "Phossphorylcholine modified chitosan: Appetent and safe material for cells," *Carbohydrate polymers,,* vol. 70, pp. 82-88, 2007.

Optimization of a Wood Plastic Composite to Produce a New Dynamic Shading System

G. Martins, F. Antunes, A. Mateus, S. Baptista and C. Malça

Abstract The shading systems of the windows may play a protagonist role in the reduction of energy consumption in lighting, artificial ventilation and air conditioning, contributing to sustainable development regarding bioclimatic architecture and improvement of the energetic efficiency in buildings. This work, resulting from a partnership between architecture and materials science, reports the development of an innovative and dynamic shading system that simultaneously fulfils various aesthetic, functional and environmental requirements. The original design of the system, derived from this collaboration, enables an industrial low cost production in virtue of the selection of technologies and materials concurrently adequate to multiple objectives. This work also aims at contributing to a sustainable construction through the use of eco-friendly materials such as biocomposites produced from recyclable polymers and vegetable fibres, using extrusion as processing technology both for blending the raw materials and for the shutter units production. The optimization of materials for this specific application involved wood plastic composites (WPC) based on high density polyethylene (HDPE), industrial residues of pine sawdust and additives such as coupling agents, pigments and UV protectors. Results showed that the optimized WPC made from recycled industrial pine wood

G. Martins (✉) · F. Antunes
University of Coimbra, Coimbra, Portugal
e-mail: agtmartins@gmail.com

F. Antunes
e-mail: filipe.antunes@ci.uc.pt

S. Baptista
Polytechnic Institute of Leiria, Leiria, Portugal
e-mail: sofia.baptista@ipleiria.pt

C. Malça
Polytechnic Institute of Coimbra, Coimbra, Portugal
e-mail: candida@isec.pt

A. Mateus · C. Malça
Centre for Rapid and Sustainable Product Development,
Polytechnic Institute of Leiria, Leiria, Portugal
e-mail: artur.mateus@ipleiria.pt

R. Fangueiro and S. Rana (eds.), *Natural Fibres: Advances in Science and Technology Towards Industrial Applications*, RILEM Bookseries 12,
DOI 10.1007/978-94-017-7515-1_26

residues presents thermal stability, mechanical strength and water absorption appropriate to be applied in the production of the new architectural solutions such as the innovative shading system presented.

Keywords Pine sawdust · HDPE · Wood plastic composite (WPC) · Shading system · Sustainable construction · Bioclimatic strategies

Introduction

The construction industry represents approximately 25 % of the industrial production in Europe and is identified as the second most responsible sector for the emission of carbon dioxide, corresponding to 40 % of energy consumption and contributing to around 30 % of the total waste produced. In this 30 % all construction stages—production, storage, transportation, application, maintenance, repair and demolition—were accounted for (Mateus and Bragança 2006, Rao and Riahi 2006). The market of construction industry is highly competitive and the investment in products perceived as environmentally friendly has been increasing in virtue of consumers demand and also as a means of differentiation and social responsibility. Ten years ago, in Portugal, 29 % of the energy consumption was attributed to buildings while the remainder corresponded to 36 % in transport, 33 % in industry and 2 % in agriculture and fisheries. Concerning the energy consumption of the residential sector, 50 % corresponded to domestic hot water; 25 % to atmospheric heating and cooling, and 25 % to lighting (Silva and Guimarães 2005). These statistics clearly illustrate the relevance of the energetic footprint represented by buildings in nowadays society.

The energetic performance of buildings is, indeed, a matter of major concern in modern architecture. In this context, the shading systems for buildings have emerged as key elements in the design of the edifices frontages, providing not only thermal comfort, controlled luminosity and protection against other environmental conditions, but also enabling various aesthetic effects regarding both the exterior and the interior spaces. In fact, shading systems are frequently made of multiple assembled parts, possessing mechanisms that enable changeable levels of exposure and simultaneously alter the outlook of the plain windows (Palhinha 2009). However the existing systems have a poor performance and available commercial solutions are characterized by limited standardized models. Therefore new developments in shading technology are required, being time to put in the market new concepts of shading and take them to the forefront of architecture and design.

The actual demand for sustainable construction has fostered the research of alternative products, namely based on composite materials resulting from the use of renewable material and/or industrial wastes, conventionally called "*green materials*" (Mantia and Morreale 2011). Wood plastic composites (WPC) constitute emblematic examples of materials that seek to address some environmental issues related to the construction sector, since they may be composed of renewable thermoplastics, such as polyethylene, and industrial wood wastes, such as pine sawdust, to generate value-added products. In fact, the reuse and recycling of materials for new applications and processes are of great importance because they allow for the closing of the production cycle to reduce energy consumption at the level of extraction and processing of raw materials, assuming these wastes and byproducts are new materials for new products, thereby initiating a new life cycle of the product (Ray 2013).

The new concept of dynamic shading systems presented in this work has two main objectives. The first one is the design an innovative architectural model that allows for high energy efficiency and the second one, the development of an adequate WPC for this application, using renewable, recyclable and industrial waste materials. Overall, the final product aims at fulfilling the current environmental concerns and thus to contribute to the use of more sustainable materials and technologies. The main raw materials selected for such WPC included HDPE from industry suppliers and pine sawdust residues from a local unit of wood transformation. Some commercially available additives, such as coupling agents based on PE-g-maleic anhydride (PE-g-MA), pigments and UV protectors, were also included.

The extensive study of optimization of a WPC for the described particular application involved a detailed characterization of the selected raw materials and the preparation, at diverse experimental conditions, of blends containing variable concentrations of pine wood and additives. The concentration of those components was optimized by torque rheometry to obtain adequate viscosities for extrusion processing. The most promising composites were characterized using SEM, FTIR, DSC-TGA, tensile testing, measurements of water contact angle and water absorption capacity. The obtained results enabled to arrive to a final recommended formulation. Only a fraction of these results is presented in this paper.

Dynamic Shading System Modeling

There is no doubt that shading systems are essential elements on buildings due to their role in climate protection by controlling the exposure to the elements (heat, cold, rain, wind, sunlight, dust) and by decreasing the use of artificial ventilation and air conditioning, enabling the reduction of energy consumption for lighting and for thermal conditions. To achieve a better performance, the shading systems must be dynamic in order to enable the regulation of the temperature, the balance between blocking of sunlight and visibility optimization, or to ensure an adequate ventilation of the space. From the point of view of the user, other characteristics related to privacy and security may also be important.

The new dynamic shading developed consists of a set of parts placed over a sliding support and connected to each other by a system that opens and closes each set at the same time, enabling the directional adjustment of the shutters by the user, according on the weather conditions. It is intended to combine the thermal control and the lighting control, with the possibility of regulating the degree of opening of the various parts that compose the new shutter. This allows creating a higher or lower level of shading, as well as different aesthetic effects as depicted in Fig. 1.

This system is projected to be fabricated using a WPC based on an adequate thermoplastic and pine wood sawdust, being the individual parts designed for extrusion processing. The structural details of the shutter parts may not be disclosed given the intention to patent the models.

This shading system will probably be a product with great marketing potential, not only in Portugal, but also in other countries with abundant sunshine, given the limitation of the current market in offering innovative, dynamic and less standardized solutions to shade the buildings windows. Furthermore, due to the sector limitations to offer passive and ecological solutions that meet sustainable construction, the new shutter here presented will contribute to sustainability because it combines the use of "ecological" materials with bioclimatic building strategies, e.g. passive solution shading.

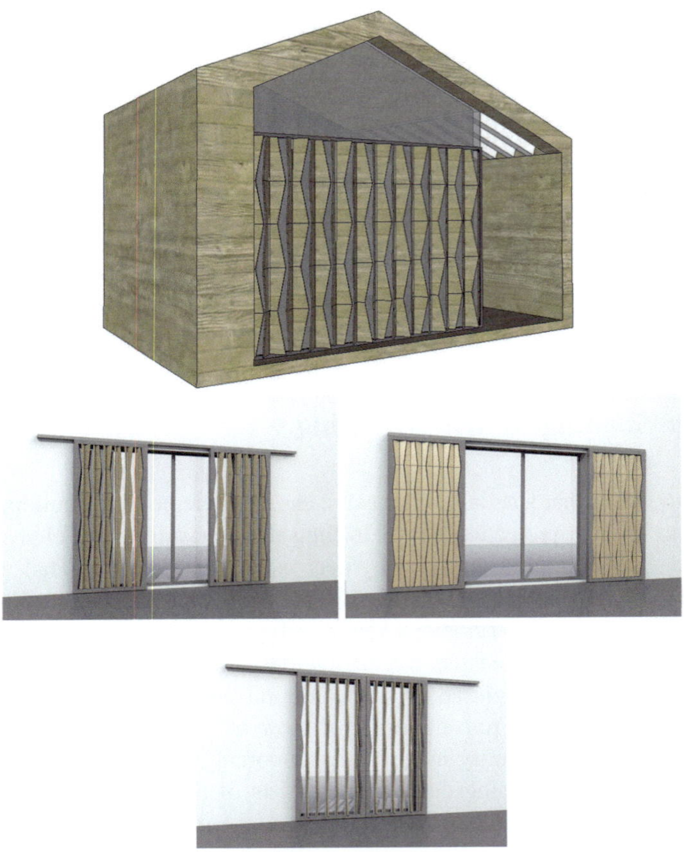

Fig. 1 Model of a dynamic shading system made of WPC

Materials and Methods

The preliminary selection of raw materials and processing methods to obtain the pretended WPC and finally, the individual shutters, needs to observe multiple requirements.

The selection of raw materials to obtain a WPC mainly includes the wood type, which acts as filler or as reinforcement, and the polymers that constitute the composite matrix. Since wood is hydrophilic and the polymers are hydrophobic, the most conventional approach to promote the interaction between these components consists of using coupling agents as additives in these composites. Chemically modified polymers with maleic anhydride, possessing a backbone compatible with the main composite polymer, become entangled in the matrix while the anhydride groups react with the hydroxide bonds of the cellulosic components of wood (Abdelaal et al. 2012). Through the use of these compatibilizers, not only the interfacial adhesion and the mechanical properties are increased, but the susceptibility to water adsorption, thickness swell and bio-deterioration may be limited to lower levels. Other additives, such as minerals as talc or calcium carbonate, are frequently used as fillers mostly for economical reasons. The addition of pigments and UV protectors may also be needed, usually in small quantities, being these additives commercially available in polymeric formulations.

Regarding the selection of the wood fibres, after deciding the type (soft wood, hardwood) that should be used, the morphology and the particle size distribution are characteristics of chief importance, as well as the thermal degradation profile. Typically, the wood fibres begin to deteriorate at temperatures around 240 °C. When wood is in the form of sawdust, as the pine wood residues of this study, a preliminary sieving process is mandatory to limit the variability of the supplied fibres.

As for the general selection criteria of polymer types for the composites, thermoplastics are usually preferred because these offer the advantage of enabling repeated melting processes, unlike thermosetting polymers which become irreversibly in the solid state after a single processing cycle that leads to polymerization after an initial increased temperature. Thermoplastics are most adequate for blending the wood fibres and obtaining the composites in a twin screw extruder, for instance, followed by the extrusion or injection molding of the composite previously obtained to achieve the intended final products. Furthermore, temperature related properties must be considered, namely the melting temperature (T_m) and the melting flow rate (MFR). The thermoplastics selected for WPCs should melt, at least, at around 50 °C less than the starting temperature of wood degradation. However, depending on the application of the final product, the melting temperature may have to be superior to a certain value. For outdoor applications of the composites, a high density polyethylene that melts at around 140 °C is more adequate than a low density polyethylene with a melting point around 110 °C. On the other hand, the melt flow rate of the polymers should comply with the processing technology of the composites. In practical terms, this parameter is mainly indicative, since the polymer/wood/additives blends are prepared under different conditions. It should be mentioned that the properties of the polymeric additives (coupling agents, pigments and others) should be close to those of the main polymer of the composite matrix.

In this work the methods applied in the optimization of composites involved first, a pre-selection of raw materials and their detailed characterization. Afterward, the study of the blending conditions using Torque Rheometry at different temperatures and using diverse wood concentrations, as well as different contents of coupling agents. Finally, the most promising combinations from the latter study were also characterized regarding chemical, physical and mechanical properties.

The results were continuously evaluated taking into consideration the following necessary upscale from laboratory to industrial production of the composites and of the shutter units.

Materials

The pine wood sawdust was supplied from a local industry, in the form of residues with random characteristics, and the polymers were acquired to industrial suppliers. Two different grades of HDPE were tested: HDPE HMA 035 from Exxon Mobile, with T_m = 134 °C and MFR = 8 g/10 min and HDPE KS 10100 UE from Dow Chemicals, with MFR = 4 g/10 min (T_m = n.d.). Two coupling agents based on HDPE-g-MA were also used: Fusabond E226, with T_m = 120 °C and MFR = 1.75 g/10 min and Fusabond E265, with T_m = 131 °C and MFR = 12 g/10 min, both from Dupond.

Methods

The experimental methods used in this study involved techniques as sieving for particle size distribution analysis, FTIR for chemical composition, DSC-TGA for thermal behavior, SEM for microstructure, Torque Rheometry for viscosity, Tensile Testing for mechanical properties, Water Contact Angle measurements for surface wettability and Water Immersion Tests for swelling.

Results

The results here presented have been selected to illustrate some of the important features of this study. The particle size distribution of different batches of wood sawdust (Fig. 2) shows that both have high percentages (around 50 %) of particles under 0.7 mm. Whereas these were classified as Fine, the other obtained groups

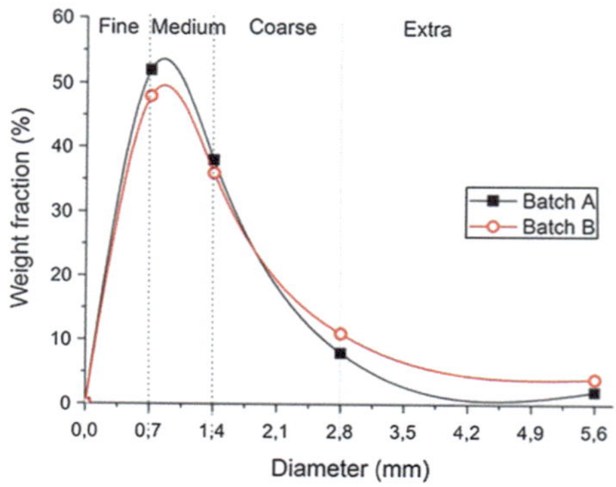

Fig. 2 Particle size distribution of different batches (A and B) of pine wood sawdust

were named as Medium (between 0.7 and 1.4 mm), Coarse (between 1.4 and 2.8 mm) and Extra (superior to 2.8 mm).

Torque rheometry performed at 160 and 190 °C with both HDPE polymers and with different batches of wood sawdust (A and B) is represented in Fig. 3. Fine (F), Medium (M) and Coarse (C) particles were used in this study that gives indirect information about the viscosity of the blends.

The SEM analysis of composites containing a high percentage (10 to 20 wt%) of a coupling agent (E 226 or E 265) showed that the interfacial adhesion between the wood particles and the polymeric matrix is enhanced, as illustrated in Fig. 4 by the physical bonding between these components.

Accordingly, the tensile strength of these composites is combined with higher strain relatively to those without PE-g-MA. Moreover, the low swelling and the high water contact angle of the surface of the composites containing high percentage of coupling agents indicate that these are adequate for outdoor applications.

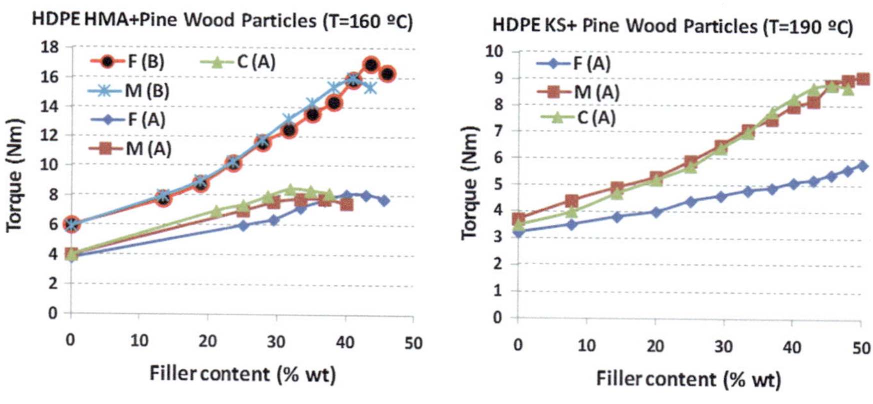

Fig. 3 Torque measurements at 160 and 190 °C of blends of HDPE (HMA and KS) with pine wood particles from batches A and B, of diverse size ranges (Fine, Medium, Coarse)

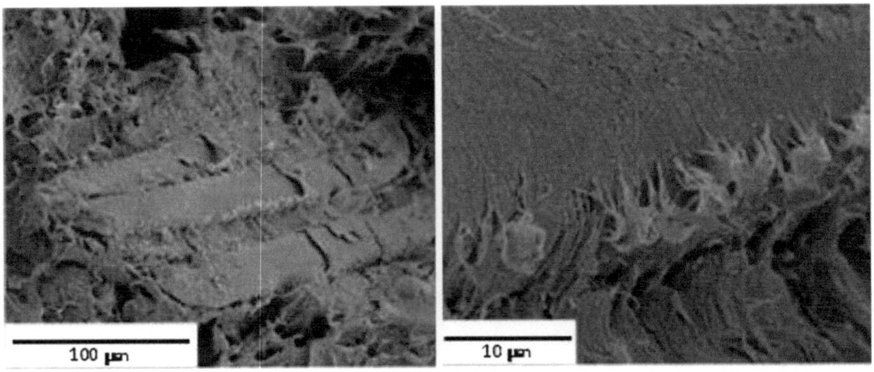

Fig. 4 Detail of the microstructure of a HDPE-pine wood composite (*left*) and the fibre-matrix interfacial bonding (*right*) resulting from the use of PE-g-MA coupling agent

Conclusions

From this study, it can be concluded that the composites which present a better performance regarding the total set of properties required to produce, in industry, the shading systems by extrusion, are composed of 50–55 wt% HDPE HMA 035, 30–35 wt%. Fine wood sawdust and 10–20 % PE-co-MA.

References

Mateus, R. and Bragança, L.: Tecnologias Construtivas para a Sustentabilidade da Construção. Edições Ecopy, Porto, (2006).

Rao, S. and Riahi, K.: The Role of Non-CO2 Greenhouse Gases in Climate Change Mitigation: Long-term scenarios for the 21st century. The Energy Journal, IAEE 27, 177–200 (2006).

Silva, S. M. and Almeida, M. G.: Avaliação do impacto energético e económico de diferentes soluções construtivas. REC 18, Universidade do Minho – Departamento de Engenharia Civil, Guimarães, 45–62 (2005).

Palhinha, M. J.: Sistemas de Sombreamento em Arquitectura: proposta de um novo método de concepção e dimensionamento. Tese de Mestrado, Instituto Superior Técnico, Lisboa, (2009).

Mantia, F.P. and Morreale, M.: Green composites: A brief review. Composites: Part A, Vol. 42, 579–588 (2011).

Ray, S.S.: Environmentally Friendly Polymer Nanocomposites: Types, Processing and Properties. Woodhead Publishing, U. K., (2013).

Abdelaal M.Y. *et al.*: Enhancement of polyolefins compatibility with natural fibers through chemical modification. American Journal of Polymer Science, 2(5), 102–108 (2012).

Biodegradation of Wool Used for the Production of Innovative Geotextiles Designed to Erosion Control

Jan Broda, Katarzyna Kobiela-Mendrek, Monika Rom, Joanna Grzybowska-Pietras, Stanislawa Przybylo and Ryszard Laszczak

Abstract Sheep wool was used for the production of innovative geotextiles designed to erosion control. The Kemafil ropes filled with wool, which are used for the production of geotextiles were buried in the ground. During ground disposal the susceptibility of wool to biodegradation was observed. The basic geometrical and mechanical parameters of wool fibres as well as their chemical composition and the surface morphology were analysed. It was observed that the ground disposal causes the decrease of breaking force of fibres and the destruction of scales on their surface. Within four months the biodegradation reaches the first stage consisting in cleavage of disulfide bonds of keratin. The rapid biodegradation is observed in fibres mechanically damaged. In the presence of the hydrogel the biodegradation is delayed. After four months, despite observed biodegradation of wool, the ropes maintain their ability to fulfil the erosion protection function.

Keywords Wool · Biodegradation · Kemafil rope · Geotextiles · Erosion control

Introduction

In recent years different attempts for utilization of sheep poor quality fibres and wool wastes were undertaken. The most common procedure of the utilization wool wastes is the open–loop recycling, which refers to using recycled fibres for the production of other products that have a lower commercial value. This method effectively extend the active life of the constituent fibre, potentially by many years beyond the first use phase (Russell et al. 2015).

J. Broda (✉) · K. Kobiela-Mendrek · M. Rom · J. Grzybowska-Pietras · S. Przybylo
R. Laszczak
Institute of Textile Engineering and Polymer Materials, University of Bielsko-Biala, Bielsko-Biala, Poland
e-mail: jbroda@ath.bielsko.pl

R. Fangueiro and S. Rana (eds.), *Natural Fibres: Advances in Science and Technology Towards Industrial Applications*, RILEM Bookseries 12,
DOI 10.1007/978-94-017-7515-1_27

For wool waste open-loop mechanical recycling is often used for the production of nonwoven fabrics, which are obtained by means of garneting or carding or airlaying of webs followed by mechanical, thermal or chemical bonding. In this way thermal and acoustic insulating materials for automotive industry or building are produced.

Another idea of utilization wool wastes consists on the production of geotextiles destined for the ground protection. Such products include woollen geomats impregnated with grass and wild flower seeds (Johnson et al. 2013). Mats provide excellent thermal protection, ensure a proper microclimate for seed germination and by gradual degradation deliver nutrients for further plant vegetation. Other example of such products are innovative geotextiles built from meandrically arranged coarse ropes connected with additional linking threads (Fig. 1).

For the production of ropes the Kemafil technology can be applied. The technology was elaborated in Germany few decades ago and still is used for the production of products for various purposes (Patent DD 110905). Over many years the technology was several times modified. After last modification the technology enables production of thick ropes with a core—mantle structure, which can be filled with various materials, covered by thin fabrics and wrapped by knitted sheath (Helbig et al. 2006).

First attempts of the application of Kemafil ropes for preventing a steep slope were performed at a motorway near Chemnitz, Germany. It was revealed that thick ropes arranged in a meander-like pattern create a network of microdams, which retain flow of water along the slope. The ropes absorb rainfall water and reduce the transport of material detached from the soil by rain drops. Due to enhanced soil and water holding capacity ropes increase seeds germination rate and promote vegetation growth (Rickson 2006; Álvarez-Mozos et al. 2014).

The Kemafil ropes can be produced from biodegradable waste materials easily accessible at local market. Ropes produced from such materials buried in the ground progressively biodegrade and therefore fulfil their erosion control function

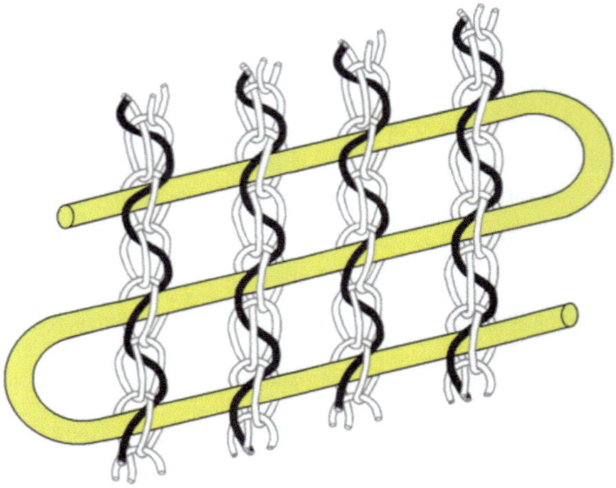

Fig. 1 The structure of innovative geogrids designed for erosion control

in limited period of time. This time must be sufficient for establishing of natural vegetation, which takes over the protective function.

The biodegradation rate of ropes depends on many factors. For wool, due to the presence of large quantities of disulphide and hydrogen bonds, salt linkages and other crosslinkings, which stabilize the wool keratin the biodegradation proceeds relatively slowly.

The biodegradation starts by the enzymatic destruction of scales on the surface of fibres. Afterwards enzymes penetrate deeper into the cortical layer and laminate it into spindle-shaped cells. In further step amorphous intrafibril keratin is removed and the splitting of cells into separate macrofibrils is observed (Pekhtasheva et al. 2012).

In the process microorganisms, bacteria and fungi, with mainly proteolytic and keratinolytic enzymes are involved. In first stage degradation involves the splitting of disulphide bridges, which is followed by the enzymatic decomposition of keratin into oligopeptides. In the next step proteolytic enzymes induce hydrolytic keratin decay by polypeptide bonds to separate amino acids, which are used in metabolic processes of oxidative deamination with the release of ammonia (Kornilowicz-Kowalska and Bohacz 2011).

In our studies the biodegradation of wool used for production of Kemafil ropes designed to soil erosion control was investigated.

Experimental

The ropes with a core—mantle structure and a diameter 12 cm were produced. The ropes were covered with woollen needle punched nonwoven and wrapped by knitted sheath made from cotton twine with a diameter of 3 mm. The core of the ropes were filled with coarse wool fibres. The ropes with and without polymeric hydrogel were obtained. The commercial hydrogel used in agriculture was mixed with wool fibres before filling the ropes.

The ropes were buried in the soil and exposed to natural outdoor weathering conditions during several months (Fig. 2).

Fig. 2 Samples of ropes; **a** before ground disposal; **b** covered by the soil

The ropes were buried at a depth of 20 cm in experimental site located on campus of the University of Bielsko-Biala (Poland). The samples were covered by neutral humus (pH 7.2). After levelling the site has been sown with grass seeds.

The ground disposal has started in the autumn at the end of the vegetation season. During first month an average daily air temperature was about 10 °C. In second and third month for few days the temperature decreased below 0 °C. Then in the fourth month the temperature was above 0 °C again (Fig. 3). During few weeks the ground was covered by snow.

The microbiological decomposition of wool fibres buried in the ground was observed. In investigations the fibres thickness and breaking force before as well as after 1–4 months of ground disposal were measured. Moreover, chemical composition and fibres morphology were analysed.

The fibre thickness was measured using optical projection microscope according to the Polish standard PN-ISO 137:2000. The breaking force of individual fibres was determined according to the Polish standard PN-P-04930-07:1993 by the tensile machine Instron 5544. The chemical composition of fibres was analysed by Fourier transform infrared spectroscopy (FTIR). For FTIR measurements the Nicolet 6700 spectrometer was used. The spectra were registered in the range from 400 to 4000 cm^{-1}. For measurements fibres were cut into 1–3 mm pieces and blended with powdered potassium bromide KBr. The blends were pressed under

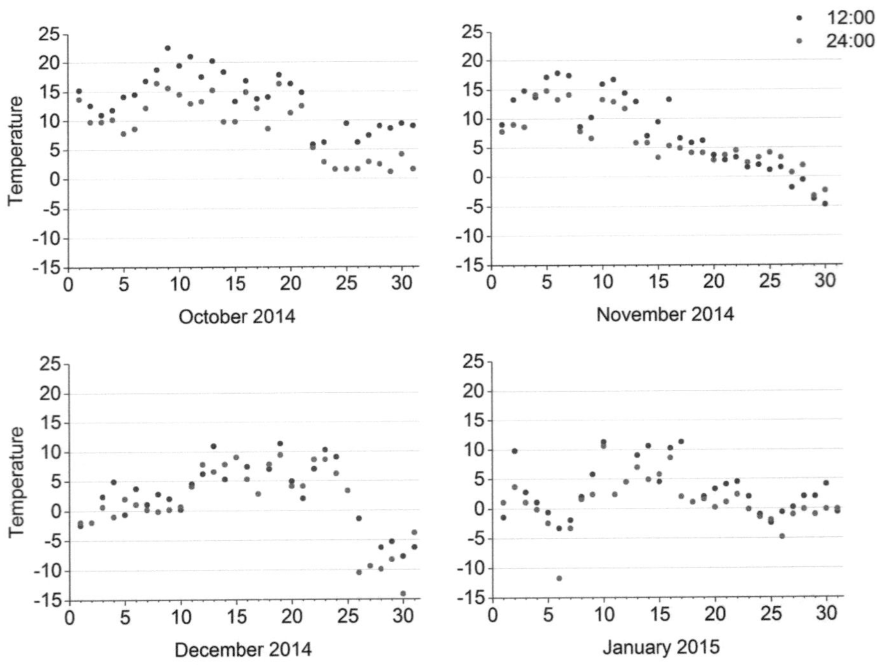

Fig. 3 Daily temperature during ground disposal

200 MPa. The tablets were placed in a measuring chamber of the spectrometer equipped with a mirror beam collimator (focused 16x). As a reference, the tablet made of KBr (without fibres) was used. The morphology studies of fibres were performed by scanning electron microscopy (SEM). The microscope JEOL JSM 5500 LV operated in backscattered electron mode was applied. The observations were carried out for fibres sputtered with gold in Jeol JFC 1200 ionic sputter.

Results and Conclusions

Mechanical Parameters

The measured thickness and breaking force of wool fibres determined before and after ground disposal are presented in Table 1. The wool fibres used for the production of ropes are thick and characterised by high thickness diversity. After ground disposal the diameter of fibres practically does not change.

The breaking force of individual fibres is highly diversified. For fibres used for the formation of the cover the breaking force is lower than the breaking force of fibres forming the core. The ground disposal does not cause the significant change of the breaking force of fibres forming the cover. For fibres used for the filling of ropes the breaking force decreases more significantly. For fibres filled the ropes produced with the hydrogel the decrease of fibres breaking force is minimally lower.

Fibres Morphology

Figure 4 shows microphotographs at magnification ×1000 of fibres before and after ground disposal. For fibres before disposal on the surface fine undamaged flat scales with well-defined scale edges are well visible. After 1 months of disposal for fibres taken from the nonwoven cover as well as for fibres from the core of ropes produced without hydrogel the first signs of destruction of the scales are observed. In many places the edges of scales are blunt and the outer layer of scales is significantly damaged.

Table 1 Parameters of wool before and after 1 and 4 months of ground disposal

Disposal months		Thickness (μm)			Breaking force (cN)		
		0	1	4	0	1	4
Wool (cover)		32 ± 9.1	31 ± 8.8	31 ± 7.9	22.2 ± 10.4	20.8 ± 9.8	22.8 ± 13.4
Wool (core)	–	27 ± 8.5	27 ± 9.2	27 ± 9.5	28.0 ± 11.7	22.9 ± 16.2	22.5 ± 15.2
	Gel		26 ± 8.9	26 ± 9.2		24.4 ± 8.4	24.4 ± 12.6

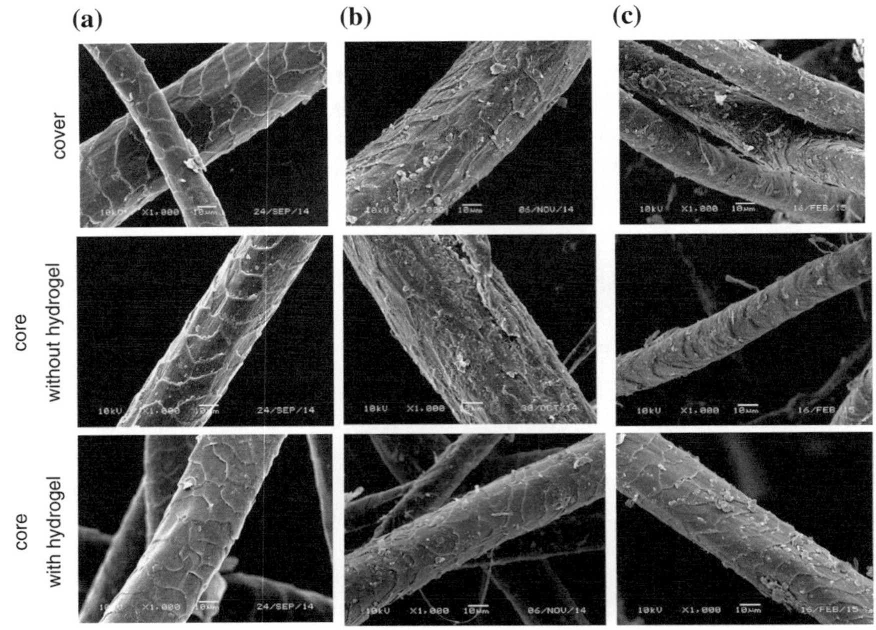

Fig. 4 Morphology of the surface of wool fibres: **a** before; **b** after 1 month; **c** after 4 months of ground disposal

After 4 months of disposal in the case of fibres taken from the covering non-woven further signs of damage are visible. For most of the fibres scales surface is heavily damaged and their edges are significantly blunted. For samples taken from the ropes, which do not contain the hydrogel a few fibres covered with scales are observed. Scales of this fibres possess strongly damaged surface and blunt edges. For the remaining fibres scales are almost invisible. In these fibres, the outer cuticle is heavily or completely damaged. For some fibre destruction goes deeper into the lower-lying layers.

In the literature two mechanisms of the wool biodegradation involving microorganism are proposed (Filipello Marchisio 2000). The first mechanism consists on surface erosion, which results in progressive degradation of the fibres from the outer cuticle to the inner cortex. The second mechanism, so called tunnelling or radial penetration, involves formation of holes in the fibre. The observed changes in the morphology of investigated fibres suggest the surface erosion mechanism.

In the case of fibres taken from ropes produced with the addition of the hydrogel on the surface of the fibres the scales are well visible. For most fibres scales exhibit only slightly damaged surface and slightly blunted edges.

The hydrogel is a cross-linked polymeric material that possesses the ability to swell and retain a significant fraction of water. The hydrogel was added to the ropes to increase their absorption capacity and improve their water storage potential.

Analysing microphotographs of wool from the ropes produced with the addition of the hydrogel one can see the swollen hydrogel located in the space between fibres as well as the hydrogel firmly adhered to the surface of the fibres (Fig. 5).

One can conclude that by absorbing large quantities of water the hydrogel sucks water from the environment and reduces the fibres humidity. In this way the hydrogel spoil conditions for the growth of microorganisms inside the fibres and delay their biodegradation. Simultaneously one can notice that the surface of scales is clean and is not covered by bacteria or other microorganism. Such picture may suggest that the hydrogel wrap microorganisms inside its network structure. By wrapping the hydrogel immobilises the enzymes and reduce their activity.

Figure 6 present microphotographs at magnification ×2000 of mechanically damaged fibres after 4 months of disposal. For all fibres both for fibres taken from the cover as for fibres from the core the widespread microbiological damage reaching splitting of the fibrillar structure is observed.

It is known that the mechanical damage of fibres results in the cracking of the outer cuticle layer. The open cracks enables easy access and penetration of enzymes to deeper layers of fibres. In such circumstances the biodegradation proceeds very fast and quickly leads to the fibres fibrillation.

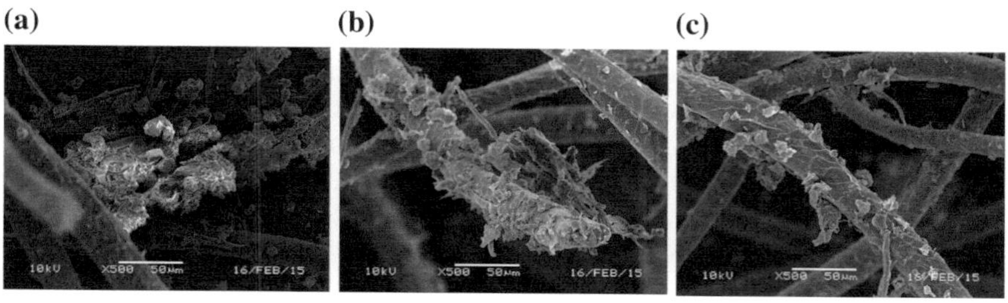

Fig. 5 Wool fibres after 4 months of deposal with the hydrogel: **a** the swollen hydrogel in the space between fibres; **b** and **c** the hydrogel adhered to the fibres surface (magnification ×500)

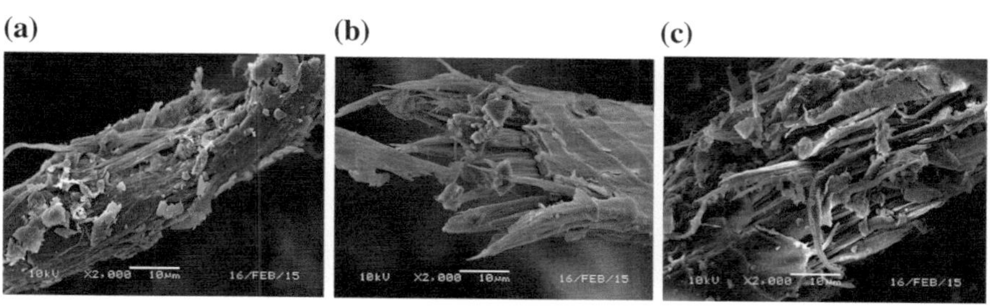

Fig. 6 Decay of mechanically damaged fibres: **a** fibres taken from the cover; **b** fibres taken from the core (without hydrogel); **c** fibres taken from the core (with hydrogel)

The splitting of the fibres structure is frequently observed in fibres taken from the cover of the ropes. The cover was formed from the woollen nonwoven, which was produced by punching technique. This technique is connected with intensive mechanical treatment, what causes a weakening the fibres manifested by a lower mechanical strength and bigger number of mechanically damages. In the soil such damages are easily attacked by microorganism, what quickly leads to fibres fibrillation.

Fibres Composition

Figure 7 presents the FTIR spectra of wool used for producing of ropes.

On the spectra typical bands, amide A, amide B and amide I–III, assigned to the peptide bonds and characteristic for wool keratin are visible. At 3292 cm^{-1} the amide A band and at 3076 cm^{-1} the amide B band connected both with the stretching of N–H bonds are observed. At 1657 cm^{-1} the band amide I related to the stretching vibration of C=O bonds is visible. At 1556 cm^{-1} the characteristic band amide II connected with the bending of N–H bonds and stretching vibrations of C–N bonds occurs. At 1244 cm^{-1} the amide III band related to the combination of C–N stretching and N–H in plane bending as well as some distribution form C–C stretching and C=O bending is observed. The amide I–III bands are attributed to the conformation of the keratin molecules. The intense amide I band corresponds to the α—helix keratin.

Figures 8 and 9 present the FTIR spectra before as well as after 1–4 months of ground disposal of fibres from the cover and the core of the ropes.

For all fibres, both from the cover and the core, the ground disposal does not affect the peak position of amide bands and only slightly influences on their intensity. The position of the amide peaks does not change also by addition of hydrogel. The same position and similar intensity of amide bands observed before and after ground

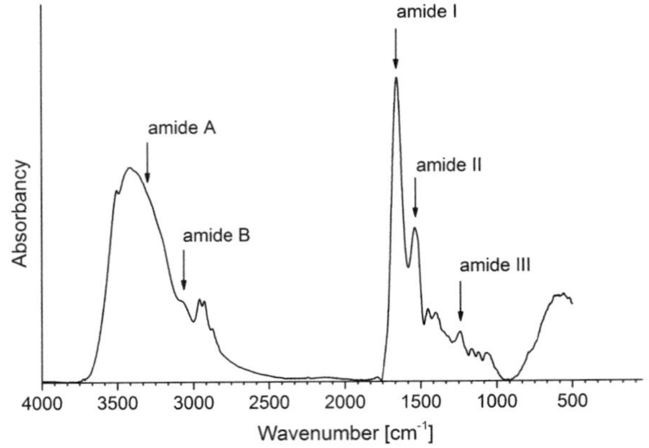

Fig. 7 FTIR spectra of wool

Fig. 8 FTIR spectra of wool fibres forming the cover of ropes before and after ground disposal

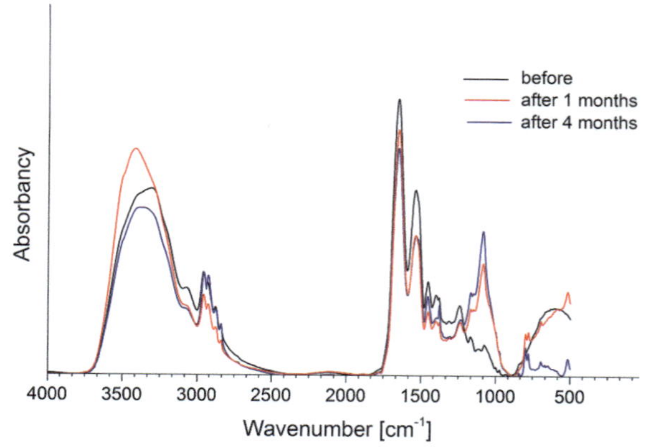

Fig. 9 FTIR spectra of wool fibres taken from the core of the ropes before and after ground disposal: **a** without gel; **b** with gel

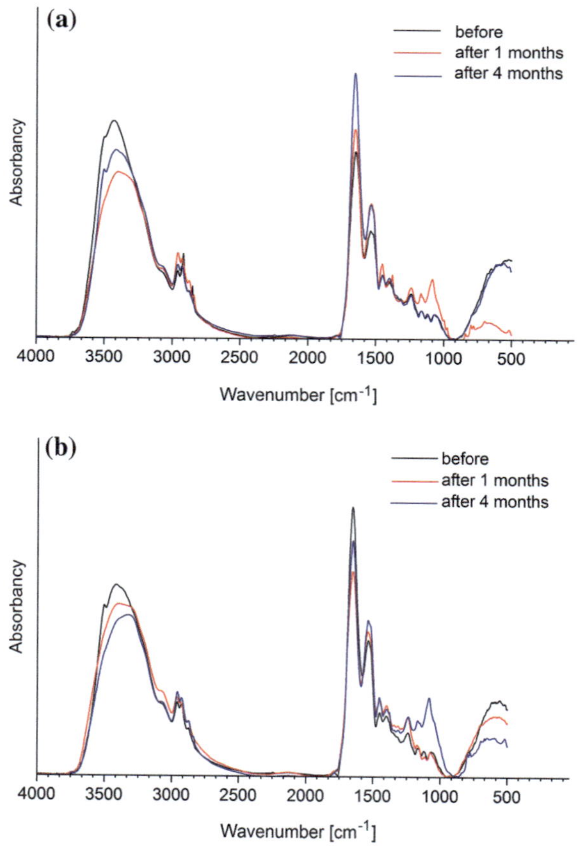

disposal testifies to the fact that the four-month biodegradation that occurs during disposal does not lead to disruption of peptide bonds. It is means that the four months disposal in autumn and winter season does not decompose the wool keratin.

It is known the biodegradation of wool keratin starts with cleavage of disulphide bonds. The bands related to such bonds are located near 500 cm^{-1} and on FTIR spectra are not well visible. The bands assigned to S–S bonds are well recognized on Raman spectra, what will be analysed in further investigations.

On FTIR spectra the cleavage of disulphide bonds may suggests the growing intensity of S–O band at 1078 cm^{-1}. The highest increase of this band is observed for fibres taken from the cover, what is connected with more intense fibrillation of this fibres.

Conslusions

During ground disposal of coarse ropes filled with wool fibres and covered with woollen nonwovens the microbiological decomposition of wool is observed. As a result, the decrease of breaking force of fibres is observed. Simultaneously the destruction of scales on the surface and changes in the chemical composition of fibres are observed. The significant changes are visible already after 1 month of the ground disposal. After 4 months the outer layer of the fibres is strongly damaged. In mechanical damaged fibres the biodegradation proceeds faster and leads to fibre fibrillation. For fibres taken from ropes, which contains the hydrogel the biodegradation is delayed.

Despite observed damages the ropes still maintain their shape and mechanical strength sufficient to fulfil their erosion protection function.

Acknowledgments The authors gratefully acknowledge the funding by ERANET-CORNET consortium under international research project PROGEO "Sustainable erosion protection by geotextiles made of renewable resources including innovative manufacturing and installation technology". DZP/CORNET-16/628/2014.

References

Russell SJ, Swan P, Trebowicz M, Ireland A. Review of wool recycling and reuse. *Proceeding of 2 nd International Conference on Natural Fibers,* Fangueiro, R (Ed), pp.1-21, ISBN:978-989-98468-4-5,Sao Miguel Azores, April 2015, Universidade of Minho, (2015).

Johnson NAG, Wood EJ, Ingham PE, McNeil SJ, McFarlane ID. Wool as a Technical Fibre. Journal of the Textile Institute 2013, 94:3-4, 26-41.

Patent DD 110 905.

Helbig R, Arnold R, Erth H, Roess T, Hevert W, Lischkowitz H., New technologies for manufacturing extra coarse rope-like biodegradable geotextiles, Technische Textilien, 2006, 49, 185-187, 244-247.

Rickson RJ. Controlling sediment at source: an evaluation of erosion control geotextiles. Earth Surf. Process. Landforms 2006, 31, p.550–560.

Álvarez-Mozos J, Abad E, Goñi M, Giménez R, Campo MA, Díez J, Casalí J, Arive M, Diego I. Evaluation of erosion control geotextiles on steep slopes. Part 2: Influence on the establishment and growth of vegetation. Catena 2014, 121, p.195–203.

Pekhtasheva E, Neverov A, Kubica S, Zaikov G. Biodegradation and biodeterioration of some natural polymers. Chemistry & Chemical Technology 2012, 6, p.263-280.
Kornilowicz-Kowalska T, Bohacz J. Biodegradation of keratin waste: Theory and practical aspects. Waste Management 2011, 31, p.1689-1701.
Filipello Marchisio, V. Keratinophilic fungi: Their role in nature and degradation of keratinic substrates. In: Kushwaha, R.K.S., Guarro, J. (Eds.), Biology of Dermatophytes and Other Keratinophilic Fungi. Bilbao 2000.

Renewable Materials for Stab Resistance

Marcus O. Weber, Susanne Aumann, Malin Obermann and Andrea Ehrmann

Abstract Different materials were investigated experimentally with respect to their stab resistance. The focus of this research is the use of renewable cellulosic materials, such as cotton and paper, in comparison with high-performance materials, such as aramid or ultra-high-molecular-weight polyethylene. The tests were carried out according to the standard of the Association of Test Laboratories for Bullet, Stab or Pike Resistant Materials and Construction Standards (VPAM). The results for common cellulosic materials achieve remarkable results in comparison with high-performance materials. Depending on the stab resistance being compared in terms of fabric thickness or mass per unit area, especially standard paper performed well, while even common cotton depicted a significant stab resistance. The results of these investigations demonstrate the advantages and limitations of renewable cellulosic materials in comparison to high-performance fibrous structures.

Keywords Cellulosic materials · Stab resistance · Knitted fabric · Structural parameters

Introduction

For a long time stab resistance is a need for human protection against arrows, swords, knifes and other weapons, and that is why the development has already started in early history. Historical protective clothing used paper, leather, wood or metal for armouring the warriors and other persons at risk. Since the development

M.O. Weber (✉) · S. Aumann · M. Obermann
Faculty of Textile and Clothing Technology, Niederrhein University of Applied Sciences, Mönchengladbach, Germany
e-mail: Marc.Weber@hs-niederrhein.de

A. Ehrmann
Faculty of Engineering Sciences and Mathematics, Bielefeld University of Applied Sciences, Bielefeld, Germany

R. Fangueiro and S. Rana (eds.), *Natural Fibres: Advances in Science and Technology Towards Industrial Applications*, RILEM Bookseries 12,
DOI 10.1007/978-94-017-7515-1_28

of man-made fibres, especially high-performance fibres, a broad variety of materials can be used for stab resistant garments.

Several groups working on stab resistant materials concentrated on woven fabrics (Firouzi 2014; Textor et al. 2006), either purely or combined with thin metal plates (Horsfall et al. 2013). Compound fabrics were investigated quite often, too (Yong 2014; Li et al. 2013; Hou et al. 2013; Stojanovic et al. 2013; Chen et al. 2008). Researchers found, e.g., that the bonding process in textile composites influenced the mechanical and stab resistance properties of the composite fabrics (Lin et al. 2008).

Additionally, the influence of different finishing procedures was investigated. Laminating aramid fabrics with polyethylene and other films of various thicknesses was found to improve the stab and puncture resistance (Mayo et al. 2009). A coating with natural rubber latex could increase the energy absorption and thus reduce the damage of the base nonwoven fabric (Hassim et al. 2012). Adding a shear thickening fluid to Kevlar and Nylon fabrics improved the puncture resistance significantly and the cut protection slightly (Decker et al. 2007). Another study correlated the stab resistance to the hardness of the dispersed particles in a shear thickening fluid and the inter-yarn friction to the puncture resistance (Gong et al. 2014).

Theoretical examinations of the stab resistance of fibrous structures have shown that puncture through a fabric occurs in different stages, including slippage of the penetrating tip into an inter-fiber spacing (Termonia 2006), which underlines the advantages of dense woven fabrics for stab resistance. Knitted fabrics, however, offer more wearing comfort due to their flexibility making them superior structures for sweaters, jackets and vests to be worn all-day. Nevertheless, research reports about stab-resistant knitted are scarce. Comparing different fibre types of diverse high-performance materials, building fabrics of different structures and combining it into multilayer fabrics consisting of fleece and jersey layers from p-aramid and similar materials exhibited the best stab resistance (Flambard 2000). In another examination of different knitted fabrics, stab resistance was correlated with cut resistance, low flexural rigidity, low elastic modulus and low inner layer friction (Flambard and Polo 2004). Comparing different structures, tuck loops have shown to improve cut and stab performance, compared with loser structures, such as plush or jersey (Alpyildiz et al. 2011). Many questions are still open, especially according to the yarns under examination, since most studies concentrated on high-performance materials which are usually not perfect in terms of wearing comfort.

This study focuses on maximum comfort while providing a significant protection capability.

Experimental

The knitted fabrics under examination are produced from ringspun cotton yarn (properties cf. Table 1) on a circular knitting machine Relanit (Mayer & Cie.) with gauge E28. The structure "Lapique" used for this study is depicted in Fig. 1 as construction pattern as well as microscopic picture.

Table 1 Cotton yarn properties

Ring spun	16.3 tex
Tensile strength	(14.96 ± 1.1) cN/tex
Max. elongation	(5.62 ± 0.29) %
Twist	921 T/m

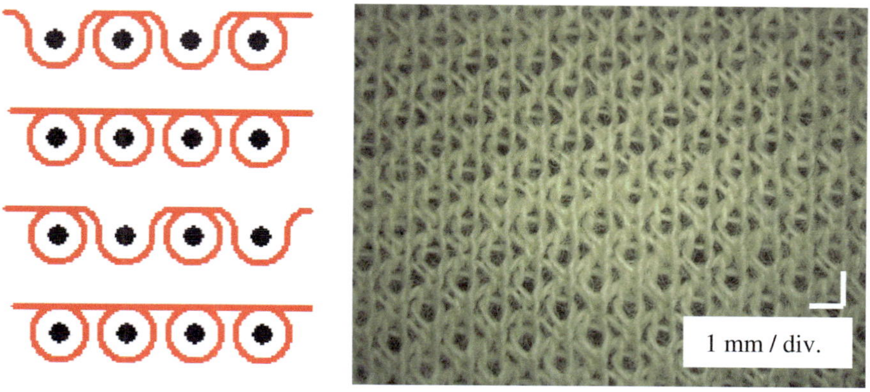

Fig. 1 Lapique knit construction (*left panel*) and microscopic picture (*right panel*)

While most measurements were performed on fabrics knitted with tightness factor 9, a series of fabrics with different tightness factors was additionally tested. The resulting fabric parameters are listed in Table 2.

Additionally, common copy paper with mass per unit area 80 g/m^2 was tested.

Finally, the tests of these natural materials are compared with the previously best stab resistant fabric, a spacer fabric knitted from ultra-high-molecular-weight polyethylene (UHMW-PE), to put the results of this study in the context of those achieved in previous tests on high-performance materials (Vidzem et al. 2013; Weber et al. 2014a, b).

The VPAM (VPAM 2004) stab resistance tests are carried out with a standard knife (Fig. 2), falling from a specific height driven by a defined weight to the samples lying on ballistic plasticine. The penetration depth in the plasticine is measured after removing the plasticine on one side of the penetration channel; opposite to the

Table 2 Parameters of Lapique knitted fabrics with different tightness factors

Tightness factor	Mass per unit area (g/m^2)	Thickness (mm)	Stitch length (mm)	Stitch width (mm)
8	119	0.81	1.59	0.96
9	108	0.82	1.62	0.98
10	105	0.83	1.71	0.99
11	103	0.83	1.73	1.01
12	96	0.81	1.74	1.04

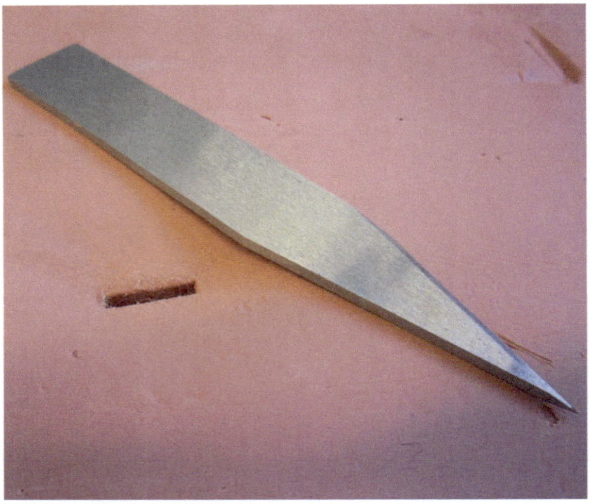

Fig. 2 Knife of specific dimensions and sharpness for stab resistance (from Weber et al. 2014a)

British HOSDB test instruction using foam rubber to support the textile (Croft and Longhurst 2007). Measurements were performed three times per experiment to decrease the measurement error. Additionally, penetration depth and cutting width were measured for each test since former investigations have shown that, depending on the material under examination; both values may be misinterpreted in case of testing knitted fabrics (Aumann et al. 2013). Measuring both values thus reduces the risk of erroneously over- or underestimating a measurement result. This is of special importance in the case of comparing fabrics with significantly differing bending and elongation properties, such as paper and knitted fabrics, which show correspondingly different indentations in the surface of the plasticine.

According to previous studies, the setting of the test unit was adapted to the natural materials (Ehrmann et al. 2013; Chung et al. 2013; Weber et al. 2014b) by reducing the falling height to 10 cm, which is correlated with a falling energy of 2.5 J.

Results

Figures 3 and 4 depict the results of the stab resistance tests for the knitted Lapique cotton fabric and paper sheets, compared with the best result of a UHMW-PE knitted fabric. In both cases, the best results can be found at the lower left side of the graph, where the penetration depths as well as the thickness or mass per unit area, respectively, have the smallest values.

Figure 3 shows the results as a function of the fabric thickness for different numbers of Lapique knitted fabric and paper layers. For both materials, the expected decrease of the penetration depth with increasing thickness of the fabric stack is visible. While the penetration depth of the knitted cotton fabric is approximately doubled by decreasing the layer depth by a factor of 2, this relation is

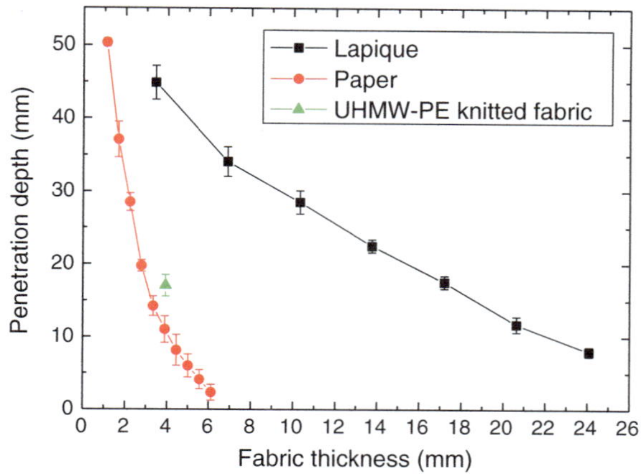

Fig. 3 Penetration depth versus fabric thickness

much stronger for paper. Here the reduction of the penetration depth with increasing number of layers is much stronger.

Additionally it is clearly visible that paper is superior to the cotton knitted fabric in terms of fabric thickness. This finding can be explained by paper being much denser than the knitted fabrics under examination. While the knitted fabrics, independent of the yarn used, are relatively open structures, with several holes where the blade can penetrate without being directly influenced by the yarn, paper has a closed surface, with the blade being decelerated directly from the first contact. Thus, each sheet of paper contributes to the stab resistant effect of the paper, slowing down the blade with each new layer.

In the knitted fabrics under examination, however, the structure is relative loose, so that the blade is only decelerated by one layer when its point has already reached the next layer or even further ones. Thus, the thickness-dependent penetration depth curve of the cotton knitted fabric has a smaller gradient than the curve obtained for paper.

Figure 4, however, shows the same results as functions of the mass per unit area. Here the qualities of the stab resistant effects of paper and the knitted cotton fabric are nearly exchanged: The penetration depth is almost always smaller for the Lapique fabric, if identical masses per unit area are compared.

This apparent change of the quality of the structures under examination can be explained by the significantly different densities of both materials. While the high density of paper is advantageous, if the results are compared with respect to the fabric thickness, is becomes a disadvantage when the mass per unit area is taken into account.

Comparing both graphs, it becomes visible that similar masses per unit area result from overall fabric thicknesses which differ approximately by a factor of 6. A mass per unit area of ~ 1000 g/m^2, e.g., is reached by ~ 1.1 mm paper or by ~ 7 mm knitted fabric. Thus the knitted fabrics use a much longer distance to

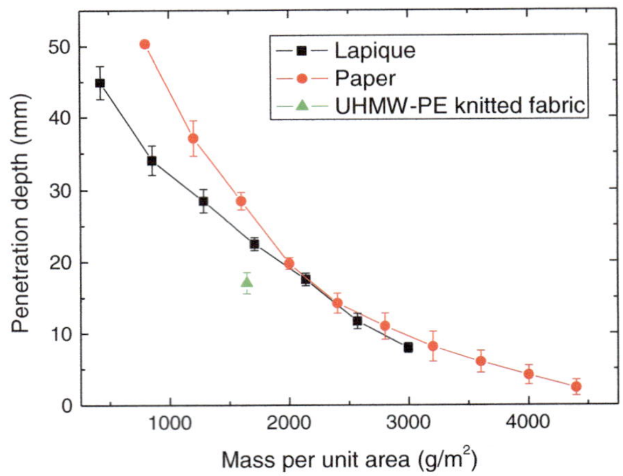

Fig. 4 Penetration depth versus mass per unit area

decelerate the knife than a pile of paper with the same mass per unit area. In this context, the disadvantage of the process of the blade penetrating the stack of knitted fabrics described before is compensated by the longer way on which the blade can be influenced.

The results display that—in terms of the thickness—smaller penetration depths can be reached with paper layers (Fig. 3), while the knitted fabric should be preferred in terms of mass per unit area (Fig. 4).

Figures 3 and 4 also contain the best result of tests performed under identical conditions of a spacer fabric knitted from UHMW-PE. In correlation to the mass per unit area (Fig. 4), the UHMW-PE fabric is clearly superior to paper and cotton fabrics. However, it should be mentioned that the mass per unit area difference between the UHMW-PE knitted fabric and the cotton knitted fabric of comparable penetration depth is only ~30 %.

With regard to the fabric thickness (Fig. 3), paper even offers a smaller penetration depth than the high-performance material. Nevertheless it must be recognized that paper has a much higher bending rigidity and lower drapability than knitted fabrics and can thus not be used to create a comfortable garment.

Comparing Figs. 3 and 4 clearly shows the influence of the density of the samples under examination. Thus in Fig. 5 the penetration depth is correlated with the tightness factor of Laqique knitted cotton fabrics, with higher tightness factors corresponding to looser fabrics. Apparently, tighter fabrics offer a better stab resistance. The increase of the penetration depth for higher tightness factors is correlated with a similar decrease of the mass per unit area (cf. Table 2), while the layer thickness is relatively similar for all tightness factors.

It can be concluded that, taking into account both the mass per unit area and the thickness of the knitted fabrics, tighter fabrics should be preferred for a better stab resistance.

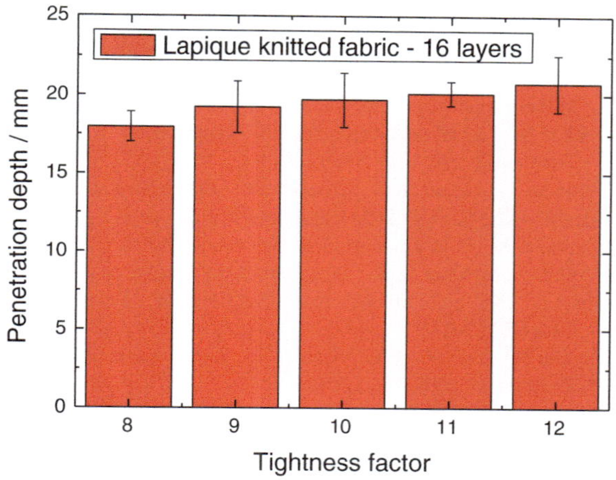

Fig. 5 Penetration depth versus tightness factor

Conclusion

To conclude, the natural cellulose materials cotton and paper were examined with respect to their stab resistant properties. Surprisingly, paper showed even superior stab resistance compared with a UHMW-PE knitted fabric as a function of the fabric thickness, and increasing the cotton fabric's mass per unit area by only ~30 % would result in a similar stab resistance as that of UHMW-PE.

This study shows that renewable cellulosic materials can have distinctive protective properties. Especially in stab resistance research, the results of this study suggest future tests of denser knitted structures, especially with higher knitting machine gauge, produced of cotton. This could combine the advantages of dense paper sheets and flexible knitted fabrics, to develop stab resistant garments which are created on a renewable base with high comfort and good protection levels.

Acknowledgements The authors gratefully acknowledge the funding by the Bundesministerium für Wirtschaft und Technologie due to a decision of the Deutscher Bundestag under grant KF 2233806.

References

Alpyildiz T., Rochery M., Kurbak A., Flambard X.; Textile Research Journal 81, 205–214 (2011).

Aumann, S.; Ehrmann, A.; Vidzem, B.; Brücken, A.; Weber, M. O.; Bache, T.: Comparison of penetration depth and cutting width in stab protection measurements according to VPAM. 7. Aachen Dresden Textile Conference, Aachen 28–29 November 2013

Chen J. M., Hsieh J. C., Lou C. W., Hsing W. H., Yang H. J., Lin J. H.; International Conference on Smart Materials – Smart/Intelligent Materials and Nanotechnology/2nd Int Workshop on Functional Materials and Nanomaterials: Advanced Materials Research 55–57, Chiang Mai, Thailand, 417–420 (2008)

Chung, S.; Ehrmann, A.; Weber, M.: Accuracy of thickness measurements on knitted fabrics. Melliand International 01 (2013) 32–33

Croft J.; Longhurst D.: HOSDB Body Armour Standards for UK Police (2007), Part 3: Knife and Spike Resistance. Publication No. 39/07/C (2007)

Decker M. J., Halbach C. J., Nam C. H., Wagner N. J., Wetzel E. D.; Composites Science and Technology 67, 565–578 (2007)

Ehrmann A., Aumann S., Brücken A., Weber M. O.; Technische Textilien 56, 198–199 (2013)

Flambard X., PhD Thesis, Université des Sciences et Tecnologies de Lille I (2000)

Flambard X., Polo J.; Journal of Advanced Materials 36, 30–35 (2004)

Firouzi D., Foucher D. A., Bougherara H.; Journal of Applied Polymer Science 131, 40350 (2014)

Gong X., Xu Y., Zhu W., Xuan S., Jiang W., Jiang W.; Journal of Composite Materials 48, 641–657 (2014)

Hassim N., Ahmad M. R., Ahmad W. Y. W., Samsuri A., Yahya M. H. M.; Journal of Industrial Textiles 42, 118–131 (2012)

Horsfall I., Watson C. H., Champion S. M.; Journal of Applied Mechanics – Transactions of the ASME 80, 031901 (2013)

Hou L., Sun B., Gu B.; Applied Composite Materials 20, 569–585 (2013)

Li T. T., Wang R., Lou C. W., Lin, J. H.; Journal of Industrial Textiles 43, 247–263 (2013)

Lin C. C., Lou C. W., Hsing W. H., Ma W. H., Lin C. M., Lin J. H.; Advanced Materials Research 55–57, 429-432 (2008)

Mayo Jr. J. B., Wetzel E. D., Hosur M. V., Jeelani S.; International Journal of Impact Engineering 36, 1095–1105 (2009)

Stojanovic D. B., Zrilic M., Jancic-Heinemann R., Zivkovic I., Kojovic A., Uskokovic P. S., Aleksic R.; Polymers for advanced technologies 24, 772–776 (2013)

Termonia Y.; International Journal of Impact Engineering 32, 1512–1520 (2006)

Textor T., Schroeter F., Schollmeyer E.; Symposium on Smart Nanotextiles held at the 2006 MRS Spring Meeting: Materials Research Society Symposium Proceedings vol. 920, San Francisco, CA, pp 1–11 (2006)

Vidzem, B.; Aumann, S.; Heimlich, F.; Werner, R.; Ehrmann, A.; Obermann, M.; Brücken, A.; Weber, M. O.; Bache, T.: Stab resistance of textile materials. 7. Aachen Dresden Textile Conference, Aachen 28–29 November 2013

(VPAM) Vereinigung der Prüfstellen für angriffshemmende Materialien und Konstruktionen: Prüfrichtlinie „Stich- und Schlagschutz", VPAM-KDIW 2004 (2011)

Weber, M. O.; Aumann, S.; Vidzem, B.; Obermann, M.; Brücken, A.; Ehrmann, A.; Bache, T.: Comparison of different yarns for stab resistant knitted fabrics. 47th International Congress IFKT, 25–26 September 2014a, Izmir, Turkey

Weber, M. O.; Aumann, S.; Brücken, A.; Ehrmann, A.; Bache, T.: Stab resistant knitted clothing. 8. Aachen Dresden Textile Conference, Dresden 27–28 November 2014b

Yong K. C.; Polymers & Polymer Composites 22, 375–380 (2014)

Hemp Fibre from Crops Grown on Reclaimed Land for the Production of Sanitary Mats

Jerzy Mańkowski, Andrzej Kubacki, Jacek Kołodziej, Maria Mackiewicz-Talarczyk, Przemysław Baraniecki and Irena Pniewska

Abstract The technology to produce biodegradable sanitary (disinfectant) mats for the creation of protection zones has been developed and implemented at the Institute of Natural Fibres and Medicinal Plants. Due to the cyclically repeated outbreaks of infectious diseases, disinfectant mats are being used increasingly. These mats may be quickly saturated with a disinfectant solution to prevent spreading of infection and to limit spreading of epidemic diseases; disinfectant solutions are applied to the wheels of cars which cross the soaked mats. Natural fibre mats consist of three layers. The surface layer is a natural nonwoven fabric based on flax and hemp fibre. The middle layer of the mat is jute fabric, which has a structure characterized by the ability to return to the original form. To form the bottom mat layer, an elastic natural resin was used providing appropriate level of insulation and preventing seepage of disinfectant solution into the ground (e.g. soil). The developed mats are completely biodegradable under the influence of external conditions and therefore, do not pollute the environment.

Keywords Hemp · Flax · Natural fibres · Nonwoven · Sanitary mats · Epidemic · Biodegradation

Introduction

The project EKOHEMPKON Life+: "Remediation Method of Degraded Land by Cultivation of Industrial Hemp in the Region of Lignite Mine Konin", led by Institute of Natural Fibres and Medicinal Plants in Poznań, was started on 01.

J. Mańkowski (✉) · J. Kołodziej · M. Mackiewicz-Talarczyk · P. Baraniecki · I. Pniewska
Institute of Natural Fibres and Medicinal Plants (INF&MP), Poznan, Poland
e-mail: jerzy.mankowski@iwnirz.pl

A. Kubacki
LENKON, Steszew, Poland

R. Fangueiro and S. Rana (eds.), *Natural Fibres: Advances in Science and Technology Towards Industrial Applications*, RILEM Bookseries 12,
DOI 10.1007/978-94-017-7515-1_29

October 2012. Location of the project is post-mining area of Lignite Mine "Konin" in Kazimierz Biskupi, wielkopolskie district.

The crop rotation system used for remediation is based on two crops: industrial hemp and alfalfa. The cultivated plants are plowed after mowing (harvesting), their biomass fertilize the soil, accelerate and replicate layer of humus of degraded lands.

Besides the process of degraded land remediation, the project will also show how remediated land would be used for cultivation of industrial crops—raw materials for production of pulp and paper, construction materials, composite materials and energy as well as for the production of natural disinfectant (sanitary) mats. The relevant research and experiments included: evaluation of the usefulness of hemp fibre for the production of sanitary mats, production of the batch of nonwovens, which were subjected to laboratory tests—determining their strength and biodegradability (Cierpucha 2013; Mańkowski et al. 2013, 2014).

Construction and the Properties of Sanitary Mats

Worldwide globalization has led to the rapid development of trade thereby contributing indirectly to the spread of various diseases transmissible by the means of transport.

Ways to reduce the spread of infectious disease of livestock are regulated by legal acts of the European Union and individual member states. The proliferation of sources of infectious animal diseases, in particular diseases which can also be dangerous to humans (e.g. Bird flu) is of particular public concern. Transmission of infectious disease occurs mainly by direct contact with sick animals, as well as with the secretions that may be present on the foods of animal origin and on contaminated equipment and means of transportation. Developed nonwoven mats on the basis of hemp and flax fibre, impregnated (sealed) with disinfecting liquids, are proposed as a barrier to bacteria and viruses, for rapid purification of footwear and wheels of the means of transport which are in direct contact with sources of diseases outbreaks.

Barrier nonwoven mats are manufactured using needling technique with application of flax and hemp fibre. Needled nonwoven meet the demanding functional requirements like high tensile strength and appropriate resilience due to the high possibility of forming the structure.

Applied needling techniques involve the binding of a longitudinal fibre structure with its own fibre by special needles. Selection of the appropriate raw material allows to obtain the desired product structure characterized by the adequate properties such as appropriate: thickness, degree of development of the edge line of the mat, surface friction coefficient, characteristic elastic properties, tensile strength and specific physico-chemical properties.

Mats made of natural fibres are composed of three layers. The surface layer is a natural non-woven fabric based on 80 % hemp and 20 % flax fibres. Bast fibres have got properties that improve the absorption coefficient of disinfectant media and cause their longer maintenance in well soaked fleece (nonwoven).

The middle layer is jute fabric, which strengthens the construction of a non-woven and has a resilient structure characterized by the ability to return to original form. The two layers are joined together by needling, using natural fibres for stitching and reinforcing nonwoven structure.

The bottom layer is covered by flexible natural resin (latex) that provides the appropriate level of insulation, prevents penetration of the disinfectant to the ground and stabilizes the position of the mat on slippery surfaces. This is important, when heavy lorries are crossing the sanitary mats sealed with disinfectants.

Materials and Methods

The raw material for the production of nonwovens is obtained on the special decoration line. The main task of the mechanical processing of fibrous materials is to extract from them the maximum amount of pure fibre by loosening tissues of the stem in the process of crushing and breaking, which take place during breaking and separating fibres from wooden parts in the process of scutching and shaking.

The technology of extracting decorticated (not retted) hemp fibre is based on the use of high-yield machines, where is obtained decorticated fibre of high purity and suitable length and thinness.

Subsequently, was specified method of forming a fleece and nonwoven stitching and was prepared the test equipment for forming and molding of mats. The appropriate reinforcing fabric, content of natural fibre blend were chosen and an insulating layer with natural resins was elaborated.

The laboratory research included quality characteristics of sanitary mats; testing of strength (resistance) of the mats to the breaking forces. This is main parameter responsible for the functional characteristics of disinfectant mats. As a result, the nonwoven fabric samples with extreme weighing, i.e. 900 and 1350 g/m^2 were selected.

Results and Discussion

Conducted tests showed that the strength of disinfectant mats is not directly affected by the thickness of nonwoven fabrics but rather by a kind of reinforcement layer constituting the mat (Fig. 1).

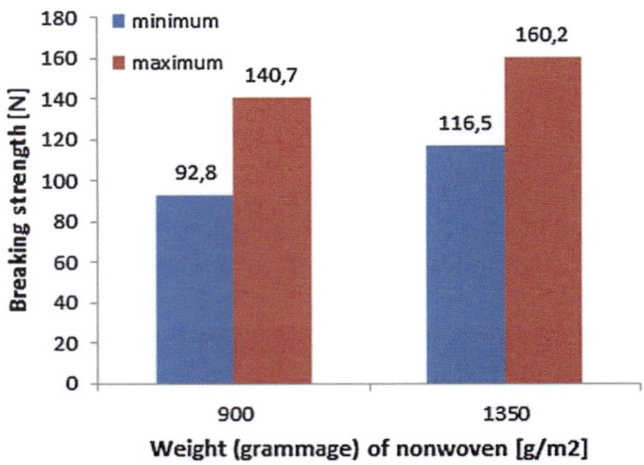

Fig. 1 The average results of tensile strength of the disinfectant mats with a weighting of 900 and 1350 (g/m^2)

Sanitary mats has also been studied with regard to the degree of biodegradability; the samples of manufactured nonwovens were subjected to the weather conditions, then visual observations (organoleptic) and strength tests within 12 months were conducted.

Biodegradation processes were evaluated on the basis of:

- sensory (organoleptic) assessment of changes in the appearance,
- changes in strength of the samples.

Measurements of mechanical and durability properties were evaluated according to the Polish standard. Organoleptic changes the appearance of the nonwovens mats evaluated at the beginning, in the 2nd month, in 3rd month, 4th month, 5th month, 6th month, 8th month, 10th month, and 12th month are shown in the pictures below (Fig. 2).

These observations confirmed the increasing biodeterioration changes in the structure of the nonwoven fabric as time goes on, so that after a period of 10–12 months, there was observed complete destruction of the nonwoven structure of the sanitary mats.

Strength tests are shown in Fig. 3. Observations of changes in the strength and durability characteristics of the mats worsen by a gradual decline in strength to the overall degradation of the structure of the mat.

The studies confirm that after a period of 10 months 90 % of nonwoven fabrics were biodeteriorated. Based on the results of sensory evaluation and analysis of the strength parameters high-vulnerability of flax/hemp fibres to biodegradation was confirmed (Fig. 3). It means that examined sanitary mats can be considered as completely safe for the natural environment (Mańkowski et al. 2009).

After unfolding the mat was evenly sprayed with disinfectant solution (liquid) properly prepared, selected by the veterinary services according to the type of animal infectious disease.

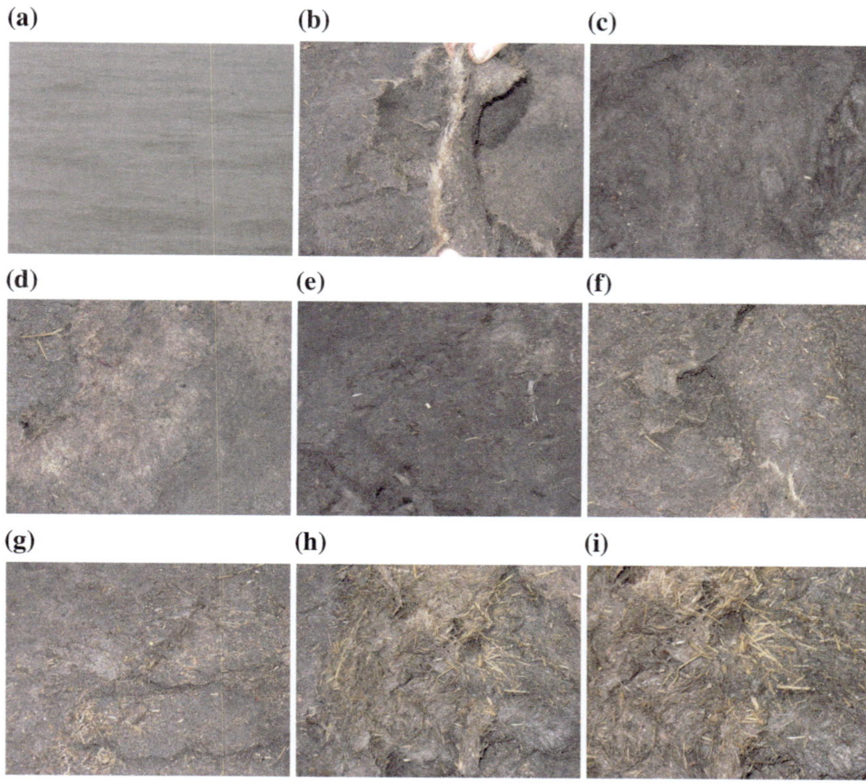

Fig. 2 The process of biodegradation (decomposition) of sanitary mats: **a** at the beginning, **b** 2nd month, **c** 3rd month, **d** 4th month, **e** 5th month, **f** 6th month, **g** 8th month, **h** 10th month, **i** 12th month

Fig. 3 The strength of the sanitary mats in the different phases of biodegradability (decomposition)

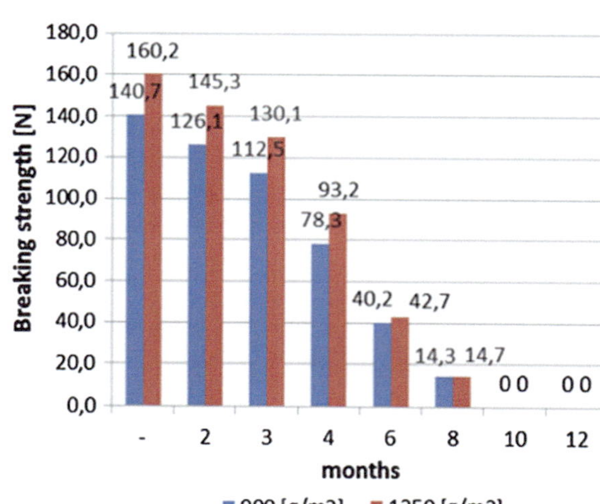

After consultation with veterinary services, the Institute of Natural Fibres and Medicinal Plants will officially prepare and propose the range of disinfectants recommended to apply why using mats made of natural fibres:

- Caustic soda NAOH—5 % solution—bactericidal and virucidal.
- Chloramine T—1 % solution—bactericidal, fungicidal and tuberculocidal viral,
- ALDEWIR—5–10 %—viral, and bactericidal
- INCIDIN Active—bactericidal,
- DESRESON AF 1 L—liquid—broad spectrum of activity including against the avian influenza virus
- FAM 30—bactericidal, viral, fungicide.

Biodegradable sanitary mats made of natural fibre, has been tested at the Department of Veterinary Hygiene and obtained a positive opinion of the Provincial Inspector for Animal Health in Poznan, combating infectious diseases and providing the protection of animals.

Conclusions

Sanitary mats developed at the Institute of Natural Fibres and Medicinal Plants in Poznan are used to create buffer zones to prevent the spread and outbreak of infectious diseases such as e.g. the bird flu (avian influenza). Their use on footpaths and vehicle crossings within the chicken coop, cowsheds, piggery, dairy plant, fodder mills, slaughterhouses, food processing plants, apiaries, livestock collection centers, mushroom producing farms, country border crossing etc. is recommended.

These mats are made from 100 % natural fibres: hemp, flax and jute. Natural fibres have significant absorbent capacity (hygroscopic abilities) compared to chemical fibres. This means that they can absorb and retain incomparably larger amounts of disinfectant solutions used for decontamination. The used disinfectant (recommended by the veterinarians), as a result of adsorption forces, drips and seeps into the ground to a lesser extent (retained by the bottom layer of mats reinforced with resins). The strength of the tested sanitary mat is not directly affected by the thickness of this nonwoven fabric, but rather by a kind of reinforcing fabric (which is the layer of the mat).

As the mats are made from natural raw materials, there exist no problems with their disposal after use, because, in a short time, they are biodegradable (decomposed) or can be burned.

The research revealed that after a period of 10 months approx. 90 % of nonwoven fabric is biodegraded.

The results of the project were implemented and covered by patent protection (international patent No. EP 09717393.4); the manufacturing of mats was carried out in the Experimental Plant of the Institute (INF&MP) in Stęszew near Poznan.

References

Cierpucha W. (editor) (2013), Technologia uprawy i przetwórstwa konopi włóknistych (Technology of cultivation and processing of fibrous hemp), IWNiRZ, Poznań.

Mańkowski J., Kubacki A., Kołodziej J., Pniewska I., Rekultywacja terenów zdegradowanych w wyniku działania odkrywkowych kopalni (Recultvation of regions degraded by the activity of the opencast mine). Biuletyn Informacyjny PILiK Len i Konopie No 20/2013, 7–10.

Mańkowski J., Kołodziej J., Pudełko K., Kubacki A., Pniewska I. (2014). Rekultywacja terenów pogórniczych węgla brunatnego za pomocą uprawy konopi włóknistych i lucerny siewnej. Remediacja, rekultywacja i rewitalizacja, ed. (Recultivation of post-mining regions of lignite mine by cultivation of industrial hemp and alfaalfa. Remediation, recultivation and revitatisation), ed. Malina G., PZIiTS Poznań.

Mańkowski J., Kubacki A., Kołodziej J.: Maty dezynfekcyjne zapobiegające rozprzestrzenianiu się chorób zakaźnych wykonane na bazie włókien naturalnych. (Disinfectant mats on the base of natural fibres protecting againts the spread of infectious deseases), Len i Konopie No 13/2009.

Part VI
Market, Opportunities, Recycling and Sustainability Aspects of Natural Fibres

Natural Fibres and the World Economy

Terry Townsend and José Sette

Abstract World natural fiber production in 2013 is estimated at 33 million tons, including 26 million tons of cotton lint, 3.3 million tons of jute, 1.2 million tons of clean wool, 900,000 tons of coir, and 1.6 million tons of other natural fibers, including abaca, flax, hemp, kapok, ramie, sisal, and silk. The farm value of natural fiber production in 2013 was around US$60 billion. A reasonable estimate of total employment in natural fiber industries, including family labor, hired labor and employment in industries providing services to agriculture, and including both full time year round employment and part time or seasonal employment is around 60 million households or 300 million people, or about 4 % of the world's population.

Keywords Natural fiber · World economy · Production

Introduction

World natural fiber production in 2013 (the latest year of complete data) is estimated at 33 million tons (Fig. 1), including 26 million tons of cotton lint, 3.3 million tons of jute, 1.2 million tons of clean wool, and 900,000 tons of coir (fibers made from coconut husks). Production of all other natural fibers, including abaca, flax, hemp, kapok, ramie, sisal, silk, and other fibers summed to approximately 1.6 million tons.

The farm value of natural fiber production in 2013 was around US$60 billion (Fig. 2), of which cotton accounted for $45 billion, wool $8–9 billion and jute

T. Townsend (✉)
Cotton Analytics, Houston, TX, USA
e-mail: Terry@CottonAnalytics.com

J. Sette
International Cotton Advisory Committee (ICAC), Intergovernmental Organization, Washington, USA

R. Fangueiro and S. Rana (eds.), *Natural Fibres: Advances in Science and Technology Towards Industrial Applications*, RILEM Bookseries 12,
DOI 10.1007/978-94-017-7515-1_30

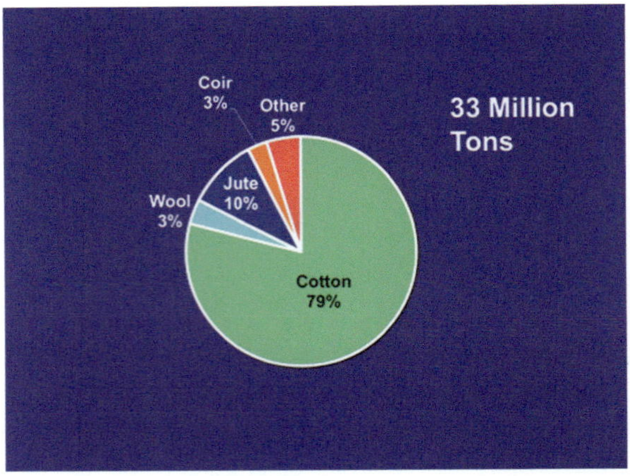

Fig. 1 World natural fibre production, 2013

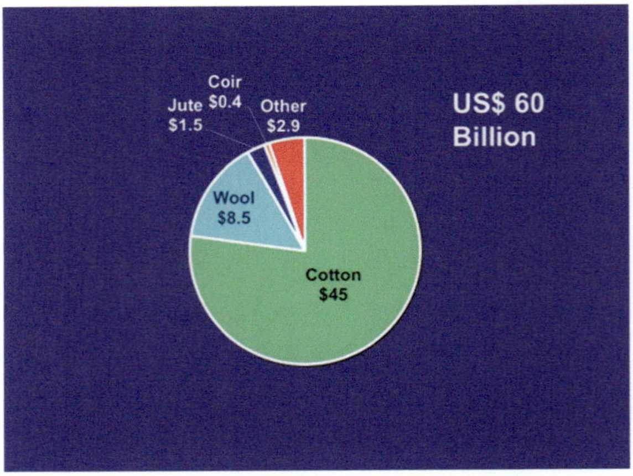

Fig. 2 Farm value of world natural fibre production, 2013

$2 billion. All other natural fibers together accounted for the balance of about $3 billion.

On average, wool was the most valuable fibre per kilogram in 2013 at more than US$7 (Table 1), even more expensive per kilogram than silk. However, there is a greater range of values for silk, with some countries reporting farm prices of more than $14/kg and others reporting much lower values. Reflecting its scarcity, kapok is also relatively expensive per kilogram, while cotton, abaca, ramie and sisal were all approximately the same value per kilogram in 2013. Jute is the least expensive natural fibre per kilogram, justifying its use in many bagging and geotextile applications.

It is difficult to estimate employment in the agricultural segments of natural fiber value chains (Table 2) because most production occurs in developing countries with weak systems of data collection, most producers are small holders and most labor is hired informally and seasonally, and because many households go in and out of fiber production from one season to the next, making it difficult to know who and

Table 1 Average farm prices of natural fibres, 2013

	U$/kg
Wool, clean	7.30
Silk, raw	4.40
Kapok	3.40
Cotton, lint	1.70
Abaca	1.60
Ramie	1.50
Sisal	1.30
Jute	0.50

Table 2 Natural fibre producing households, 2013

	Millions
Cotton	45
Jute	6
Wool	5
Coir and bast	1
Silk	0.3
Ramie	0.1
Sisal	0.1
Other	*0.5*
Total	60

how many are employed in any one year. Nevertheless, a reasonable estimate of total employment in natural fiber industries, including family labor, hired labor and employment in industries providing services to agriculture, and including both full time year round employment and part time or seasonal employment is around 60 million households or 300 million people in total, or about 4 % of the world's population.

Wool

Global wool production was 1.2 million tons, clean basis, in 2013 (International Wool Textile Organization (IWTO) 2014). Apparel accounts for 51 % of wool use (Fig. 3). Australia is the leading producer of wool, which is mostly from Merino sheep. China is the second largest producer, and New Zealand is the third largest and the largest producer of crossbred wool.

The value of world wool production in 2013 is estimated at US$8–9 billion. The wool industry produces wool on several million small-hold and commercial farms worldwide and employs millions of people in wool production, harvesting and throughout its many processing stages.

The structure of wool production varies enormously from country to country. For example, the American sheep industry (ASI), where sheep are raised primarily

Fig. 3 World wool industry

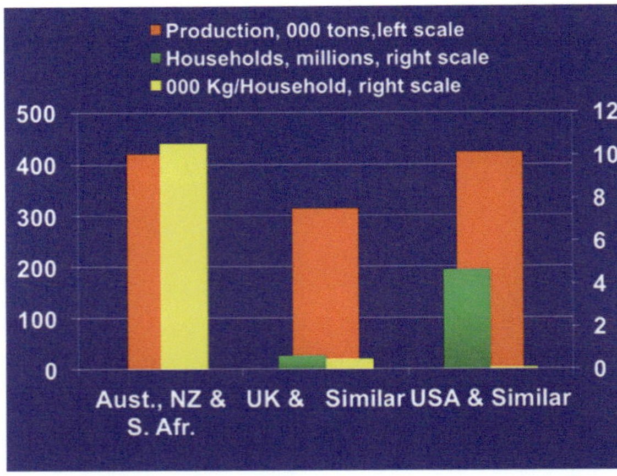

for meat rather than wool, has 79,500 sheep producers through the United States producing about 7230 tons of wool, or about 90 kg clean basis per producer. On the other hand, Australian Wool Innovation reports that in Australia (where wool is the primary product) 25,000 wool growers produced 270,610 tons of clean wool in 2013, or about 10.8 tons per producer. The British Wool Marketing Board has 46,000 registered wool producers accounting for 23,400 tons of wool, or about 500 kg per producer.

Assuming that productivity per producer is the same in New Zealand and South Africa as in Australia, that productivity in European countries is similar to that in the UK, and that wool production per producer in other countries where meat is the primary output is similar to that in the United States, there might be about 5 million households producing wool around the world.

Sisal

Sisal derives its name from a small port in the Yucatan peninsula of Mexico through which the earliest supplies of agave fibres, locally known as Henequen, were exported.

Mexico had practically a monopoly of agave fibre production, mainly for cordage, until the early 1900s. However, demand stimulated cultivation elsewhere during the twentieth century, and plantations appeared in many parts of the world based on different species of agave that had been transplanted from its original home. Agave sisalana (sisal) proved the most successful of these species.

Cultivation of agave sisalana developed into a major industry in East Africa, particularly in Tanzania and Kenya, prior to World War Two. Hybrid varieties of the plant were developed that gave extended production life and higher fibre content to the leaves. During the post Second World War years, rapid growth occurred in Brazil, and Brazil is the largest producer and exporter today, followed by China.

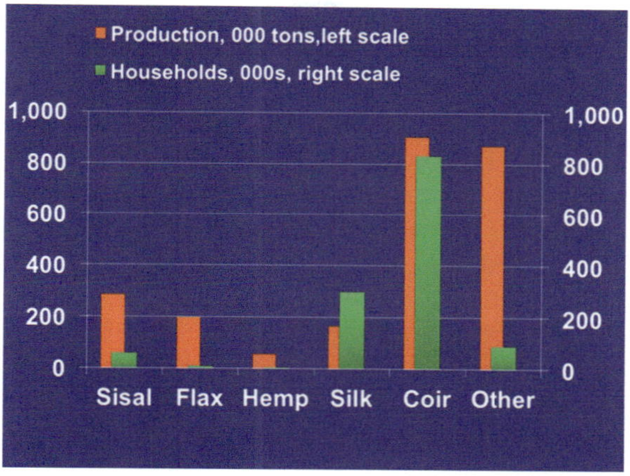

Fig. 4 World natural fibre industries

Sisal is produced by smallholders in Brazil, while sisal in other countries is commercially produced on medium to large sized plantations.

Sisal is unique among fibre crops in that it is a perennial, and the first harvest comes only three years after planting. Sisal is labor intensive, and employment in sisal production is estimated at one person per ton, or about 300,000 persons in between 50,000 and 100,000 households (Fig. 4).

Over many years the main products made from sisal were binder twines and cordage used mainly in baling agricultural products. However, demand for sisal baler twine has declined over the last 30 years due to competition from twine made from polypropylene. Nowadays, apart from the ongoing production of agricultural twines and cordage, sisal is increasingly used in weaving yarns for natural carpets, the cores of elevator sisal ropes and as buffing/polishing cloths for a wide range of metal surfaces, such as stainless steel cutlery and car body parts.

The end uses of sisal fibre have diversified in recent years, and it is now used in composite materials, as a replacement for glass fibres and the strengthening of plastics. It is also being used in various components in the automobile industry and in commercial aircraft, in the geotextiles sector for land reclamation schemes and the stabilization of slopes in road construction. There are applications in plaster reinforcement in the construction of domestic property in certain parts of the world, and sisal is also used as padding for mattresses and domestic furniture, and in the handicraft sector for handbags, placemats, animal figures and other products.

Flax/Linen

Fibrous flax is cultivated mainly in Europe in France, Belgium, Netherlands, Poland, and Lithuania, as well as in Ukraine, Russia, Belarus, Egypt and China (CELC/CIPALIN). European production of flax fibers ex scutching mills (a process of separating flax fibers from the woody parts of the plant) was 160,000 metric tons in 2013, including 115,000 tons of long fibers and 45,000 tons of short fibers. European production accounted for approximately 80 % of world production, which is estimated at about 200,000 tons.

Approximately 10,000 companies in 14 European Union countries are involved in the linen industry to produce finished fabrics, from planting and harvesting flax, to scutchers, to spinners, weavers, knitters, finishers and traders. These companies are gathered in the European Confederation of Flax and Hemp (CELC).

Most growers in Europe devote about 11 ha to flax each year, and flax is usually less than 15 % of total area on each farm. Flax is grown without irrigation, and all commercial varieties are developed conventionally. There are around 7000 flax growers in Europe alone, and worldwide there are probably between 8000 and 10,000 producers.

Flax fibers show very good mechanical properties (especially stiffness and absorption of sound and vibrations) and low density, which predisposes them for use as reinforcement of composites, giving the latter a more eco-friendly character. The markets for flax fibre are 90 % textiles—fashion, home and lifestyle, and 10 % for new composite goods such as sports equipment and materials for the automotive industry.

Hemp

About 100,000 ha of hemp were grown worldwide in 2014, including 40,000 ha in China, 25,000 in Canada and 17,000 in Europe (FAO, Intergovernmental Group on Hard Fibres and Jute; FAOSTAT3). (Production in Canada is for seeds only.) Assuming that average farm size is similar to flax, about 3000 households are involved in commercial hemp production.

Hemp fibres are mainly used for specialty pulp and paper, insulation material and for bio-composites in automotive applications. As of 2014, 55 % of hemp was still used in specialty pulp and paper, mostly supplied by French producers. Insulation material, together with other non-woven applications, like mulch fleeces, which are used in gardens and agriculture, water cress growing fleeces, and mats for farrowing pigs, account together for 25 % of use. Biocomposites account for 20 % of hemp applications.

The only really established biocomposite market is compression moulding in automotive interior applications. Hemp biocomposites account for 7500 tons, including 3700 tons of hemp fibre and 3800 tons of other fibers. Automotive biocomposites for interior applications are a growing market and are used for door panels/inserts, trunk liners, spare wheel covers, parcel trays, headliners, A-B-C columns and much more. Chinese hemp fiber is mainly processed into textiles and exported around the world.

Cotton

Cotton is grown commercially in about 80 countries on approximately 2.5 % of the world's arable land; making is one of the most significant of all cash crops (International Cotton Advisory Committee). Cotton is also a highly-traded commodity with about 150 countries involved in exports or imports of lint. Cotton connects people to markets because it is storable, durable, has a high ratio of value to cost of transportation and because it can be grown in arid regions. Therefore, cotton is grown mostly in land locked countries and interior regions of continents.

An estimated 40–45 million households are involved in cotton production around the world each season (Fig. 5), and including seasonal labor an estimated 250 million people are employed in cotton production during some parts of each season. By far, the largest number of people involved in cotton is in China, where an estimated 30 million households are cotton producers. Average farm size in Eastern China is only about one-tenth of one hectare. Another 9 million households are involved in cotton production in the Indian subcontinent, and about 3.5 million African households are producing cotton each season. All other cotton producing regions, including Central Asia, the Middle East, Australia, Europe, South America and North America account for less than one million households together.

The estimated average farm price of cotton in 2013/14 was about $1.70/kg, and with production of 26 million tons, the farm value of cotton production was about $45 billion.

The biggest challenges facing natural fiber industries, including the cotton industry, are economic in nature caused by competition with polyester and other oil-based fibers. Natural fibers accounted for all fiber use a century ago, and as late as the 1980s, cotton alone accounted for 50 % of world apparel fiber use. However, as of 2013, cotton's share of world fiber consumption was less than 30 %, and all natural fibers together accounted for just over one-third of world fiber use. Synthetic filament (manmade continuous fibers, like fishing line) and synthetic staple fibers

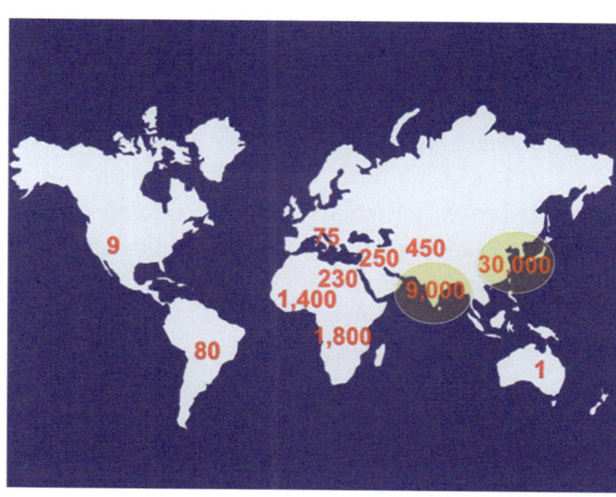

Fig. 5 Households producing cotton, Thousands

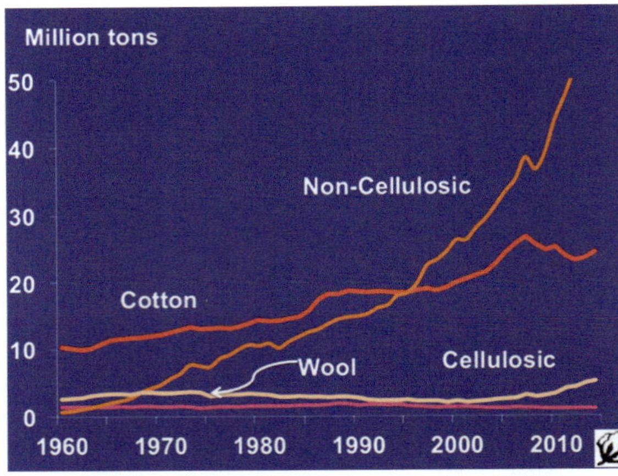

Fig. 6 World fibre production

(manmade fibers that have been cut into shorter segments) accounted for 58 % of total fiber use in 2013, and cellulosic fiber (manmade fibers made from cellulose, usually extracted from wood) accounted for 6 % of the total. The loss of market share to polyester, and especially to polyester filament, is a threat to the survival of the cotton industry and the entire staple fiber spinning industrial chain (Fig. 6).

Jute and Kenaf

Jute and kenaf (kenaf and jute are different plants but have similar fiber characteristics) are cultivated almost exclusively in developing countries of East Asia and in some parts of Latin America (FAO, Intergovernmental Group on Hard Fibres and Jute). Bangladesh and India accounted for 97 % of world production during 2013/14, with most production occurring in the Ganges Delta shared by the two countries. Until the late 1990s, world production of jute fluctuated between 3 million and 3.7 million tons, with the notable exception of a record crop of over 6.0 million tons in 1985. Between 1998 and 2000, world production exhibited a marked decline to an average level of 2.6 million tons because of competition with polypropylene. However, world jute and Kenaf production reached 3.3 million tons in 2013/14, and the farm value of production was about $1.5 billion.

Jute is processed mainly in the producing countries themselves and is used for the manufacturing of traditional products such as hessian cloth, food grade bags, carpet backing and other floor covering. Because of a national law requiring the use of jute in packaging material, India is the largest jute consumer. Diversified jute products, such as geo-textiles and composites are manufactured in relatively small quantities. Jute cultivation and processing is labor-intensive and therefore provides a livelihood and is an important source of food security for many farmers and their families in Asia.

Fig. 7 Jute production is concentrated in the ganges delta

Jute is the major cash crop for over 3 million farm households in Bangladesh (Bangladesh Jute Textile Mills Corporation). Assuming productivity is the same in India and other producing countries, 6–7 million households worldwide, meaning around 30 million people, are involved in jute cultivation. When the full value chain, from agriculture, to marketing, transportation, manufacturing and trading are considered, 25 million people in Bangladesh alone, one-fifth of the population are dependent on jute (Fig. 7).

Silk

World silk production amounts to less than 200,000 tons of filament, but with an average farm value of more than $4/kg, the total value is about $700 million. Silk is extraordinarily labor intensive because of the need to tend the silk worms, harvest the cocoons and unwind silk filament from cocoons; an estimated 800,000 people worldwide are involved in production of silk filament fiber.

India is the second largest producer of silk in the world after China. Silk is produced year round in Thailand. Most production is after the rice harvest in the southern and northeastern parts of the country.

Other Natural Fibers

World production of abaca, ramie, which is often blended with cotton in apparel fabrics, and sisal, an industrial fiber, totaled about 500,000 tons in 2013 (FAOSTAT3). World abaca production had a value of about $100 million, ramie about $200 million and sisal about $400 million. Together, about one million people are involved in the production of these crops.

Brown coir fibre is obtained from mature coconuts, while white coir fibre, which is finer, is extracted from immature green coconuts. Coir fibre is elastic enough to twist without breaking and it holds a curl as though permanently waved. Twisted or curled fibre is used in mattresses and bristles that are twisted to obtain a "spring/curling" effect. Total coir production was more than 900,000 tons worth about $400 million in 2013, and employment in the coir industry totals several million people.

Conclusions

In a world economy measured in trillions of dollars, natural fiber industries can easily be overlooked by policy makers. However, natural fibers provide employment to 4 % of the world's population and serve as a basis for industrialization and value added employment that benefits millions more people and amounts to billions of dollars of added economic activity.

Natural fiber industries contribute to food security and poverty alleviation. The greatest threat to the sustainability of natural fiber industries is competition with oil-based synthetic fibers.

Acknowledgments The authors gratefully acknowledge the support of the Discover Natural Fibre Initiative (DNFI) and its members and the Food and Agriculture Organization of the Untied Nations (FAO) in gathering the data used in this report.

References

Bastfibres, other, Fiber Crops not specified elsewhere, Hemp tow waste, Kapok, Ramie and Silk: http://faostat3.fao.org/

Abaca, Coir, Jute, Kenaf & Allied Fibres, and Sisal, Henequen and similar fibres: FAO Statistics, December 2014

Cotton: International Cotton Advisory Committee (ICAC). www.ICAC.org

Flax: CELC/CIPALIN for European statistics; world total estimated. www.mastersoflinen.com.

Wool and Other Natural Fibres, Animal Origin: International Wool Textile Organization (IWTO). www.IWTO.org

Wool as an Heirloom: How Natural Fibres Can Reinvent Value in Terms of Money, Life-Span and Love

Ingun Grimstad Klepp, Tone Skårdal Tobiasson and Kirsi Laitala

Abstract This paper addresses a main challenge for natural fibres; falling prices and increased focus on quantity versus quality. This is a challenge not only related to economic issues and profit, but is also unsustainable in an environmental perspective and in light of the challenges the textile sector and the world face. The paper uses wool as an example and in a surprising approach links the history and century-old traditions of natural fibers to an environmental thinking which supplements the traditional thinking around circular economy and LCA. Fabrics with a long life are the ones that have the lowest environmental impact (Fletcher and Tham 2015; Laitala 2014). Longevity or lifespan is a complex phenomenon in which both technical and social, or aesthetic aspects, are intertwined.

Keywords Wool fiber · LCA · Life span · Monetary value

Imported Silk and Billowing Sails

The Vikings, our most famous and infamous ancestors were expert sailors and textile producers. The two are of course, inter-twined. To sail one needs sails, and to sail in open boats across large bodies of water in the northern hemisphere, one needs warm clothing. The Vikings traveled extensively. They travelled westward across the Atlantic to the American continent and to the east and south into Russia

I.G. Klepp (✉) · K. Laitala
SIFO - Consumption Institute Norway, Oslo, Norway
e-mail: ingun.g.klepp@sifo.hioa.no

K. Laitala
e-mail: kirsi.laitala@sifo.hioa.no

T.S. Tobiasson
NICE Fashion - Nordic Initiative Clean and Ethical Fashion, Oslo, Norway
e-mail: tone@nicefashion.no
URL: http://nicefashion.org

R. Fangueiro and S. Rana (eds.), *Natural Fibres: Advances in Science and Technology Towards Industrial Applications*, RILEM Bookseries 12,
DOI 10.1007/978-94-017-7515-1_31

and to Constantinople and the Holy Land. In order to produce one single sail, spinning the wool would have taken 4800 h while weaving one sail is estimated to 3200 h (Jørgensen 2014). But they also needed sheep: 200 in order to supply the amount of wool for one single sail. In order to ready a fleet with 30–60 men (with sails, clothes and woolen cloaks that doubled as sleeping mats): 35–65 full-time equivalents, and 850–1600 kg of wool (Ibid.).

The most important export article from Norway throughout hundreds of years has been dried cod, which in Southern Europe is and has been transformed into tasty dishes like bacalhau. For this activity, it took more time to prepare the textiles than the labor associated with the boats, fishing and transport.

We have thus far described the practical textiles, but the Vikings also enjoyed beauty and luxury. Silks and dyestuff were treasured items and the basis for trade with the Byzantine Empire, through today's Istanbul. The rich Viking graves are filled with textiles in wool, silk and linen. But more important is that the Vikings mastered advanced and beautiful techniques for exploiting and maximizing the raw beauty and performance of the fibers (Vedeler 2006).

Clothing was the highest-priced assets most people owned, and stay important in written records until the end of the 1800s. Young people built up a stock of apparel that was meant to last a lifetime, and clothing was passed on when no longer used by the original owner. But cheap imports of cotton from slave plantations in America, and eventually finished goods from the growing textile industry led to a fall in price, functionality and durability, and thus created greater opportunity for replacement and fashion changes (Ulväng 2012). These changes have increasingly accelerated and created today's idea of "fast fashion", with constantly falling prices and growth in quantity, environmental impact and strain on our ecosystems, animals and humans.

In this paper we will discuss how natural fibers can contribute in a positive manner. We will first discuss some important parameters for the discussion on sustainable issues in textiles. We will then offer an alternative approach to a sustainable development. This will first be discussed theoretically, followed by some concrete examples. But as most of our examples are from Norway, we will include some facts about wool sourcing and the current market for Norwegian wool and wool products.

Growth Spiral and Circular Ideology

Today the situation for natural fibers is under pressure due to both economic and ecological conditions. Over the past 50 years, the amount of textiles produced has exploded, and prices have dropped dramatically (Fig. 1). Just since 2001 the import of clothing to Norway has almost doubled and it now amounts to 16.6 kg/capita (Statistics Norway 2013) while the current prices for clothing are the same as they were in 1980 (Statistics Norway 2011). Both the high level of production and consumption of textiles and clothing, and the way in which the textiles and clothing

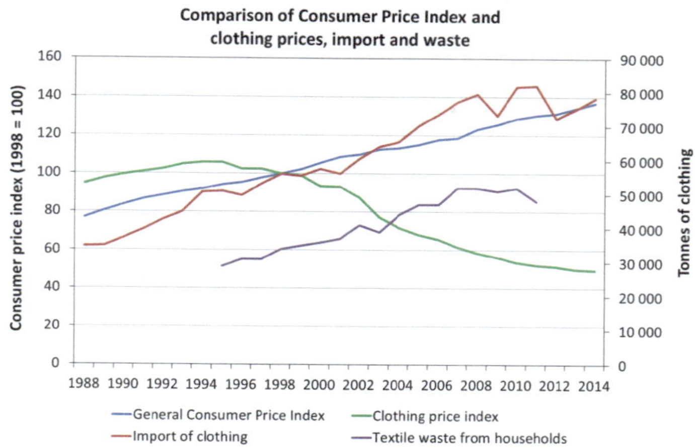

Fig. 1 Comparison of consumer price index, clothing price index, and the amounts of clothing import to Norway and textile waste from household (Statistics Norway 2015)

are produced, are connected with significant environmental problems (Fletcher and Tham 2015).

A lot of work has been initiated to address these issues. On behalf of the Nordic Council of Ministers there has been conducted a survey of initiatives in the Nordic countries and around the world. This survey shows that most larger and established initiative are addressing the problems by focusing on improving waste management, and incremental improvements in the value chain (Klepp et al. 2015). This is also what has progressed furthest as part of the international initiatives on textiles and environment, such as Mistra's Future Fashion, WRAP's work in the UK, EU's Circular Economy push and Launch Nordic. The idea of a circular economy has received much attention and has interpreted narrowly as reuse of materials and recycling of materials (Klepp et al. 2015). But the production of waste is not the cause, but rather the result of the problem we face. Growth in the amount of consumed textiles and the low utilization of textiles' potential lifetime are not addressed through better waste management. Thus the focus on recycling legitimizes consumption, and can contribute to a less environmental-friendly pricing of recycled materials over new. The emphasis on recycling also provides synthetic fibers with an apparent leg up on natural fibers because the synthetic fibers are most suitable for recycling, not only down-cycling (Fletcher 2008).

The rapid growth in the quantity and the fall in prices are also particularly challenging for wool, silk, linen and other natural fibers at higher price points. The focus on quantity over quality leads to cheap fibers such as viscose, synthetic fibers and cotton being preferred. The same focus contributes to systematic quality work and environmental improvements being difficult.

An important and much used analysis tool is so-called LCA, Life Cycle Assessment. LCA is used to obtain information about the total environmental impact of a product or activity. The aim is to include the entire life cycle from raw material production to disposal, including transport and use. It is common to use this type of assessment to evaluate which phases of the production chain have the greatest environmental impact, so companies can make improvements in areas where changes have significant effect. LCAs are also used in other comparisons where less suited. Using this type of assessment tools in the comparison of different products or fibers should be done with care, as the various parameters and measurements have a big impact. Many believe that in comparisons between fibers, the industry-based parameters contribute to the low ranking of for example wool (and other natural fibers) in assessments and comparison studies (Kviseth and Tobiasson 2011).

The expression LCA is based on the idea of looking at the whole life from cradle-to-grave or cradle-to-cradle. A circular economy is the dominant mind-set within much of the on-going environmental work, the goal being a closed loop system where everything is reused and recycled. Yet the assessments performed only concentrate on the production methods. They neither focus on the growth in amount of products nor use and disposal.

One way to look at the entire life cycle is through the expression *functional unit*. The functional unit may be provided for a given period of time or to obtain a quantified performance. An example for paint is 1 m^2 wall maintained for 10 years. In this way emissions are directly linked to the product's function and can easily be compared with other products that have similar function (Rønning et al. 2011).

LCA studies of clothes seldom use of functional units. In a survey of sheets, "a year of protective covering of a bed" has been used as the unit of analysis (De Saxce et al. 2012), and recently also in a Swedish study of clothing "one use that refers to the use occurring within a 24 h time period" (Roos et al. 2015). However, for apparel the most common analysis is to choose a standard garment and look at the production and possibly some very simple and imprecise estimates of consumption and disposal (Allwood et al. 2006; Labouze et al. 2006; Steinberger et al. 2009). Another common method is the cradle-to-gate principle. This is used to compare fibers and only includes production of the raw material as finished fiber, but excludes steps such as dyeing, weaving and sewing. It is common that such reasoning is behind systems that compare fibers against each other, such as Made-By's assessment (Made-By 2013). When LCA studies are based on a small part of the life cycle, one is in principle comparing disposable products with items that can be used 100 or 1000 times.

The result of the lacking focus on quantity is that despite the efforts to reduce environmental impacts the environmental impacts of increased production exceed the reductions made (Fletcher and Tham 2015). For natural fibers this is a commercial problem because the fibers are wrongfully given a lower score in LCAs when their strongest card; use and longevity, is not counted. Continued growth, persistent low prices and an environmental political focus that undermines wool and other natural fibers strongest card, is both an environmental and profitability challenge. An alternative is therefore necessary, both on the basis of an economic and an environmental mind-set.

In the newly launched project KRUS (financed by the Norwegian Research Council) we are exploring the increased value, both economically and perceived, and reduction in amount. Within Norwegian wool-production we will look at new ways of increasing value. At the core of the project is how the local affects the economic and emotional value. Slow and local apparel are thus integral terms in KRUS (Nordic Fashion Association 2015a).

Life Span and Love

Norway's oldest and surprisingly well-preserved item of clothing is a 1700 years old tunic (Guhnfeldt 2011). It has not been in active use, but was for many years frozen underneath a glacier. We do have fabrics that have had an active life for many centuries, such as the Baldishol tapestry, which is 1000 years old (The Norwegian National Museum 2012). Apparel generally wears out quicker, but 100 year old clothes are still in use, items from the period between the two World Wars are not entirely unusual. Technically, there is nothing stopping clothing from becoming really old, and many Norwegians have older garments in their wardrobe, for example a suit, their national costume or knitted sweaters.

When these clothes are still usable, it is not just because they are of good technical quality, but also that they are and have been socially accepted for prolonged use. We might call this a type of social durability. Garments used less also wear less, and thus can last long. An example of such clothing is christening gowns (Klepp and Laitala Unpublished). The reason is that christening gowns are used for only one type of occasion, and still are considered valuable after they are used. Looking at a very ordinary garment such as jeans, the number of days in use is much higher. How often you use a trouser additionally depends on how many pairs you have. Years are a common measurement stick in LCAs, but how much use a year represents is highly dependent on the amount of clothes owned. The increase in amount of clothing should therefore result in longer clothing life—and not shorter. This has major implications for the meaning of heritage and the value attached to used clothing. But an estimated number of years (alone) says very little about how much a garment is used. The amount of repeated use will thus be a better measure, and perhaps a device that is more suitable as a unit in a comparative analysis. However, the number of years is not without significance, and goes to the relationship between technical and social lifespan.

Until now we have only written about technical quality, i.e. that the garments are worn and potentially only withstand a certain amount of use. This is the main reason (49 %) for disposal of clothing for adults in Norway, but clothes are also thrown away for other reasons (Laitala and Boks 2012). Clothes go out of use because they do not fit the user's body (19 %), taste (11 %), and lack of space or changes in life conditions (8 %). They can also be disposed of because they do not work as expected or there are new and better alternatives on the market (6 %), changes in fashion or taste of the owner (5 %). The term social lifespan covers several of these. Some garments have an extremely short social lifespan, f. ex. wedding dresses. But fashion as such plays a much smaller role in disposal of clothing than most imagine (Laitala and Klepp 2013). The chance that something will become un-fashionable before it is worn out increases if the amount clothing increases. If we own a lot, we do not use 'up' what we have until the social lifespan catches up with the clothes. In the material we have collected on the relationship between fashion and technical life, we can see this clearly. The clothes that go out of use due to fashion are somewhat older than the clothes disposed of for other reasons, as they have extended their stay in the closet unused (about 2.8 years) (Laitala and Klepp 2013).

This means that if we produced clothing for both a long technical and social lifespan, we will be able to reduce production and therefore lower the environmental impact considerably. Such adjustments are dependent on higher prices to maintain current levels of employment and profitability.

Monetary Value and Tradition—Elements of New BMs

An environmental focus on longer lifespan and increased economic, aesthetic and emotional value must be wagered to match the way textiles are produced and used in order to gain practical significance. In such a process, both firms working for new business models (BM) and consumers who are willing to change practices, are necessary. In such a restructuring process, we need to answer questions such as what can help increase the love of items, and hence the willingness to pay and to care for fabrics and apparel? Tradition and good raw materials, as well as knowledge about "where the clothes come from" are possible answers we are looking at in KRUS. We will now give some examples from consumption, production and marketing that can contribute to a change in this direction.

Norwegian Wool

In Norway we have a declining stock of sheep, which is currently around 1 million winter-fed sheep and 2.5 million in the summer. They produce around 4.5 million kilograms of wool (Hebrok et al. 2012). There are several older breeds such as the Norwegian spæl sheep (Fig. 2), but the most typical is a modern crossbreed, called "Norwegian White Sheep". The wool is shorn either on the farms or at the wool stations and slaughterhouses spread around the country; the wool is then classified and sent to the UK for scouring. We have several large and smaller woolen mills and a varied production of knitting yarns of high quality based on both Norwegian and imported wool (Ibid.) Norway is also a world-leader on use of wool underwear and next-to-skin wool products, but this industry is based on imported merino; knitted and produced mainly abroad by Norwegian or foreign companies. A few companies produce this type of apparel in Norway. Production of ready-to-wear apparel in Norwegian wool was very limited until a few years ago,

Fig. 2 Norwegian spæl sheep. *Photo* Grethe Ringdal

and the knowledge-level of what was or not produced from local resources has been very low (Klepp and Tobiasson 2013).

Merino and alpaca are popular raw materials and much used. In spite of many textile factories closing down and out-sourcing to low-cost countries with low or lacking environmental standards, there are still textile mills and factories in Norway. In addition to wool mills, we find weaving and knitting factories. The weaving factories have survived based on quality materials for our national costumes, furniture and throws as their main products. While the knitted products, mainly Norwegian sweaters, are in use both in Norway and as export and souvenir products (Klepp and Tobiasson 2013; Hebrok et al. 2012). There is, however, not a uniform labelling of what is "Norwegian", neither for the raw material (wool) nor for the yarn or other product.

Scandinavian Business Seating—New Flexibility for Norwegian Wool and Local Production

The Norwegian furniture company is a major player in the office and home-office chair market (Fig. 3). They own three different brands, HÅG, RH and RBM, and their production has been mainly in Røros in Norway and Nässjö in Sweden. Their

Fig. 3 Close up of Scandinavian business seating chair. *Photo* SBS

main supplier for textiles has been Gudbrandsdalen Uldvarefabrik (GU), a historic and fully integrated weaving mill based in Lillehammer, Norway. However, GU had not sourced Norwegian wool since the 1950s (Klepp and Tobiasson 2013). The reasons cited were vegetable matter, price point and whiteness of wool, in comparison to New Zealand wool. In addition, New Zealand wool was Eco-Flower approved (Ibid.).

The materials used for furniture textiles is generally—if wool—a crossbreed type of wool and not merino. This is because durability and crimp are more important than softness. In interiors, synthetic materials have been on the increase based on price. If flammability and durability are considered, wool is perhaps a much better choice. This was a main concern for SBS. But they had not until they came across the research from the project Valuing Norwegian Wool, considered that the wool could also be Norwegian, not only spun and woven in Norway. "The story of the Norwegian Wool project is about returning to our local resources and bringing the advantages and potential of Norwegian wool back to life. It is about origin, about appreciating the value of an important resource and making sure it is being used to its full potential," the advertising material for the project explains. All three brands have now developed one chair where the textile is 100 % Norwegian (Scandinavian Business Seating 2015).

The wool is the top classified crossbreed type (C1), it was scoured and spun at the SandnesGarn mill and woven at GU un-dyed. The dyeing is based on demand, which ensures less environmental impact. The textile has also been made available to designers, and will be suited for outer-wear. "What surprised us the most, was that the claim of more vegetable matter in the Norwegian wool, turned out to be un-founded, but we also found that the exceptional crimp and luster of the Norwegian wool surpassed the New Zealand wool. This made it so much more suitable for our purposes," said Pernille Stoltze who had initiated the project during an information meeting. The project has already generated international attention.

Sølv—New Twist on an Old Business Model

"Driving slow in the fast lane of fashion" is the catch-phrase of this company. The Norwegian womenswear label's mission is reducing consumption through delivering timeless high-quality garments for the owner to love and care for over time (Sølv 2014). They are passionate about local craftsmanship and work closely with their Norwegian suppliers; the sheep-farmers, the yarn-mill, the weavers and knitting-factory. They sell their collections directly to their customers through pre-order only. This enables the customer the opportunity to follow the garment journey from raw material to sewing table, while it is being made. Over-production is therefore not an issue.

The name "Silver" plays on the idea of heirloom. In Norwegian, this is "arvesølv"—the inherited silver or other valuables being handed down. The fact that every item is pre-ordered and made to fit to the customer, goes back to a tailoring tradition that pre-empts mass-production. This also opens up for the customer's possibility to customize their purchases later on. In the tradition of our national costumes, this is integrated in the design.

The close cooperation with the Hillesvåg woolen mill has ensured the commercial development of some older breeds with pigmented wool and with a double-coated fleece; two types of wool which offer challenges for a stream-lined industry production.

Graveniid—New Use of Lapp Traditions, Northern Norway

Lapps have a minority indigenous status in Norway. They have a strong tradition in the use of their traditional costumes both as party garments and in daily life. The new thing with this business is that it assimilates local understanding of clothing as well as color and pattern, but produces a more trendy design based on imported raw materials. They have established a knitting factory in the Finnmark region of Norway, the northern-most province.

> The year was 1985 when Graveniid Duojit in Karasjok, a village in the heart of Norway's northernmost and most hardy county was established. Marit Guttorm Graven, who belongs to a reindeer Sami family, saw a need for both warm and robust clothes, while she wanted to offer clothes one could dress up in, even in the cold north. It turned out that there was a need for this type of clothing and there still is. The first customers were local, including the

> reindeer herders, and they quickly became our regular customers. Their need for functional, warm and durable clothes laid the foundation for the collection you can find in our online store today. As the years passed, the products spread to a larger market. In line with this philosophy, we have constantly developed rough products for different needs and conditions (Graveniid 2010).

The products are for the most part based on imported merino yarns.

Maxhosa by Laduma—New Use of Xhosa Traditions, South Africa

In our South African case, tradition is used as the means to renegotiate prices and value. Local value chains are important in the education process of consumers, and to reestablish an understanding for textiles' fair value. The designer, Laduma Ngxokolo, was concerned that in the practice of his generation of young men. When entering "manhood" and going through the rituals and educational process, they were required to replace their old boyhood wardrobe with a "man's wardrobe" (Laduma 2013). "But all the clothes that were available were imported," he told an audience at Campaign for Wool's October meeting in Oslo. The lack of ethical designs and raw materials frustrated him to the point of making a difference. In cooperation with among others Cape Wool, his label Maxhosa is based on 80 % local merino wool and 20 % mohair blend.

His approach was to completely reinterpret his heritage in the setting of this "confirmation", which could be compared on some level to that Norwegian girls when "entering adulthood" also are offered new clothing, but in this case only one set: a national costume ("bunad"). However, as the focus on value, quality, heritage and durability were part of Laduma's areas of prioritization—there are definite points that tangent.

Laduma also told the story of how he started to knit at the age of 16 (when he bought a knitting machine), and that it was the beading techniques and not knitting techniques that led to the bold patterns. But aside from tapping into the local heritage and value chain, Maxhosa has been able to ensure a focus on natural, locally sourced fibers instead of synthetic and perishable textiles. The focus has been on quality and longevity, which is important for such a change of wardrobe.

Selbu Mittens—New Thinking About Fair Price for Quality

We are accustomed to thinking of "underpaid" workers as a problem in poorer parts of the world. But in the textile industry, this is also something that persists in the west, for designers and craftsmen. The low price is an issue because for the industry as a whole, it lowers prices and the understanding of the value of garments.

Fig. 4 Selbu mittens (*Photo* Anne Bårdsgård)

Knitting is a popular activity, and certainly for the traditional two-tone patterns linked to geographical areas and major sports events. Selbu-mittens are something most people in Norway own, or at least have an understanding of (Fig. 4). They are also not unusual as gifts, f. ex to foreign guests. Hand knitted Selbu-mittens are sold in the handicraft and many tourist stores, but they are also commonly hand-knitted and recipes abound. A pair of hand-knitted mittens in Norwegian spun wool cost between 200 and 400 NOK (ca 30–60 €). Which makes it obvious that this is not the basis for a so-called "living wage".

The Selbu mill has developed yarn packages of naturally pigmented dark wool and natural light wool (Selbu spinneri 2015). The pattern they sell is reconstructed from old mittens. As production costs are high, the yarn costs between NOK 1500 and 2400/kg. Thus yarn packages are costly, NOK 500 per kit, including yarn and pattern, and not including knitting needles. The mill discussed knitting mittens for sale, which would mean a cost of NOK 2500 pr. pair, calculated with an hourly wage of 30 NOK per hour for knitting (approximately 80 h). The minimum wage in Norway is around NOK 160/h. A sober pricing of mittens based on one-fifth of the minimum wage still provides a price 10 times higher than normal priced pair of Selbu-mittens in Norway. Some of the price difference here is that the yarn is a niche product with high production costs, but also ordinary Selbu-mittens take time to knit. So far, this project is very young and we have therefore limited opportunities to say something about the market for these products. However, we will follow up and find it interesting in relation to the development towards quality rather than quantity in the textile sector.

VikingGold—Reconquering the Vikings Textiles

Our last example brings us back to the Vikings and their textiles, their focus on beauty and quality. The example is from an innovation project we have worked on, financed by KreaNord (Nordic Fashion Association 2015b). The ambition is to combine archeological knowledge with modern design and today's value chain. The concrete aim: to show-case how our older Nordic textile history can inspire—and more abstractly—recapture the associations and connotations to "Viking"—not their power, rape and pillage factor; but their artistry and artisan standard for textiles. Sadly, today's association for "Viking" is bad quality, ugly and cheap—and the result is a tourist market we would like to ignore.

The project has been a cooperation between Iceland, Demark and Norway; looking at both business models and technical solutions. But we are also trying to understand how heritage can be reinvented as a currency in an international setting. Binding the story of origin to the product and finding an integral way of communicating this link, is vital—something we are seeing in food. In VikingGold we have followed the wool from the old sheep breeds (Old Norwegian Spæl) to the actual sorting, classifying, scouring—and all the way through the value chain (Fig. 5). This has been done more closely than with any other Norwegian wool, except for the small-scale mini-mill handling. The result—at the weaving level—is surprisingly exquisite. This, once again shows the lack of knowhow is not a hinder, but a possibility!

The Vikings' focus on quality of textiles is inspirational. The 200 m of fabric we have produced is perhaps the most expensive diamond twill material ever woven, but so what? This luxury material is not based on over-pricing, branding etc., but based on show-casing a textile with the esthetical qualities the Vikings could identify with.

The unique aspect of this project is that it combines tradition conveyed through archaeological knowledge with an innovative and conscious design process. Parallel with this, a reconstruction of the same fabric we used as inspiration has been produced.

The project has had many positive side effects, such as greater collaboration between Norwegian spinning mills and weaving mills, contact between archaeological and design hubs, and the current value chain reconnecting. Judging from the purpose of this paper, we are prepared to put forward both the focus on tradition, production and high pricing as aspects of interest. Thus this is an opportunity to follow some products to market or to the catwalk as such, which differs considerably from the normal emphasis on price and volume. The background and documentation for the project is available on the website nicefashion.org.

Fig. 5 Viking gold yarn. *Photo* Tone S. Tobiasson

Conclusion: From Nature with Love

The focus on recycling and "sustainable fibers" has one winner, and that is the synthetic fibers industry. This is because their products can be recycled forever and be produced without the challenges of animal welfare and land-use. But seen from a consumer perspective, these products quickly become less attractive.

If we instead focus on lifespan and value, the higher priced natural fibers have many benefits. To the extent, textiles still hold their position as heirlooms; the natural fibers such as silk, linen, wool represent this status. Based on these fibers we still find clothing and textiles commanding high price and value. **Increased knowledge, increased focus on slow clothes and local products can contribute to an actual change. This can happen when textiles recapture some of their lost**

economic, emotional and esthetical value. In such a shift, it is imperative to find good examples both within production and consumption. We have offered examples from production which focus on new business models where raw materials, crafts, tradition and local production play an increasingly important role. They represent small niche producers, but if we look to what has happened within the food sector we see that niche has an important role for future change. A sustainable future scenario could be a wardrobe with only favorite clothes, long-lasting and much loved; and an industry that no longer needs to settle for a decreasing share of the consumer income.

Acknowledgements The authors gratefully acknowledge the funding by IWTO, KreaNord, and the Norwegian Research Council, BIONÆR program.

References

Allwood JM, Laursen SE, Malvido de Rodríquez C, Bocken NMP (2006) Well dressed? The present and future sustainability of clothing and textiles in the United Kingdom. University of Cambridge, Institute for Manufacturing, Cambridge.

De Saxce M, Pesnel S, Perwuelz A (2012) LCA of bed sheets - some relevant parameters for lifetime assessment. Journal of Cleaner Production 37 (0):221-228. doi:10.1016/j.jclepro.2012.07.012.

Fletcher K (2008) Sustainable fashion & textiles: Design Journeys. Earthscan, London.

Fletcher K, Tham M (eds) (2015) The Routledge Handbook of Sustainability and Fashion. Routledge, London.

Graveniid (2010) Om Graveniid. http://graveniid.no/blogg/om-graveniid/. Accessed 23 June 2015.

Guhnfeldt C (2011) Norges eldste klesplagg. Aftenposten.

Hebrok M, Klepp IG, Tobiasson TS, Laitala K, Vestvik M, Buck M (2012) Valuing Norwegian Wool. Professional report no 5-2012. SIFO, Oslo.

Jørgensen LB (2014) Tekstilfremstilling i det forhistoriske Norden. Osterøy museum, Lonevåg, 4th October.

Klepp IG, Laitala K (Unpublished) Klærs levetid -LCA på liv og død In: Vittersø G (ed) Dematerialisering som miljøstrategi – mulige veier til redusert forbruk. Oslo.

Klepp IG, Laitala K, Schragger M, Follér A, Paulander E, Tobiasson TS, Eder-Hansen J, Palm D, Elander M, Rydberg T, Watson D, Kiørboe N (2015) Mapping sustainable textile initiatives and a potential roadmap for a Nordic actionplan. TemaNord 2015:545. Nordisk Ministerråd, Copenhagen.

Klepp IG, Tobiasson TS (2013) Ren ull. Aschehoug, Oslo.

Kviseth K, Tobiasson TS Pulling Wool over our Eyes: The Dirty Business of LCAs. In: KEA Conference Towards sustainability in Textiles and Fashion industry, Copenhagen, 26th – 27th of April 2011 2011.

Labouze E, Guern YL, Abbayes Cd (2006) Analyse de Cycle de Vie d'un Pantalon en Jean. ADEME and Bio Intelligence Service S.A.S., France.

Laduma (2013) MaXhose by Laduma http://www.maxhosa.co.za/about/laduma-ngxokolos-profile. Accessed 23 June 2015.

Laitala K (2014) Clothing consumption - An interdisciplinary approach to design for environmental improvement. Norwegian University of Science and Technology, Trondheim.

Laitala K, Boks C (2012) Sustainable clothing design: Use matters Journal of design research 10 (1/2):121-139. doi:10.1504/JDR.2012.046142.

Laitala K, Klepp IG (2013) Bare mote? Materialitetens betydning for klærs levetid. In: Strandbakken P, Heidenstrøm N (eds) Hinsides symbolverdi - Materialiteten i forbruket. Novus, Oslo, pp 145-167.

Made-By (2013) Environmental benchmark for fibres (Condensed Version). vol December, 2.1 edn. MADE-BY, Santa Barbara.

Nordic Fashion Association (2015a) KRUS. http://nordicfashionassociation.com/project/krus. Accessed 23 June 2015.

Nordic Fashion Association (2015b) Viking Gold. http://nordicfashionassociation.com/project/vikinggold. Accessed 23 June 2015.

Roos S, Sandin G, Zamani B, Peters G (2015) Environmental assessment of Swedish fashion consumption. Five garments – sustainable futures. Mistra Future Fashion.

Rønning A, Lyng K-A, Vold M (2011) Kunnskapsplattform for beregning av klimabelastning fra bygg og byggematerialer - Litteraturstudie. OR.02.11. Østfoldforskning, Kråkerøy.

Scandinavian Business Seating (2015) New Norwegian wool production - a collaboration of design, quality and sustainability. http://www.sbseating.com/newsroom/news/norwegian-wool. Accessed 23 June 2015.

Selbu spinneri (2015) Selbu Spinneri – rett fra sauen. http://selbuspinneri.com/. Accessed 23 June 2015.

Statistics Norway (2011) Statistisk årbok. Statistisk Sentralbyrå, Oslo/Kongsvinger.

Statistics Norway (2013) Table 08801: External trade in goods, by commodity number (HS) and country SSB. Accessed 30 December 2013.

Statistics Norway (2015) Consumer price index. SSB. https://www.ssb.no/en/kpi.

Steinberger J, Friot D, Jolliet O, Erkman S (2009) A spatially explicit life cycle inventory of the global textile chain. The International Journal of Life Cycle Assessment 14 (5):443-455.

Sølv (2014) SØLV - Brand philosophy. www.solv-studioet.com. Accessed 23 June 2015.

The Norwegian National Museum (2012) Baldisholteppet. Nasjonalmuseet for kunst, arkitektur og design. http://dks.nasjonalmuseet.no/kunstverk/baldisholteppet. Accessed 25 November 2014.

Ulväng M (2012) Klädekonomi och klädkultur: böndernas kläder i Härjedalen under 1800-talet. Gidlunds förlag, Möklinta.

Vedeler M (2006) Klær og formspråk i norsk middelalder. Kulturhistorisk museum, Universitet i Oslo, Oslo.

Hemp Cultivation Opportunities and Perspectives in Lithuania

V. Jonaitienė, Z. Jankauskienė and I. Stuogė

Abstract Hemp is one of the oldest cultivated crops. Hemp (*Cannabis sativa* L.) is environmentally friendly and helpful; it is especially suitable for use in crop rotation, not exhausting the soil, reducing its pollution by weeds. Their production meets the environmental principles. Not only the growth, but also the products from hemp are important for protection of the environment. Lithuania legalized hemp cultivation in 2014. The demand for hemp and their products is still developing in Lithuania. Hemp cultivation and processing is profitable, but all received potential income must be expanded into stalk processing to fibers; the processing and production to finished products would create higher added value. The plant for stalk processing into fiber and the possibilities of hemp final products use in production will create additional jobs in the country. The aim of this research is to analyse the possibilities and perspectives of hemp cultivation in Lithuania.

Keywords Hemp · Cultivation · Fibre · Seed

V. Jonaitienė (✉)
Faculty of Mechanical Engineering and Design, Kaunas University of Technology,
Studentų str. 56, 51424 Kaunas, Lithuania
e-mail: vaida.jonaitiene@ktu.lt

Z. Jankauskienė
Upytė Experimental Station, Lithuanian Research Centre for Agriculture and Forestry,
Linininkų 3, 38294 Upytė, Panevėžys District, Lithuania
e-mail: soja@upyte.lzi.lt

I. Stuogė
Department of Ecology, Institute of Animal Sciences, Lithuanian University
of Health Sciences, R. Žebenkos str. 12, 82317 Baisiogala,
Radviliškio District, Lithuania
e-mail: ina.stuoge@gmail.com

R. Fangueiro and S. Rana (eds.), *Natural Fibres: Advances in Science and Technology Towards Industrial Applications*, RILEM Bookseries 12,
DOI 10.1007/978-94-017-7515-1_32

Introduction

Hemp is an annual plant, valued for grown fibers and seeds. Hemp (Lat. *Cannabis*) is flowering plants (*Magnoliophyta*) tribe, belonging to hemp (*Cannabaceae*) family. This is one of the oldest cultivated plants. Few distinct types of hemp could be defined: industrial hemp (*Cannabis sativa*), Indian hemp (*Cannabis indica*) and *Cannabis ruderalis* (Karus and Vogt 2004).

Molecular components of hemp, known as cannabinoides, consist of two main specific components, i.e., THC (Tetrahydrocannabinol) influencing human psyche, and CBD (Cannabidiol), reducing this impact. Industrial hemp has a small THC, but significant CBD content.

True hemp are the plants with amazing vitality and ability not only to kill and displace other tiny crops or weeds, but to improve soil characteristics also (Karus and Vogt 2004).

Results and Discussion

Hemp Cultivation Technology Opportunities

Hemp has been grown in Lithuania some time ago for fiber and food, so it is no wonder that there are created many words in Lithuanian folk songs about them and they are referred in folklore. Hemp fibers and seeds are found in ancient burial places in Lithuania. Although hemp was grown in Lithuania for centuries (it was mainly used in the textile industry, for cord cocoons, and food), currently we do not have enough raw materials and no longer traditions (Butėnaitė 2012).

Hemp cultivation technology is described by employees of Upytė experimental station from Lithuanian Research Centre for Agriculture and Forestry (LRCAF) who have examined the facilities of acclimatization of derived hemp varieties from five different countries in 2008–2009, and found that all of them are suitable for the cultivation in Lithuania, only the harvest depends on species and growing purpose.

Hemp is growing fast, just over 100 days it can reach 3.5–4 m in height. High hemp forms large shadows and thus inhibits the growth of weeds, so cannabis cultivation is environmentally friendly, the pests do not like them, and cultivation does not require any chemical agents. The cannabis is suitable for growing in ecological farms. The studies performed in Lithuania have also shown that soil pollution with weeds decline after cannabis inclusion in the rotation. Cannabis is valuable proceeding in crop rotation. Its deep root system positively effects soil structure and reduces the amount of nematodes and fungi in the soil. Hemp does not emaciate soil, so you need less fertilizer for after succeeding crops. The soil remains with optimal properties after their growing since it is estimated that about 10–20 % higher yield of wheat is obtained when they are cultivated in the soil where earlier the hemp was grown. Hemp is one of the few cultures in Europe, which is grown not in organic farms, but without the use of chemicals. The plant can be used for land cultivation since hemp is growing rapidly. Hemp is capable of returning the polluted abandoned fields to agricultural production cultivation, as they have the properties absorbing heavy metals (Carus et al. 2013).

Cannabis needs well-manured land. Usually hectare of cannabis needs 100–125 kg of nitrogen, 40–60 kg of phosphate fertilizer, and 65–90 kg of potassium. Large quantity of biomass can be obtained in a short period of growth of the cannabis plant. Substantially big part of this biomass remains in the field after harvesting, and so comes back into the soil in the form of organic fertilizers, thus enriching the soil. Hemp is a highly effective weed suppressant. Typically, 200–300 seedlings per square meter completely overshadow weeds and they are ceased to exist before the harvest. Hemp varieties with less than 0.3 % of tetrahydrocannabinol (THC) content are already bred. These varieties are certified for commercial cultivation in the European Union. Their range is very different, as the flowering time, plant height (biomass), fiber content in the stems, seed yield, fatty acid composition, and the THC content differ (Poiša et al. 2010). The choice of species mostly depends on hemp growth purpose: fiber, seed or fiber and seed. Most new varieties are still derived in Europe and Ukraine.

Hemp sowing time is chosen when the soil temperature reaches minimum 5.5–8 °C. Seeds germinate within 24–48 h and spring up within 5–7 days with adequate moisture and warm temperature. Sowing should start as early as possible in order to maximize stalk yield fiber. It can be sown later, if hemp is grown for seeds, thus reducing plant height and for easier harvest.

Crop density can vary within very large limits; 60–90 kg of seeds per hectare is taken for hemp grown for fiber, depending on soil type, fertility, and sort type. 44–56 kg of seeds per hectare is taken for production of hemp for seeds.

Hemp harvest will depend on the purpose of cannabis production and the products we want to get. Hemp harvest should begin as soon as the last pollen drops off in order to obtain high-quality fiber (usually they slough 70–90 days after sowing). Seeds are harvested after about a month after 60 % of the seeds ripen. When cutting hemps for fibers, they should be cut into approximately 60 cm parts, soaked, then tied, and storaged or processed. The adapted combine harvester is used to harvest hemp seed: the seeds are brushed slowing down the speed of the harvester blade, and then the unnecessary stems are cut. The combine harvester of German production is designed to remove the seeds and fiber yields at the same time. Typically, the adapted combine harvesters are used for hemp harvesting, thus protecting against long hemp fiber entanglement in rotating mechanism.

The harvest time varies in the countries where industrial hemp is breeding. In Poland the cultivation of cannabis is produced only for seeds; it is harvested when the seeds are grown up in panicles located in the middle of the stem and later blooming is rising up: at first it blooms in the middle section, still later—on tops. Hemp in Italy is cultivated only for fiber; so they are harvested in full blooming stage, while the cultivation of cannabis for seed and fiber is about 1 month later than about 70 % of panicles bloom. Hemp for fiber in France is harvested at the end of blooming and growing for fiber and seed they are harvested after 1–2 weeks after full bloom. Hemp grown for fiber in Ukraine is harvested when in most plants are found mature individual seeds, while growing for fiber and seed they are harvested when 75–80 % of seeds reach biological maturity. Dutch cannabis growers experience show that the best cannabis hemp fiber is during hemp blooming and about 1 week after blooming. Later fiber lignifies and its quality is deteriorating.

Hemp so far has been evaluated for its strong fiber, and its seeds for nutrition, as they contain more protein and amino acids than in cereal grains. Cannabis has high energy content; it contains 18–28 % of protein, 35–38 % of fat, 25 % of carbohydrates, a wide variety of minerals. Hemp seeds contain proteins without trypsin inhibitor, which impedes the absorption of proteins in the human body. Oil containing omega 3 and omega 6 fatty acids (ratio 2.5:1) required for the human body can be pressed from the seeds, and it is associated with a healthier lifestyle (Kolarikova et al. 2013).

Hemp Cultivation Perspectives in Lithuania

Hemp is useful plant, suitable for both food and cosmetics. Hemp is environmentally friendly building material, suitable for the construction of houses. Building blocks from cannabis have A1 non-combustibility class, they are produced not only using cannabis boon, but chopped fibers also. These blocks can be called 100 % organic, because only lime and other natural supplements are used as a binding agent. Cannabis has no non-recyclable parts, it is fully recyclable.

Hemp oil is one of the most famous products and it is extracted since ancient times. This is a very valuable product, having a high iodine, polyunsaturated (omega-6 and omega-3), and polyunsaturated fatty acids and other very important, necessary for the human organism, materials. Two types of materials are produced from hemp stems that can be used in industry: hemp boon is obtained after crushing hemp woody core of the stem and hemp fiber, obtained from the separation of the stem fiber. Cannabis stem, depending on the species, contains approximately 20–40 % of fiber which is on the surface of the stem and 60–80 % of wood (boon). Hemp fiber is very strong compared with other natural fibers (cotton, flax, and nettle), so it has long been widely used in ropes, net, and sail production, also in agrotextile and geotextile. Hemp boon is useful for thermal insulation composite production due to their small-pored porous structure. High-quality building blocks, ideal for eco-friendly passive houses are produced by blending hemp boon with clay or lime. They have good heat resistance in winter time and cool in summer, resistant to mold, moisture-proof, and highly breathable. Cannabis has a lot of cellulose, and is therefore used in papermaking. Cannabis has a lot of cellulose and is used for the paper production processes. Cannabis paper is of a high-quality and durable for long term storage for forms, cash, and cigarettes production) (Jankauskienė and Gruzdevienė 2010).

Lithuania legalized hemp cultivation in 2014. The experience in hemp growing and processing is rather poor. Hemp cultivation and processing in Lithuania is very relevant in growing processing sections, such as: textile, paper, and others. There is a possibility of hemp appliance in production of textile, furniture, and building construction products taking into account its eco-friendliness, harmony and increasing production in Lithuania. The effective manufacturing, including cultivation and processing, is required for the successful development of the sector.

Lithuanian hemp growers must comply with the following requirements:

- To grow hemp only outdoors in unified field;
- To sow only hemp varieties legal in EU;
- To use only certified hemp seed;
- mandatory declare hemp crops;
- inform about the cultivation, areas, blossoming;
- declare warehouses, storage, retting, wintering sites;
- report on hemp grown for fiber yield and the use of purchased seed.

The State Plant Service under the Ministry of Agriculture provides hemp growth for fibre production supervision. 53 varieties of hemps were included in the European Union (EU) common catalogue of agricultural plant species in 2013. The seeds for the market are not intended for sowing and their products, raw or soaked hemps, and their stalks must have an accredited laboratory certificate attesting that the THC content in hemp from which these products were received, do not exceed 0.2 %. The importers must obtain the status of approved importer in order to bring the seeds of hemp grown for fibre to Lithuania. Only legal or natural persons of good reputation might become the importers.

The hemp stalks, pellets from the chaff, dust, due to higher biomass than of many other plants and good incandescent properties are supposed to be used as fuel. Currently, the supply and demand of hemp production is still developing in Lithuania. The introduction of existing products (hempseed oil, hemp salt, clothes from hemp, etc.) and the process of the search for the markets, new product manufacturing, processing perspectives, expanding areas for crops are going on (Poiša et al. 2010).

Lithuania Hemp allowed growing only from 2014, still not sufficient to supply a wider use, because the country is not yet in hemp fiber processing factories. Lithuania has already started to produce building blocks of cannabis and has already built houses using hemp fiber (about 100 m^2 of construction of a house requires about 1 ha of cannabis). Organic hemp blocks—an alternative construction material made from hemp shives and binder can change light concrete blocks used in the construction of buildings (Balčiūnas 2013).

Recently the processing of hemp stalks is not developed in Lithuania yet; their transportation to processing plants, located in other countries (e.g., Latvia) is going on. But in the future it is planned to make the processing in the country, creating higher added-value while growing hemp not only for hempseeds and hemp oil extraction, but also for stalk production for the fiber.

The interest in hemp, their cultivation, processing and use is growing in Lithuania. Hemp areas increased rapidly in Lithuania in 2011–2014 (Fig. 1). Their growing areas were very small till legalization, but now they are increasing each year. By 3.8 times larger crops areas were declared in 2014 than in 2013. Industrial hemp was grown on 54 ha in 2011, and already 1063 ha of hemp crops were declared in 2014, where 755 ha were grown for hempseed and 308 ha for hemp

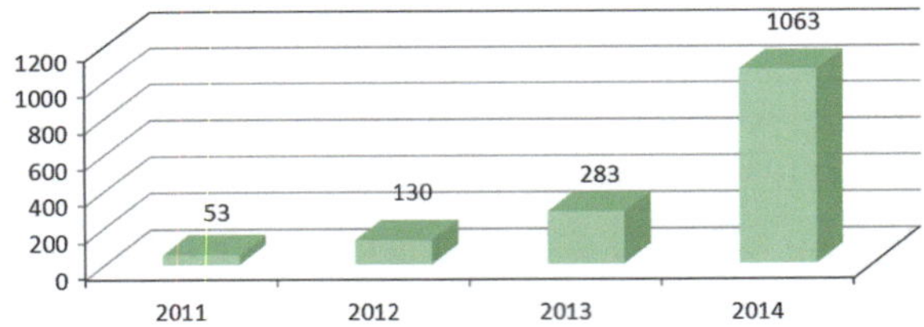

Fig. 1 Hemp area (ha) in Lithuania in 2011–2014

fiber. Therefore, the cultivation of hemp in Lithuania was oriented to seed yield in 2014 as 71 % of all industrial hemp crops were accounted for seed crops (Direct payment 2013; Direct payment 2014).

The factories using flax fiber can restructure into hemp production under the opportunity for processing them in Lithuania, thereby reducing the dependence on external markets and increasing local production and use. The possibility of hemp fiber use in the field of textile for the production of linen could be of very small-scale; higher application could be in carpets and technical textiles production. Their production has increased over the last five years.

The cultivation of hemp in Lithuania is oriented to seed growing. The crops for seed accounted 71 % of all grown hemp in 2014. The farmers are facing problems, typical for any newly produced agricultural culture, although the interest on hemp is growing, their production and processing in Lithuania is increasing. In particular, seed selection in Lithuania is not carried out, and the entire seed which must be certified, is imported. This means the dependence on foreign markets and higher costs.

The demand of renewable energy from the agricultural resources in Lithuania is increasing and hemp stalks or processed into pellets products for their good heat properties could be successfully applied for the fuel.

The paper in Lithuania in the olden times was made from hemp and nettle. Paper from hemp can be recycled 7 times, while the paper from wood can be recycled only 3 times. The perspective of cannabis use in the processing of paper and cardboard, taking into account ecology, is highly relevant, as paper and paperboard production increased in Lithuania (paper and paperboard production for 39 %, corrugated cardboard, boxes and cardboard packaging for 93 %).

Textile production increased in Lithuania in 2009–2013. However, the increase resulted in a growing fabric of artificial and synthetic fibers and yarns from artificial fibers, which accounted for 80 % of all produced textile in 2013. Woven fabrics of artificial and synthetic fibers and yarns from artificial fiber production have increased by 3.6 times over the past five years, while from flax it increased by 31 %. Synthetic materials demand is considerably greater, so it is likely that the production of materials from hemp, like flax production demand, would not make rapid rise in Lithuania.

Despite claims that natural fibres demand will increase because of limited cotton processing capability, most interested manufacturers are optimistic about prospects of fiber in the hemp market. There are also concerns about damage to the environment caused by the cultivation of cotton and looking to the cannabis growing as an environmentally friendly and even useful plant. However, manufacture from hemp fibre in the field of textile fabrics can be of very small volumes, the higher use adaptation could be in the carpet production, where the production increased by 6.7 times over the last five years (Jankauskienė and Gruzdevienė 2009).

The co-operation of fibre production and processing must be developed in order to have viable development of cannabis subsector with regard of grown hemp stalks crushing and ploughing. The development of the sector requires investments in processing, because low volume of cannabis growing is a barrier to invest in modern technologies, but the sector is still developing, and farmers master production technologies, application in production has just started, and still searching for opportunities. The sector looks promising, but it all depends on the competitiveness of raw material prices, supply and processing capacity in the future.

Conclusions

The cultivation of hemp grown for fibre was permitted in Lithuania in 2014. The farmers master their growing technologies and the processing has not been expanded. In the short term, the cultivation of hemp is oriented for seed production and their processing: oil pressure because it does not require large investments as in processing the stalks into fibre. Small cannabis growing volumes is a barrier to the investment in modern technology.

At the present time in Lithuania, not only the areas of hemp grown for the seed have upward trend. 1063 ha were used for hemp growing in 2014, 308 ha of which for hemp growth for fibre. These areas are still insufficient for development of effective processing.

Stem processing to fibre must be developed with regard to hemp growing and processing; the processing to final products must be created in order to obtain greater added value.

The demand for hemps and their products is still developing in Lithuania. Currently, oil producers are limited to small quantities. Consumers are familiarizing with the already manufactured products; hemp is already used not only in food, cosmetics, but also in the furniture and construction industry production. Use of hemp is extensive and the potential processing is provided in Lithuania in expanding processing segments, such as: paper and cardboard boxes, especially the expansion of online shopping network, and growing popularity of parcel shipping. Cannabis cultivation and processing are very important in order to increase the use of renewable resources for thermal energy generation in Lithuania, because they have good heat characteristics and can be applied to biofuel production. The possibility of using cannabis in Lithuania is visible in the construction and furniture production and for the manufacturing of other products. Hemp use in the textile industry to manufacture fabrics is less potential.

References

Butėnaitė L, Jonaitienė V, Stanys S, Jankauskienė Z, Gruzdevienė E (2012) Bast Crops Cultivation Possibilities and Perspectives in Lithuania // Product Design and Technology : Proceedings of Conference / Kaunas University of technology. Kaunas : Technologija. ISSN 1822-492X., p. 234-236.

Direct payment. 2013 Crop areas in accordance with the declaration of Statistics declared areas [online]. NPA.

Direct payment. 2014 Crop areas in accordance with the declaration of Statistics declared areas [online]. NPA.

Karus M, Vogt D (2004) European hemp industry: Cultivation, processing and product lines. Euphytica 140:7–12.

Kolarikova M, Havrland B, Ivanova T (2013) Energy balance of hemp (*Cannabis sativa* L.) grown for energy purposes. Agricultura Tropica Subtropica 46(1):10–5.

Jankauskienė Z, Gruzdevienė E (2013) Physical parameters of dew retted and water retted hemp (*Cannabis sativa* L.) fibres. Zemdirbyste-Agriculture 100(1):71–80.

Balčiūnas G, Vėjelis S, Vaitkus S, Kairytė A (2013) Physical properties and structure of composite made by using hemp hurds and different binding materials. Procedia Engineering 57:159–166.

Jankauskienė Z., Gruzdevienė E (2010) Evaluation of Cannabis sativa cultivars in Lithuania // Žemdirbystė / Zemdirbyste-Agriculture, vol 97, No 3, p. 87-96. – Akademija, Kėdainiai, Lithuania. (ISSN 1392-3196).

Poiša L., Adamovicšs A, Jankauskiene Z., Gruzdeviene E (2010). Industrial hemp (*Cannabis sativa* L.) as a biomass crop // Treatment and use of organic residues in agriculture: challenges and opportunities towards sustainable management. Proceedings of the 14th Ramiran International Conference of the FAO ESCORENA Network on the Recycling in Agricultural, Municipal and Industrial Residues in Agriculture. Lisboa, Portugal, 12-15 September, 2010, p. 326-330 (ISBN: 978-972-8669-47-8).

Jankauskienė Z., Gruzdevičienė E (2009) The investigation of industrial hemp acclimatization in Lithuania // Scientific bulletin of ESCORENA, Arad, vol. 1, p. 10-13. (FAO-ESCORENA international conference „Week of natural fibres“, „Aurel Vlaicu“ university, Arad, Romania, 21-24 June, 2009) (ISSN 2066-5687).

Carus M., Karst S., Kauffmann A., Hobson J. and Bertucelli S. (2013) The European Hemp Industry: Cultivation, processing and applications for fibres, shivs and seeds. European Industrial Hemp Association (EIHA) information.

Review of Wool Recycling and Reuse

Stephen Russell, Paul Swan, Mariell Trebowicz and Angus Ireland

Abstract The clothing and textile industry forms a considerable part of the world's economy. Of the more than $1 trillion sales of clothing worldwide, roughly two thirds is consumed in Western Europe and North America (University of Cambridge 2007). In addition to high-street products purchased directly by consumers, clothing is also consumed in the form of workwear, uniforms and corporate apparel distributed by brand owners and employers. Once this clothing is discarded by its user the main End of Life (EoL) options are in descending order of resource efficiency: (a) *Reuse*—garments worn again after donation and/or re-sale; (b) *Closed loop recycling*—garments used as raw materials for the manufacture of new products of similar value; (c) *Open loop recycling*—garments used as raw materials to manufacture industrial products of lower value; (d) *Incineration*—thermal energy generation; (e) *Landfill*. This chapter focuses on the collection, recycling and reuse of wool garments with particular emphasis on mechanical recycling in open and closed loop systems.

Keywords Wool fibre · Clothing · End of life · Recycling

The original version of this chapter was revised: The erratum to this chapter is available at DOI 10.1007/978-94-017-7515-1_36

S. Russell (✉) · M. Trebowicz
Nonwovens Innovation and Research Institute, Leeds, UK
e-mail: s.j.russell@leeds.ac.uk

P. Swan · A. Ireland
Australian Wool Innovation Limited (AWI), Sydney, Australia

P. Swan · A. Ireland
International Wool Textile Organisation (IWTO), Brussels, Belgium

R. Fangueiro and S. Rana (eds.), *Natural Fibres: Advances in Science and Technology Towards Industrial Applications*, RILEM Bookseries 12,
DOI 10.1007/978-94-017-7515-1_33

Introduction

Collection and Sorting of Post-consumer Clothing

Reuse and recycling of clothing can only practically take place if it is disposed of in such a way that items can be easily collected and diverted from municipal waste. Post-consumer clothing enters the reuse and recycling supply chain by mechanisms such as:

(a) Doorstep collections,
(b) Clothing bank donations,
(c) Charity shop donations or,
(d) Take-back schemes.

The majority of items from household or doorstep collections consist of clothing. Waste textiles also arise from commercial laundries that dispose of bed sheets, pillowcases and tablecloths, when their life is exhausted.

Companies controlled by charitable organisations commonly operate the collection infrastructure and the incentive for collection is based upon donated garments having a commercial value to the collector. For corporate clothing, brand owners or their agents often control the end of life disposal of their garments, which can involve mechanically shredding items to remove any identifying emblems or labels even though they may be perfectly wearable and not damaged. Recently, methods of quickly removing brand labels have been developed that enable corporate brand owners to redirect garments for reuse rather than recycling (Philpot et al. 2013).

Collection rates of post-consumer clothing vary by geographical region. The collection infrastructure is established in developed countries where the consumption of clothing per capita and the quality of clothing is relatively high. In these high consumption countries, there is a growing awareness of the environmental impact of clothing manufacture and disposal, particularly amongst those in the LOHAS (Lifestyles of Health and Sustainability) demographic. Europeans generate around 14 million tonnes of textile waste of which 5 million tonnes are recovered: three quarters of this is reused, typically for wear in other countries, or recycled, mainly as wipes or used in industrial applications (Morley 2014). Efforts are underway to increase collection rates in the USA, where only ca. 15 % of post-consumer textile waste is donated or recycled, with the rest going to landfill (MassSEP Offices 2011). In China, collection rates are relatively low, although there is growing recognition of the need to address the end of life disposal issues of clothing, such that it was included in China's 12th Five-Year Plan for textiles, to reduce energy consumption and to improve environmental awareness. This is particularly noteworthy given that China currently manufactures more than a quarter of the world's clothing and textiles (University of Cambridge 2007), and is expected to rise to over 50 % (Ethical Fashion Forum 2014).

In 2015, despite millions of tonnes of clothing being diverted from landfill, collection rates across all developed countries could be substantially increased. This has potential to release a large volume of valuable textile fibre, representing a major step forward in resource efficiency. In terms of increasing the donation of used clothing for recycling or reuse, issues of perception and education can be important. For example, some consumers believe that they should only donate clothing that they consider will be suitable for reuse (Oakdene Hollins 2013) whereas in fact, worn out clothing is still a valuable resource for open loop recycling and the manufacture of numerous industrial products.

Destinations of Post-consumer Clothing

After collection, garments are sorted to identify reusable (re-wearable) items suitable for resale, as well as garments that are too worn or damaged for reuse that can be recycled (Fig. 1). The estimated value of the total global trade in second-hand garments more than doubled from $1.26 bn in 2001 to $2.97 bn in 2010, based on UN Commodity Trade Statistics.

The financial returns generated by the sale of reusable garments, effectively subsidies the collection and sorting of recycling grades (Oakdene and Hollins 2013). Collected post-consumer clothing is treated like any other commercial commodity. In general, garments suitable for reuse have a higher economic value than recycling grades, i.e. low quality clothing, or clothing that is worn out, stained or damaged.

Historically, sorting was carried out near to where the clothing was first collected but the proportion exported to lower labour cost countries such as India, Pakistan or North Africa (mainly Morocco and Tunisia), Eastern Europe and special economic zones has increased. Sorted clothing is often repacked and re-exported under different classifications until it reaches its destination markets. Globally, the USA, Germany, UK, South Korea and Canada account for more than half of all post-consumer clothing exports (Crang et al. 2013). India is reported to have imported >220,000 tonnes p.a. of used clothing from the global consumer markets of the USA, Europe, the Far East and elsewhere for sorting and reprocessing (Norris 2012, 2012).

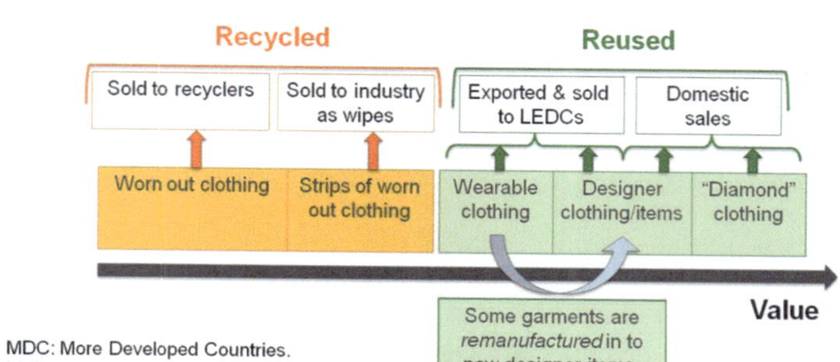

Fig. 1 Process flow for recycled and reused clothing (domestic sales also includes sales to MDCs)

High value and 'diamond' items (designer products) are identified during sorting and then sold by the charities and clothing merchants directly to consumers or retailers in developed countries or in Eastern European markets and other more developed countries (MDCs). Diamond items only account for approximately 1 % of the total volume of collected clothing, however they also account for the largest profit because they can be resold in developed countries (Dreier 2011).

A potentially disruptive development for collectors has been the trend for consumers to sell high value garments using the internet rather than donating via doorstep collections or clothing banks. Although internet sales extend the life of a garment, there is a risk of economically destabilising the industrial collection infrastructure because of the cost that is incurred for every tonne of clothing that is collected and sorted.

The price received for recycling grades of waste clothing can be much less than the cost of collection and sorting, but it is preferable to paying for disposal, which may be the only other option. As new applications for the recycling grades develop, demand can be expected to increase.

The bulk of re-usable clothing is consumed by less economically developed countries (LEDCs), most commonly in Africa, where in some locations >80 % of the population wear second-hand clothing (Bureau of International Recycling 2014), but this is not the only destination. Because of the warm climate, garments such as wool knitwear and jackets that are not generally suitable for reuse in Africa are frequently directed to Eastern Europe instead. Over ninety countries around the world consume used clothing that is deemed suitable for reuse (Bureau of International Recycling 2014).

Heirloom Clothing Items

In addition to the industrial collection, sorting, recycling and reuse infrastructure, it is known that consumers tend to hold on to high-value articles of clothing for many years before passing them on to friends and relatives. Such garments therefore remain in circulation for many years, passing between different generations in some cases. The proportion of garments treated in this way, as well as their fibre composition, is not known based on currently available data, but it is reasonable to assume that many will consist of high value garments including those containing wool.

Composition of Fibre Available for Recycling and Reuse

The relative proportions of fibre types entering the waste stream for recycling and reuse might be expected to mirror relative virgin fibre volumes consumed in clothing production. However, certain garment types are retained longer than others depending on their perceived value, serviceability and frequency of use. Consider for example, the expected overall life span of a cotton T-shirt versus a wool overcoat.

Few national studies of the fibre composition of post-consumer garment waste have been reported. A US study reported the relative fibre content of a used clothing sample donated for recycling and reuse by weight as follows: cotton, 59.29 %; polyester, 21.82 %; acrylic, 5.72 %; wool, 5.14 %; nylon, 5.09 %; and silk, 2.94 % (Chang et al. 1999). A larger sample of 35,000 garments donated to the Salvation Army in the UK indicated that the wool content was approximately 5 % by weight of the total (Ward et al. 2013). Therefore, in both the USA and UK studies, the proportion of wool was roughly 5 %, which is substantially higher than wool's share of the virgin fibre supply of about 1.5 %.

There is also a scarcity of comparative data on the international incidence of garment recycling and reuse, however a 2012 survey (The Nielsen Company 2012) of the wardrobe composition of 467 individuals across seven major wool-consuming nations, is informative. Key findings of this study are summarised in Fig. 2 and revealed the following:

Fabric Type:

- Cotton and cotton blends dominated male and female wardrobes (68 and 67 %), however the proportion of wool and wool-rich garments was relatively high for males at 13 % and for females 8 % (see Fig. 2).
- Chinese wardrobes had the largest wool content (males 19 % and females 13 %) followed by Italian wardrobes (males 17 % and females 13 %).
- Wool content was highest for accessories (25 %) followed by formal wear (17 %), underwear (11 %) and casual wear (10 %).

Garment Age:

- The average age of clothing in all wardrobes was 2.4 years.
- Wool was likely to be the oldest fabric in the wardrobe. Cotton garments were typically aged 1.9 years while wool garments were typically aged 2.8 years.

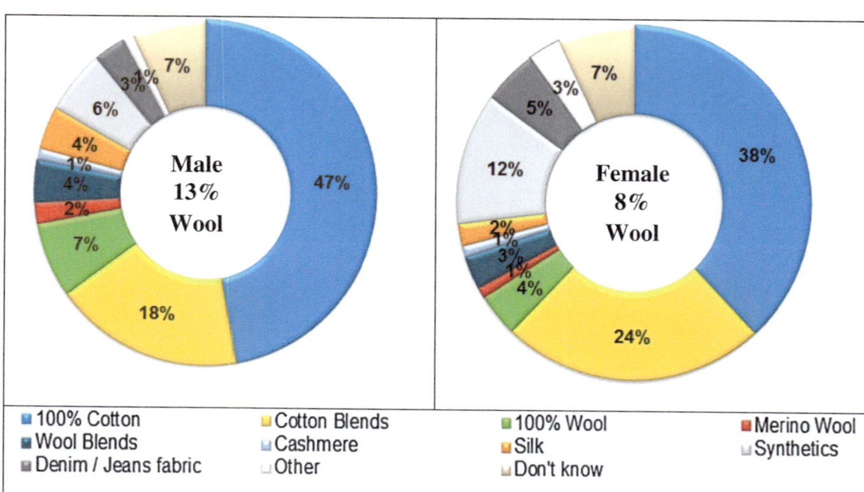

Fig. 2 Global consumer wardrobe survey: fibre composition (The Nielsen Company 2012)

Disposal:

- Donation to charities and family was highest for formal (57 %) and casual (54 %) clothes followed by sportswear (47 %) and accessories (41 %).
- A higher proportion of wool garments (47 %) was donated than synthetic (41 %) or cotton (40 %).
- Females were more likely to donate than males (53 % vs. 39 %).
- Underwear (61 %), and socks (58 %) and to a lesser degree nightwear (34 %) were typically binned. The next most common outcome for these categories was recycling at home (e.g. for use as wipers).

This study suggested that the life of garments containing wool was longer than those composed of cotton or synthetics and the respondents indicated an intention to preferentially donate wool-rich garments towards a second life.

Wool Fibre Recycling

Wool textile waste can be classified into two distinct categories: *post-industrial* (pre-consumer) and *post-consumer* waste (Fig. 3). The wool processing pipeline is essentially made up of a series of closed loop recycling steps which continuously feeds waste wool fibre from one processing step to another during production.

Post-industrial waste consists of fibre, yarn and fabric waste generated during production, and cutting waste from garment manufacture, whereas post-consumer waste is made up of clothing discarded by the user due to damage, being worn out or simply out of fashion (Modint 2010). For reasons of economic efficiency,

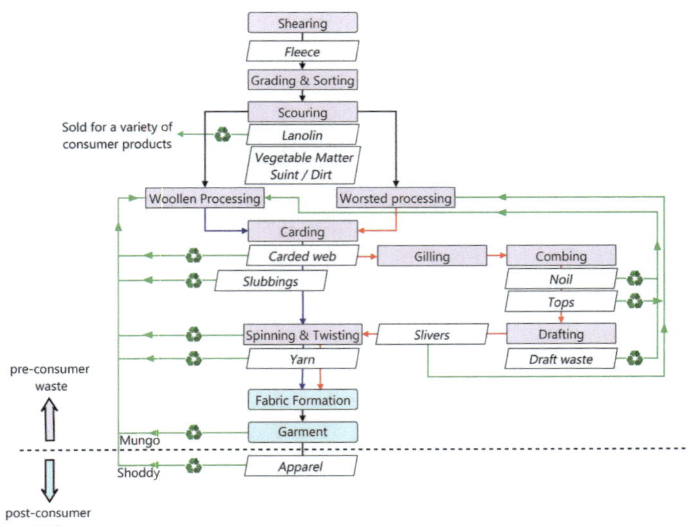

Fig. 3 Simplified wool fibre processing steps showing recycling flows of post-industrial (pre-consumer) and post-consumer wastes

post-industrial wool processing waste is routinely recycled back into the manufacturing process flow.

Many worn garments unsuitable for reuse are recyclable and wool fibre recyclate is an important, cost-effective raw material that provides an alternative to using virgin fibre either alone or in blends.

Post-consumer wool clothing has long been compatible with both open loop and closed loop mechanical recycling processes, and has been commercially exploited as a raw material for at least two hundred years. The practice traces its roots to Benjamin Law in the UK whose work led to the development of the wool shoddy industry in ca. 1813—one of the longest standing examples of post-consumer product recycling in the textile industry. The compatibility of wool fibre recyclate with open and closed loop recycling systems is partly dependent upon minimising fibre breakage and maximising residual fibre length after mechanical pulling processes. The level of yarn twist and the type of fabric construction influences the amount of fibre breakage during mechanical recycling.

Shortly after shoddy was introduced, mungo was developed as a way of incorporating recycled tailors' clippings into the woollen processing pipeline. Historically, only woollen and worsted fabrics were used to make shoddy and mungo products.

Open Loop Recycling of Wool Garments

Open loop recycling of wool involves mechanically pulling garments back into a fibrous state and using this material as a feedstock to make new products. Frequently nonwoven fabrics are produced using such materials by means of garnetting, carding or airlaying to produce webs followed by mechanical, thermal or chemical bonding.

Industrially developed countries within Europe, the USA and Asia purchase bales of clothing that have been pre-sorted into different grades. Grading is typically based upon fibre composition and colour shade. The commercial value of this material is a function of fibre composition, the homogeneity (i.e. mixed waste is less valuable), colour/shade (which can affect any subsequent dyeing operations) and whether it is knitwear or woven fabric waste, which influences the residual fibre length that can be obtained after the pulling operation.

Some manufacturers specialize in pulling the waste and selling the resulting pre-processed fibre recyclate to nonwoven fabric or yarn manufacturers. Alternatively, manufacturers purchase the baled clothing directly, and use their own recycling facility to mechanically pull the clothing back into fibre, before exploiting it as a raw material to manufacture new products. The latter approach is common in the production of insulator pads supplied to mattress/furniture and automotive product manufacturers, Fig. 4.

Woven fabrics such as worsteds (jackets/suits, coats) yield a shorter fibre after mechanical recycling (pulling), which means it is more difficult to use as a raw material for yarn spinning. This recyclate, mixed with other recycled fibre types can

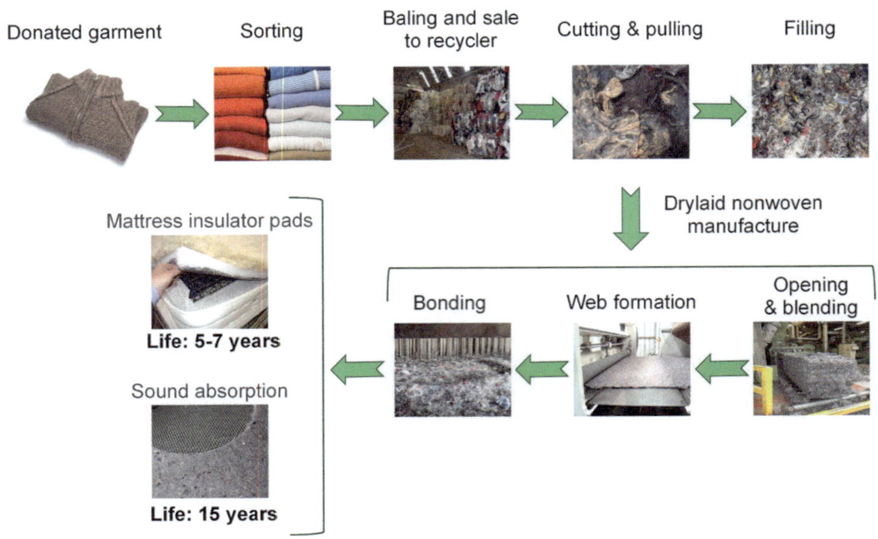

Fig. 4 Open loop mechanical recycling of wool for use in industrial products

still be used in nonwoven fabrics to make products such as mattress insulator pads, where the wool content assists in meeting fire retardancy requirements. In other products, there may be no absolute specification in terms of relative fibre content. Industrial products containing post-consumer fibres that can include a proportion of wool recyclate include:

- Nonwoven mattress insulator pads (placed directly over the springs in mattresses to protect the insulation layers that are laid above (Fig. 5);
- Nonwoven sound insulating materials for the automotive industry (e.g. body shell liners for sound insulation, backings for face materials including the floor carpets, boot/trunk liners);

Fig. 5 Mattress insulator pad (*grey layer*) produced from a blend of recycled wool and mixed fibres to protect against spring penetration

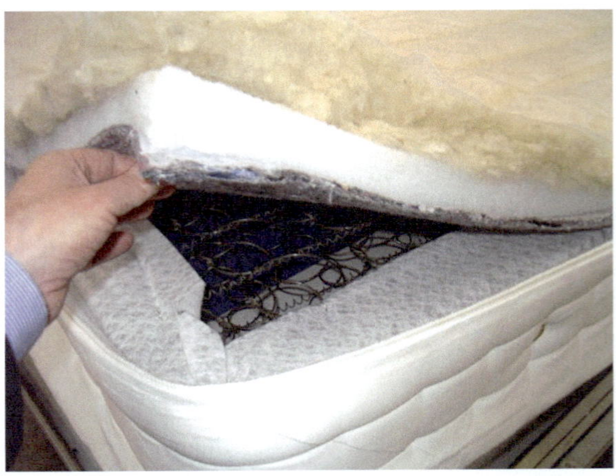

- Nonwoven building insulation (thermal);
- Protective sheeting fabrics for painting;
- Oil sorbents;
- Piano felts;
- Blankets (including emergency relief products);
- Rugs and clothing accessories.

Open loop recycling means the wool fibre is successfully diverted into a second life, away from landfill. This can be significant in terms of overall environmental sustainability, because the second operational life may be substantially longer than the first. For example, the operational life of mattress insulator pads covering the spring unit in mattresses can be up to seven years. Furthermore, since the nonwoven processes that manufacture such products are less sensitive to short fibre length compared to spinning processes, there is potential to recycle multiple times.

Closed Loop Recycling of Wool Garments

The basic steps in a closed loop recycling process for wool are summarised in Fig. 6. Yarns containing recycled wool are converted into other relatively high-value clothing products using conventional processes. Examples of a closed

Fig. 6 Example of closed loop mechanical recycling of wool garments

loop approach for wool include the Cardato initiative in Italy, and the M&S Shwop coat. In addition to improving resource utilisation, incorporating recycled wool as a blend component is an established means of reducing raw material costs in yarn manufacture.

Ideally, the recycled wool waste should be compatible with the same carding, spinning and fabric manufacturing operations used for processing the virgin fibre during woollen yarn production. Wool extracted from post-consumer garments, particularly knitwear, is a particularly useful raw material because of its compatibility with woollen yarn manufacture. A bale of mixed colour wool sweaters in Europe is usually more valuable than one comprising mixed synthetics. For wool knitwear, sorting of items into uniform colours can assist or sometimes eliminate the need for subsequent dyeing processes. Sorting to produce finer diameter wool batches can also yield premium prices.

Recycling of Corporate Clothing Containing Wool

A large proportion of the fibre content in corporate clothing comprises polyester, but wool is also extensively utilised, particularly in uniforms and knitwear. In knitwear, blends with acrylic are popular, and a large fraction is supplied in grey, black or navy. Wool is found in men's and ladies' corporate suit jackets and trousers, sometimes in blends with polyester and elastane (e.g. Lycra). Similarly, wool-rich or polyester-wool blends are common for ladies corporate skirts and dresses. The suiting fabrics containing wool are frequently treated with stain-resist chemistry.

Corporate clothing often carries a visual identity associated with a brand. Corporate clothing and workwear frequently carry embroidered motifs, labels, printed emblems, logos, transfers or prints that ideally need to be removed or destroyed to facilitate recycling. Corporate garments in territories such as the UK also carry a tax-tab, which enables recipients to avoid being taxed for receiving a benefit in kind from their employer.

Because of the visual appearance, aesthetics, presence of tax-tabs, emblems or logos on corporate garments, as well as the perceived security risk of allowing their reuse by unknown users in the future, the bulk of corporate clothing (>90 %) is landfilled or incinerated at the end of life. Some companies will permit recycling, provided the garments are first mechanically shredded to remove all identifying features. Initiatives that could potentially increase rates of wool recycling in corporate garments, thereby minimising landfill and incineration include:

- Uniforms that are designed for the end-of-life as well as the use phase e.g. homogeneous fibre composition, inbuilt de-labelling technology, garments that yield a longer fibre length after shredding;
- Working in collaboration with recyclers to secure larger volumes of garment waste to enable cost effective batch processing and to maximise economic viability for the recycler.

- Encouraging models that permit closed loop recycling of wool-rich products e.g. fibre recyclates that can be used as a blend component in yarns to make new uniforms;
- Developing convenient systems to maximise the return and collection of corporate garments from staff, cost-effectively (to improve collection rates).

Case Studies

Numerous clothing retailers and industrial groups have initiated schemes to promote recycling and reuse of textile and related products – many of which are using wool. Some of these initiatives encourage consumers to donate used clothing or to purchase garments containing recycled fibre. At the very least, these initiatives are raising consumer awareness about life cycle issues and resource efficiency, including the longevity and end of life destiny of the products they buy. In addition it is instructive to be aware of the manufacturers registered with the Textile Exchange that have developed expertise in the manufacture of yarns and fabrics containing either post-industrial or post-consumer fibre recyclates as blend components, including recycled wool.

Retailers and other organisations are fostering improvements in the environmental impact of clothing using several different models:

- *Endurance guarantees* (life time guarantees on clothing products);
- *In-store clothing deposits* (making it easier to dispose of clothing responsibly);
- *Take or buy back schemes* (financial incentives to encourage donation of clothing);
- *Eco product ranges* (clothing consisting of recycled fibres and other recovered materials).

A brief overview of a selection of case studies is now given that illustrates the mechanisms that have been developed to improve the environmental profile of clothing products. The examples presented first, relate to the recycling and reuse of wool garments, but non-wool specific schemes are also included to highlight approaches that could inform developments in the future.

Cardato; Prato, Italy

- Certifies member companies that utilise a high proportion of regenerated or recycled wool (>70 %) in their yarns.
- Claimed to be the first certified 'zero emissions' textile product.

Wool Again; USA

- Woollen yarns made from post-consumer recycled wool content (up to 80 %).

BENU Wool; Christian Fischbacher Co. AG, Switzerland

- Premium woven wool home furnishing fabrics manufactured from both post-industrial and post-consumer recycled wool.
- No additional fibre dyeing is undertaken, reducing water, waste, dyes and chemicals, and energy consumption.

Shwopping; Marks and Spencer (M&S), UK

- Take-back scheme incentivising customers to bring back used garments.
- Facilitates use of fibre recyclate in M&S product ranges.
- Workplace recycling—providing clothing deposit bins in workplaces.

H&M; Sweden

- Accept donated clothing of any brand and any condition.
- Customers receive a voucher that can be redeemed in-store.

Jack & Jones; Norway

- Take-back scheme incentivising customers to bring back used items.
- Customers receive a voucher that can be redeemed in-store.
- Post-industrial waste is used to create low impact denim.

Klättermusen; Sweden

- rECOver clothing scheme—the label in the garment carries a redeemable value (this is removed by the consumer and returned to trigger the voucher).
- Manufacture clothing using recycled fibre.

NorrØna; Switzerland

- Upcycled clothing range manufactured using recycled Polyethylene terephthalate (PET).
- About forty plastic bottles are used to make each upcycled jacket.
- Estimates that 20 % less energy is required to manufacture the product than a fleece made of virgin fibre, without sacrificing performance.

Mammut; Switzerland

- In-store recycling for Mammut clothing only.
- Produce clothing using recycled PET (REPREVE® 100 yarn).

Boardroom; Canada

- Recycled clothing—closed loop PET—the Eco® mark.
- Customers return ECO-apparel 100 % PET for recycling and the money raised is donated to a non-profit organization.

Rekixx; USA

- World's first completely landfill-free sneaker.

- Raw material inputs are lab-certified as 100 % recyclable materials.
- The product does not require separation or disassembly—all parts can be recycled together.

Patagonia; USA

- Sustainable fibres used in many products and minimisation of packaging and transportation waste.
- Returned Patagonia clothing is repaired at no cost depending on condition.
- Buy back clothing (good condition) for resale.
- Take back clothing—recycle into new clothing or make into new products.

Puma; Germany

- In-store recycling—take back any brand shoes, clothing & accessories.
- InCycle collection—manufactured clothing that contains either biodegradable or recyclable materials (cradle to cradle basic certification).
- 'Clever Little Bag' introduced in 2010 to replace traditional shoeboxes.
- Goal for 2015 is to have 50 % of its international collections made of more sustainable materials.

Nike; USA

- In-store recycling: clothing deposit bin for shoes.
- 'reuse a shoe'—collect worn out athletic shoes and grind down to create a new material to make high-quality sports surfaces including courts, turf fields, tracks and more.
- Produce many items of clothing from recycled PET—the 2014 world cup Brazilian National Team kit was made from 100 % recycled polyester (knitted fabric composed of cotton and PET).

Teijin Limited, Onward Holdings Co., Ltd. And Fuji Xerox Co; China

- First closed loop recycling system for corporate clothing (uniforms) in China.
- Based on ECO CIRCLE scheme, which involves the chemical recycling of PET.

References

University of Cambridge (2007) Well dressed? http://www.cam.ac.uk/research/news/well-dressed.

Philpot B, Pye A, Stevens G (2013) De-labelling Branded Corporate-wear for Re-use. Project Number: MPD007-007, WRAP Final Report, December 2013, pp.1-71

Morley N (2014) Textile & Clothing. http://oakdenehollins.co.uk/textiles-clothing.php. Accessed 1-10-2014

MassSEP Offices (2011) Textile Recycling Meeting Summary.

Ethical Fashion Forum (2014). http://www.ethicalfashionforum.com/the-issues/china. Accessed 1-10-2014

Oakdene Hollins (2013) Closed loop fibre recycling - current status and future challenges. http://www.oakdenehollins.co.uk/media/Closed%20Loop/Briefing_note-closed_loop_clothing_recycling.pdf Accessed 1-10-2014

Crang MA et al (2013) Rethinking governance and value in commodity chains through global recycling networks. Transactions of the Institute of British Geographers 38: 12-24.
Norris L (2012) 'Shoddy rags and relief blankets: Perceptions of textile recycling in north India' in Alexander, Catherine and Reno, Josh (Eds.). Economies of Recycling: The global transformation of materials, values and social relations. London: Zed Books. pp 35-58.
Norris L (2012) Recycling imported secondhand textiles in the shoddy mills in Panipat, India: an overview of the industry, its local impact and implications for the UK trade http://www.wornclothing.co.uk/wp-content/uploads/2011/01/SummaryShoddy.pdf. Accessed 18-3- 2014
Dreier W (2011-2012) Cleaning out the closet,Vanguard, College of Human Environmental Sciences, University of Missouri. pp 18-22. http://hes.missouri.edu/documents/vanguard/2012/cover.pdf Accessed 20-7-2015
Bureau of International Recycling (2014). http://www.bir.org/industry/textiles/?locale=en_US. Accessed 1-10-2014
Chang Y, Chen HL, Francis S (1999) Market Applications for Recycled Postconsumer Fibres. Family and Consumer Science 27(3), pp. 320-340.
Ward GD, Hewitt AD, Russell SJ (2013) Fibre composition of donated post-consumer clothing in the UK. Proceedings of the ICE - Waste and Resource Management, Volume 166, Issue 1, 01 February 2013, pp. 29 –37
The Nielsen Company (2012) Global Wardrobe Audit All Countries. Report Prepared for Australian Wool Innovation
Modint (2010) Recycled fibres from virgin to eternity, CSR factsheet. pp 1-8.

Brazilian Scope of Management and Recycling of Textile Wastes

Welton Fernando Zonatti, Júlia Baruque-Ramos and Wânia Duleba

Abstract Brazil is one of the largest world producers of textiles and clothing. It is an important producer of cotton fiber, yarn, plain and knitted fabrics. This stimulates the country's GDP generating millions of direct and indirect jobs in this sector. However, such activity generates various environmental problems, such as the generation of solid wastes from textile and clothing industrial processes, as well as the post-consumer clothing, which are discarded commonly in landfills. Thus, the present study presents considerations about mechanical processes for textile recycling, recycled cotton yarn and jeans characteristics, the generation of solid wastes in textile and clothing Brazilian industry and the actions directed toward sustainability. The textile reuse and recycling may contribute to the reduction of environmental problems, in addition to being economically viable source of income and social benefits for all involved in this chain, from street collectors to the industrial sector.

Keywords Textile recycling · Mechanical processes · Cotton fibers · Reverse logistics · Brazilian scope

Introduction

Besides being a major producer of goods and services in sectors such as agriculture, mining and petroleum (Mattos 2006), Brazil is also one of the major world manufacturer of textiles: the fifth largest producer, with a production of about

W.F. Zonatti · J. Baruque-Ramos (✉) · W. Duleba
School of Arts, Sciences and Humanities, University of São Paulo, São Paulo, SP, Brazil
e-mail: jbaruque@usp.br

W.F. Zonatti
e-mail: welton@usp.br

W. Duleba
e-mail: wduleba@usp.br

R. Fangueiro and S. Rana (eds.), *Natural Fibres: Advances in Science and Technology Towards Industrial Applications*, RILEM Bookseries 12,
DOI 10.1007/978-94-017-7515-1_34

1.8 million tons of made-up articles (IEMI 2014) and also the 5th largest producer of cotton (ABIT 2011), one of the most employed natural fiber in the world, with which are made 60 % of the garments in the country (Mello et al. 2007; ABIT 2011). The country ranks 7th in the production of yarns and woven fabrics and the 3rd in the production of knitted fabrics (Finkler et al. 2005), as well as generating 8 million of direct and indirect jobs (Neuls 2012).

According to IEMI—Institute of Studies on Industrial Marketing (2014), specialized in research and analysis of textile and clothing sectors, in 2013 the textile and clothing chain produced about R$ 58.2 billion (near 20 billion €), equivalent to 5.7 % of total Brazilian industrial production, disregarding the activities of mining and construction, which complement the secondary sector of Brazilian economy. In 2013, the textile and clothing sector had 1.6 million of jobs or the equivalent of 16.4 % of the workers at the industrial production in that year, proving to be a strong social impact segment—by employing a huge number of women and workers with lower qualification—and economic, since their production centers are extended nationwide.

São Paulo State is the main industrial producer and employer in Brazil, presenting all the segments of textile chain. The companies located in this State represent 29 % of the national total and employ about 30 % of people (more than 500 thousand) in comparison to whole Brazilian textile chain in 2010 (out of 1.6 million people) (IEMI 2014).

Even bringing economic benefits to the country, on the other hand the textile and clothing sectors creates many environmental problems. Along the textile and clothing chain there are several operations generating solid, liquid and gaseous wastes. For instance, since the cotton ginning up to generation of sludge in biological treatment of liquid effluents (Sinditêxtil-SP 2009). However, the main focus of this paper is about solid wastes generated by mainly the Brazilian clothing industrial sector.

Mechanical Processes for Textile Recycling

The textile recycling can be classified depending on raw materials to be reprocessed, resulting in the use of different production lines and specific processes to obtain the final products. They can be divided into: mechanical recycling, chemical recycling; thermal recycling and mix of technologies (Groupe CTT and ITS 2014).

Mechanical recycling is the most found, because it can process textile wastes of various compositions. Both virgin and recycled fibers can further employed in known yarning or nonwoven processes. However, textiles wastes, consolidated in the form of articles or fabrics, should be previously separated by composition and color (or not, depending on the final product to be obtained) and then be shredded in a specific machinery (Zonatti 2013; Zonatti et al. 2015).

The activity of the textile recycling emerged in the Prato city (Italy) in the mid-nineteenth century with the recovery of wastes of wool (Camera di Commercio

Prato 2014). It is characterized as the practice or technique in which wastes and leftovers from industrial processes, clothing manufacturers or from disposal of post consumption items can be reprocessed with a treatment to change their physical and/or chemical characteristics. They can be classified into (Sinditêxtil-SP 2009):

(i) recycling within the process: it allows the reuse of the wastes as raw material in the process that caused their generation (for instance, in the industry of home-wear articles);
(ii) recycling out of the process: it allows the reuse of the wastes as raw material in a different process from which that caused their generation(for instance, the employment of textile fibers for composite production in automotive industry).

Wang (2006) stated that the recycled fibers are produced in a secondary cycle processing. To obtain them, there are mechanical processes on the textile wastes. Conventionally, the waste is pre-treated by means of cutting or separation and then transported to undergoing the proper recycling, transforming the consolidated material into individual fibers.

Generally the mechanical textile recycling process is as follows: pecking of textile wastes (to decrease the size and maximize the uniformity of the scraps) and subsequent shredding this material by pricked cylinders (generally 3–6, wherein the number of needles increases in each next cylinder in the process). Subsequently, at the industrial level, connected to these first steps it can be, for example, nonwoven continuous production lines, production of composites by airlay process or others.

The mechanical recycled fibers have lower lengths in comparison to virgin ones, but it is difficult to clearly define their other characteristics compared to materials produced from primary fibers. According to Gulich (2006a, b), processes and equipment conventionally used for measuring the characteristics of textiles are not very helpful when applied to recycled fibers. This is due to the mixture of recycled fibers, which are not homogeneous and/or they can be of the same composition (for example, 100 % cotton wastes), but with different origins and also different qualities.

Recycled Cotton Yarn and Jeans

Regarding yarn characteristics, it is known that the elongation, tenacity and irregularities are the most important parameters indicating its quality (Araújo and Mello e Castro 1986). Halimi et al. (2008) studied comparatively recycled cotton and virgin cotton yarns. This study demonstrated that the tenacity of recycled yarn decreased 26.3 % when 100 % of waste as raw material was employed (Fig. 1). The tenacity decreased only 11.6 % when only 25 % of waste mixed with 75 % virgin fiber was employed. In addition the yarn regularity was not affected until the proportion of 25 % of waste material. However, when the cotton waste content was above 25 %, the effect on the yarn irregularity was considerable. According to the literature (Klein 1993; Jackowsk et al. 2002), the yarn elongation is mainly related

Fig. 1 Visual appearance of denim fabrics: **a** made from 100 % virgin cotton fibers; **b** made from 100 % recycled cotton fibers. *Source* Photographs taken by authors

to the properties of the raw material. According to Halimi et al. (2008), when the waste material in the yarn is 25 %, its elongation decreased 1.6 % in relation to the yarn composed exclusively of virgin raw material. In this way, it can be concluded that the introduction of 15–25 % of cotton waste material in yarn does not affect significantly the tenacity, regularity and stretching, fundamental properties regarding quality, reliability, comfort, dyeing, among others.

Duarte (2013) evaluated the characteristics of fiber jeans mixed cotton and recycled polyester compared to CO/PET conventional jeans, as the environmental impact in textile production, as well as the durability of its materials. To evaluate the durability of jeans physical tests of resistance to elongation and tensile strength; tear resistance; and pilling, according to NBR 11912, ASTM D 2261-11 and ISO 12945-1, respectively were conducted. It was also evaluated two prototypes of pants, with each type of jeans, subjected to intensive wear and washes every 30 days. Regarding the results obtained, the environmental impact assessment showed that the production of jeans with recycled fiber consumes fewer resources, such as water, energy and chemical inputs than the production of conventional jeans. On the other hand, mechanical strength tests and wear test showed that the CO/PET recycled jeans is less durable than the jeans made by conventional manufacturing. The materials and prototypes evaluated were contextualized in scenarios of sustainability and fashion. This author concludes that the jeans with recycled fibers is suitable for hyper technological scenario and fast fashion, while the conventional jeans is appropriate for the scenario of slow fashion and hyper culture.

The Generation of Textile Wastes in Brazil

Countries as China, Bangladesh and India, among others, are major exporters of apparel, mainly to developed countries of the European Union and United States, which in their turn do not need to hardly deal with the environmental problems arising from industrial textile and clothing production, but with the problems related to the disposal of garments by individuals after their use. This justifies the fact that there are more scientific reports and international bibliographic references focusing on the reuse of clothing items than the textile physical recycling, since the textile waste generation is a more important issue for developing or underdeveloped countries, which has the largest textile industries worldwide.

Thus, one of the main environmental problems of the textile chain in Brazil is the generation of solids wastes from industrials processes, clothing manufacturers and also the post-consumption, being the waste generation of clothing industries the most significant of all (Zonatti 2013; Zonatti et al. 2015).

Only one reference was found (Sinditêxtil-SP 2013), which states that Brazil produces 175,000 tons per year of solid wastes from its textile and clothing industries, being mostly scraps from clothing industries. More than 90 % of these scraps was discarded incorrectly in landfills. This reference also report the situation in neighborhood of Bom Retiro, located in São Paulo city. In this neighborhood are installed 1200 clothing enterprises (60 % large generators), which dispose 16 tons per day of waste textiles and this amount represents 2 % of scraps generated annually in Brazil.

In the case of clothing industry, the generation of solids wastes comes from the spreading and cutting processes. According to Ribeiro and Andrade Filho (1987), the spreading is characterized by the provision of the fabric in superposed layers, according to a predetermined schedule. For spreading the fabric, the operator lays it successively on the cutting table and over all place the risk marker paper. Then the cutting machine work follows the lines present in this risk marker paper. Between the pieces destined to be sewn, scraps left over and can overcome the proportion of 10 % in relation to the surface of all fabric layers (Fig. 2a and b).

Representing this Brazilian problematic the neighborhoods of Bom Retiro and Brás (São Paulo city) are important textile centers in São Paulo and large generators of solid waste, which is incorrectly discarded (Zonatti 2013; Zonatti et al. 2015).

(a) **(b)**

Fig. 2 Textile wastes samples: **a** textiles residues from spreading and cutting of fabric layers of sorted colors and **b** mixed textile scrapes from clothing industry. *Source* Photographs taken by authors

Only in Bom Retiro 25 % of all Brazilian garment production is generated (Yázigi 2006). There are so few cooperatives working with this industry (IPT and SEBRAE 2003) and the wastes are minimally availed by issues of logistics, transportation and due to the poor infrastructure of these cooperatives, which do not have physical space to store large amounts of scraps. Within this perimeter are estimated per day, originated by the clothing manufacturer sector, 16 tons of textile waste from Bom Retiro area and 10 tons near Republic of Korea Plaza in Brás according to the enterprise LOGA—Environmental Logistics (2014), which collects solid wastes from São Paulo city. From a total of 100 % of trash solid residues, 50 % are estimated to be scraps of fabric, and from this, about 10 % are not synthetic and 90 % of synthetic fabrics such as polyester and minority polyamide (Sinditêxtil-SP 2013). Still, according to LOGA (2011), all collected fabric scraps end up in landfills. Informally in the metropolitan region of São Paulo city 50 tons per day of textile scraps and discarded garments were estimated.

Despite the large amount of textile wastes not employed and discarded in landfills, on the other hand, in 2012 in Brazil 9,829,928 Liq. kg of textile scraps and residues composed by silk, wool, cotton, man-made fibers, were imported. This is equivalent to US $ 11,421,644 (FOB—Free on Board). This fact shows that there is a great demand for textile waste aiming recycling and for various purposes, such as: production of blankets (for industrial and geotextile applications), coatings, composites for industrial and civil construction, fillers, production of yarns and strings among others (Sinditêxtil-SP 2013).

It is observed that the Brazilian market involved with textile reuse and recycling prefers to import instead to use the national textile waste available in abundance, due to a number of problems related to mismanagement of this material, such as: mixture of different raw materials and its difficult separation (Fig. 3a–c), various types of dirt mixed to the waste (Fig. 3d), lack of tax incentives related to the textile recycling, little qualified and specialized workers in this segment, few references and research that propagate the knowledge in this area, among others. Despite of the existence of an incipient little national market for textile wastes (mainly focused in scraps made of cotton—for mechanical recycling aiming automotive applications;

(a) (b) (c) (d)

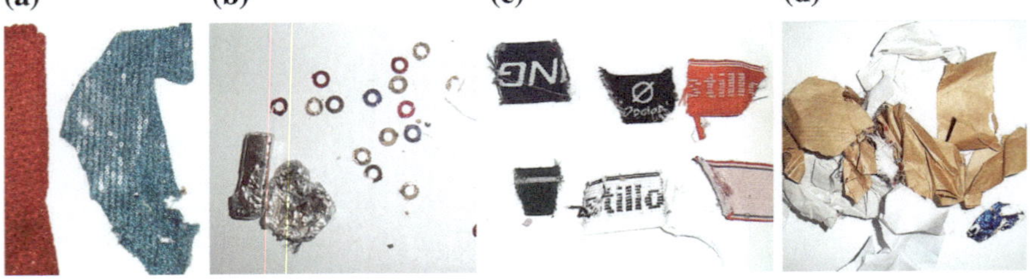

Fig. 3 Problems encountered in the textile recycling: **a** special fabrics with metallic threads, **b** mix of metal or metallic materials, **c** mixed of elastic materials and **d** different types of papers and dirt mixed with textile residues. *Source* Photographs taken by authors

and polyamide—for thermal recycling aiming automotive and technical applications), it is emphasized that besides to still disregard the domestic textile waste from textile and clothing industries, the Brazilian market for textile reuse and recycling also disregards the potential integration of discarded garments by individuals after consumption, due to the lack of studies and data aiming technical possibilities, for example, either the absence of a strongly reverse logistics established for this specific case.

Incipient Brazilian Actions Toward the Sustainability

In 2010 the National Policy on Solid Wastes (NPSW)—Law 12,305—August 2, 2010—was approved. From there Brazil can regulate issues related to solid wastes. According this regulation the trash is classified as: (i) wastes with potential of being availed through reuse or recycling; and (ii) rejects, which cannot be reused or recycled. The Law content ranges from conceptualizations about solid wastes, the instruments of legal interference, their management, company responsibilities and other issues that allow intervention in the matter of solid wastes (Brazil 2010; Brazilian Environment Ministry 2015).

An example of application of NPSW in the industrial clothing sector was in the case of Londrina city (Paraná State, Brazil), notorious polo of clothing manufacturers in the south part of the country. In this city textile scraps were disposed indiscriminately in public spaces, "passing" the responsibility for these wastes (and the consequences of environmental degradation) to the population. Since 2011 the clothing manufactures were heavily audited, held accountable and penalized, in reason of the prosecution proposed by "Protection of Environment of Londrina" and "Environmental Institute of Paraná" resulted in a joint recommendation to regulate the management of solid wastes in that city (Public Prosecution of Brazilian State of Paraná 2011).

As pointed before, the neighborhoods of Bom Retiro and Brás, located in São Paulo city, represent important textile centers. Only in Bom Retiro, the revenues of the district reaches, approximately, R$ 6 billion (near 2 billion €) a year. The wholesaler clients are originated from all over Brazil, Latin America and Africa—approximately 70,000 shoppers a day (Yázigi 2006). However, these clothing manufacturers are also major generators of solid wastes (Zonatti 2013). Routinely, employees of thousands of clothing manufactures discards bags mixing ordinary trash with fabric scraps on the streets (LOGA 2011; Sinditêxtil-SP 2013). Almost the totality ends up in landfills, but before the trash collection performed by enterprise LOGA, it is common street collectors to open these bags, select the textile wastes with higher market value (cotton fabric and jeans scraps followed by scraps of other compositions and larger dimensions) and leave spread the rest of the garbage/waste on public streets, generating pollution, clogged drains, floods and other environmental and social impacts (Sinditêxtil-SP 2013).

Directed to the clothing industrial sector there is no governmental institutionalized policies regarding Reverse Logistics, cooperatives, social inclusion of street collectors or other relevant actions regarding the industrial clothing segment neither in São Paulo city or in any other part of the country. In addition, there is no concern or Brazilian public policy about the issues related to harmful chemical elements or biological agents in fabrics that can be hazardous to human or animal health due the contact of textile with the body skin or about the contamination of the soil after their discard.

As pointed before, there is an incipient and increasing private and increasing recycling sector in Brazil, mainly focused in scraps made of cotton—for mechanical recycling aiming automotive and other applications; and polyamide—for thermal recycling aiming automotive and technical applications. Regarding post consumption garments, exception some private thrift shops or few non-governmental organizations dedicated to reuse and donation of garments, there is no institutionalized actions and the most part of apparels ends up in landfills as ordinary solid trash.

In the neighborhood of Bom Retiro (São Paulo city) a pioneer project called "Retalho Fashion" is still in implementation. The "Retalho Fashion" is headed by "*Sinditêxtil-SP*—Union of Industries of Spinning and Weaving of São Paulo State" together with "*ABIT*—Brazilian Association of Textile and Apparel Industries", "Francisco Matarazzo school—*SENAI*—National Industrial Learning Service", the "Chamber of Shopkeepers of Bom Retiro" and the "Municipality of the São Paulo city". This project aims to develop a waste management plan, organizing and facilitating the collection of textiles wastes in the neighborhood, seeking to preserve the environmental and to generate skilled occupation, creating socially fair working conditions for workers who depend on this means of livelihood and restoring the environmental and cleaning conditions of the area involved (ABIT 2012; Sinditêxtil-SP 2013; São Paulo City Hall 2014).

Thus, the "Retalho Fashion" was divided into three stages (Sinditêxtil-SP 2013):

1st Stage Diagnosis of the region; mobilizing the business community; planning of project execution;

2nd Stage Determination of the necessary infrastructure for implementation of the project and the demand for textile waste;

3rd Stage Project Implementation.

Currently the project is in the 2nd Stage, in which is scheduled to determine the minimum needed infrastructure for its consolidation, including the price and rental of physical space, quotation and purchase of machinery such as forklifts, presses and industrial scales, among others. At this stage, the project is primarily focused on manual separation of textile wastes (by color, size, composition, for example) generated in the Bom Retiro neighborhood. Thus, it is intended to skill street collectors already in the region, improving the process and, hence, to allocate in a way more sustainable way the textile wastes to private recycling companies, in order to preserve the cleaning and environmental conditions (Sinditêxtil-SP 2013).

Thus, despite current actions are important, it is necessary for the feasibility of proposals related to NPSW, Reverse Logistics, clear plans and objectives for managing solid wastes, as well as specific legislation about toxicity and correct disposal of textiles, to account and declare how much is generated, the shared liability for goods, to encourage the creation and operation of cooperatives, public policies for social inclusion of street collectors, environmental monitoring and development of technical and scientific knowledge that will benefit all sectors of society.

Conclusion

Despite of the absence of Brazilian governmental policies related to textile recycling and reuse, excepted NPSW, there is an incipient and increasing private recycling industry in Brazil employing textile scraps from clothing manufacturers. Regarding post consumption garments the most part of apparels ends up in landfills. The implantation of beneficial actions, such as "Retalho Fashion", are also noticed. Thus, it is concluded the textile reuse and recycling can contribute to the reduction of environmental problems, in addition to being economically viable source of income and social benefits for all involved in this chain, from street collectors to the industrial sector.

Acknowledgments We gratefully acknowledge **CAPES** (Coordination for the Improvement of Higher Education Personnel of Brazilian Education Ministry) and**s** (University of São Paulo).

References

ABIT – ASSOCIAÇÃO BRASILEIRA DA INDÚSTRIA TÊXTIL E DE CONFECÇÃO. **Guia Têxtil 2011.** Available in <http://www.guiatextil.com/site/noticias/industria/abit_esclarece_principais_duvidas_das_normas_de_vestibilidade>Accessed Set. 8, 2011.

ABIT – ASSOCIAÇÃO BRASILEIRA DA INDÚSTRIA TÊXTIL E DE CONFECÇÃO. **ABIT participa de projeto de Formação e Inclusão Social**. Available in<http://www.abtt.org.br/boletim/boletim12/abit3.htm>. Accessed Jan. 9, 2012.

ARAÚJO, M. de; MELLO E CASTRO, M. **Manual de Engenharia Têxtil** – Vol. I. Portugal: Ed. Calouste Gulbenkian, 1986. 694p.

BRAZIL. Política Nacional de Resíduos Sólidos ("National Policy on Solid Wastes") (2010). **Brazilian Law n° 12.305, Aug. 2, 2010**. Establishes the National Policy on Solid Wastes; modifies the Brazilian Law n° 9.605, Feb. 12, 1998; and other matters.

BRAZILIAN ENVIRONMENTAL MINISTRY. **Relatório técnico de caracterização de resíduos sólidos**. Available in<http://www.mma.gov.br/port/conama/processos/18018FE8/Relat_LixiviacaoAmiantoFibrocimento_Rel031E09.pdf>Accessed Feb. 10, 2015.

CAMARA DI COMMERCIO PRATO. **Cardato Regenerated CO_2 Neutral**. Available in < file:///C:/Users/W/Downloads/Prato_Chamber_of_Commerce_project_description.pdf > . Accessed Dec. 23, 2014.

DUARTE, L. S. **Estudo comparativo do impacto ambiental do jeans CO/PET convencional e de jeans reciclado**. 2013. 89 p. Dissertation (Master in Production Engineering). Engineering School. Federal University of Minas Gerais. Belo Horizonte/MG, 2013.

FINKLER, M.; SCAPINI, P.; FREIRE, E.; ZATTERA, A. J.; ZENI, M. Compósitos de HDPE com Resíduos de Fibras Têxteis. Parte I: Caracterização Mecânica. **Polímeros: Ciência e Tecnologia**, vol. 15, n. 3, p. 171-175, 2005.

GROUPE CTT – CENTRE MULTISERVICES POUR L'INDUSTRIE TEXTILE; ITS – INSTITUTE OF TEXTILE SCIENCE. **Textile recycling: an overview on technologies and tendencies applications**. Available in <http://www.textilescience.ca/downloads/april12_2013/Recycling_Claire_Guignier_apr12_2013.pdf>. Accessed Dec. 10, 2014.

GULICH, B. Designing textile products that are easy to recycle. In: **Recycling in textiles** (Edited by Youjiang Wang). Cambridge (England): Woodhead Publishing Limited in association with The Textile Institute, 2006a. 230 p.

GULICH, B. Development of products made of reclaimed fibres. In: **Recycling in textiles** (Edited by Youjiang Wang). Cambridge (England): Woodhead Publishing Limited in association with The Textile Institute, 2006b. 230 p.

HALIMI, M. T.; HASSEN, M. B.; SAKLI, F. Cotton waste recycling: Quantitative and qualitative assessment. **Resources, Conservation and Recycling**, vol. 52, p. 785–791, 2008.

IEMI - INSTITUTO DE ESTUDOS E MARKETING INDUSTRIAL. **Relatório Setorial da Indústria Têxtil Brasileira** - Brasil Têxtil 2014. 14ª edição. São Paulo, 2014.

IPT - INSTITUTO DE PESQUISAS TECNOLÓGICAS; SEBRAE – SERVIÇO BRASILEIRO DE APOIO ÀS MICRO E PEQUENAS EMPRESAS. **Cooperativa de Catadores de Materiais Recicláveis – Guia para Implantação**. São Paulo: Publicação IPT, 2003.

JACKOWSKI, T.; CHYLEWSKA, B.; CYNIAK, D. Influenceofspinningprocessparametersonstrengthcharacteristicsofcottonyarns. **Fibres& Textiles in Eastern Europe**, v.10, n. 3, p. 27–31, 2002.

KLEIN, W. New spinning systems, short-staple spinning series. **The Textile Institute**, v. 5, p. 40-41, 1993.

LOGA – LOGÍSTICA AMBIENTAL DE SÃO PAULO. **Coleta domiciliar**. Available in<www.loga.com.br> Accessed Feb. 26, 2014.

LOGA – LOGÍSTICA AMBIENTAL DE SÃO PAULO. **Interview with Mr. Francisco Vianna** on Aug. 18, 2011.

MATTOS, P. T. L. **O Novo Estado Regulador no Brasil**: Eficiência e Legitimidade. São Paulo: Ed. Singular, 2006. 331 p.

MELLO, M. B. C.; CAVALCANTI, G. A.; GONÇALVES, H. S.; DUARTE, S. T. V. G. Inovações Tecnológicas na Cadeia produtiva Têxtil: Análise e estudo de caso em indústria do Nordeste do Brasil. **Revista Produção OnLine**, v. 7, n. 2, p. 99-117, 2007.

NEULS, G. Economia movida a CO_2. **Página 22**. Escola de Administração de Empresas de São Paulo da Fundação Getúlio Vargas, n. 68. São Paulo: Ed. Vox, 2012.

PUBLIC PROSECUTION OF BRAZILIAN STATE OF PARANÁ. Oct. 18, 2011. **LONDRINA - MP-PR expede recomendação para regularizar gestão de resíduos sólidos**. Available in <http://www.mppr.mp.br/modules/noticias/article.php?storyid=1955>Accessed Nov. 5, 2014.

RIBEIRO, L.G.; ANDRADE FILHO, J. **Introdução à Engenharia Têxtil**. v. 1 (208 p.), v. 2 (214 p.), v.3. (196 p.) Rio de Janeiro: SENAI, 1987.

SÃO PAULO CITY HALL. **"Retalho Fashion" dará destinação correta aos resíduos de confecção**. Apr. 25, 2014. Available in <http://www.prefeitura.sp.gov.br/cidade/secretarias/servicos/noticias/?p=170087> Accessed Nov. 29, 2014.

SINDITÊXTIL-SP – SINDICATO DAS INDÚSTRIAS DE FIAÇÃO E TECELAGEM DO ESTADO DE SÃO PAULO. **Guia técnico ambiental da indústria têxtil – Série P + L**. São Paulo: CETESB: SINDITÊXTIL-SP, 2009.

SINDITÊXTIL-SP – SINDICATO DAS INDÚSTRIAS DE FIAÇÃO E TECELAGEM DO ESTADO DE SÃO PAULO. **Retalho Fashion: Inclusão social e preservação ambiental por meio da reciclagem de resíduos têxteis**. São Paulo /SP, 2013. 15 p.

WANG, Y. (Ed.). **Recycling in textiles**. Cambridge (UK): WoodheadPublishing, 2006. 248p.

YÁZIGI, E. **Esse estranho amor dos paulistanos: requalificação urbana, cultura e turismo**. São Paulo: Ed. Global, 2006. 376p.

ZONATTI, W. F. **Estudo interdisciplinar entre reciclagem têxtil e o design: avaliação de compósitos produzidos com fibras de algodão**. 2013. 177 p. Dissertation (Master in Textile and Fashion). School of Arts, Science and Humanities. University of São Paulo. São Paulo/SP, 2013.

ZONATTI, W.; GUIMARÃES, B.M.G.; DULEBA, W.; BARUQUE-RAMOS, R. Thermoset composites reinforced with recycled cotton textile residues. **Textiles and Clothing Sustainability**, v. 1, n. 1, p. 1-12, 2015.

Cotton Dyeing with Extract from Renewable Agro Industrial Bio-resources: A Step Towards Sustainability

Teresa Linhares and M.T. Pessoa de Amorim

Abstract The massive consumption of synthetic dyes and subsequent discharge into the aquatic ecosystem is a major environmental issue. Consumer's demand for natural textile products is driving research and development efforts to partially replace synthetic dye products for those based in renewable materials. In the growing concerns for environmental and human health issues, natural dyes are now being seen as promoters of "green chemistry" in textile dyeing, due to its non-toxic and biocompatibility properties. Colour-fastness of natural dyes may not always fulfil the usual standard requirements, but it can be improved by an inorganic or organic mordant. However, as defended by several authors, high standards on colour-fastness may not to be a top priority, mainly for environmentally aware consumers. The aim of this work was to obtain natural dyeing extracts for cellulosic fibres from *Olea europaea* pruning wastes, a native species from the Mediterranean basin, and from *Acacia dealbata* tree, a species native to south-eastern Australia introduced in Europe as ornamental, that become invasive in Mediterranean-climate regions. Dyeing conditions were studied in terms of pH, electrolyte and mordants. Colours of dyed cotton samples were measured for CIELab and LCh colour space values; colour-fastness of dyed samples was assessed by wet rubbing and washing tests according to ISO standards.

Keywords Cotton fiber · Natural dyes · Dyeing parameters · Sustainability

Introduction

General Approach

Development of science has lead to the so far best-informed generation about the risks related to the unsustainable activities, which can irreversibly damage Earth's

T. Linhares · M.T.P. de Amorim (✉)
Department of Textile Engineering (DET), University of Minho, Guimarães, Portugal
e-mail: mtamorim@det.uminho.pt

R. Fangueiro and S. Rana (eds.), *Natural Fibres: Advances in Science and Technology Towards Industrial Applications*, RILEM Bookseries 12,
DOI 10.1007/978-94-017-7515-1_35

balance and regeneration ability. Greater awareness regarding these facts tends to increase environmental consciousness of societies and organizations. Polluting industries, such as the textile industry, have been subjected to severe censorship due to its responsibility in environmental degradation. Textile Industry consumes nearby of 10^6 tons of synthetic dyes per year, and it is estimated that 15 % of those chemical products are released into the environment during application processes.

Synthetic Dyes

Manufacturing of synthetic dyes from petrochemical sources and its use in the textile industry are now facing serious challenges, since they generate large volumes of wastewater with detrimental effects to aquatic organisms and to human life, such as allergens or even carcinogens, amongst other harmful effects that may affect nature's balance (Shahid-ul-Islam and Mohammad 2014). Moreover, synthetic dyes are generally stable compounds that persist in nature (Pereira and Alves 2012). Presently, consumer's and industry demand for natural products is driving research and development efforts to partially replace synthetic products for those based on renewable materials.

Natural Dyes

Until the discovery of synthetic dyes by the end of the nineteenth century, all colouring matter for textile dyeing came from natural products. Since then, natural dye's application slumped sharply due to the improved properties of colour-fastness and colour strength of synthetic dyes, as well as its easier repeatability and reproducibility of colours. However, in the growing concerns for environmental and health issues, natural dyes are now being seen as promoters of "green chemistry" in textile dyeing, due to its non-toxic and biocompatibility properties. Research and development on natural dyes, mainly derived from non-food crops has increased substantially over the last years.

Natural Dyes Chemistry

Natural dyes can be described as direct dyes, mordant dyes or vat dyes, depending on their application and bonding with fibres to dye.

Direct dyes are water-soluble compounds particularly adapted to dye cellulosic fibres, a linear polymer with numerous polar hydroxyl side groups. The molecular planar shape of these compounds allows an aligned adherence to the fibres; direct dyes are also named as substantive dyes due to the huge affinity with cellulose.

Despite its easier application techniques, natural dyes often exhibit insufficient colour-fastness to washing and light.

Colour-fastness standards of natural vat dyes are excellent but the high cost, as well as its complicated extraction processes, tends to limit their use.

The mordant dyes are the largest group of natural dyes (Hill 1997). However, they have a reduced affinity to the cellulosic fibres. These compounds are also designated as adjective dyes since they need an auxiliary chemical, the so-called mordant, to establish the bond between the dyeing molecule and the lateral groups of the fibre.

Mordants

Mordants are additional acid or basic chemicals, with affinity for both textile fibres and dyes. They act as a link between fibre and dye's molecules, by forming an insoluble colouring complex, which improves material's colour-fastness against light and washing. Acid mordants, that bond basic dyes have generally been derived from tannin; basic mordants, which are used to bond acid dyes come from the salts of various metals, particularly aluminium, chromium, iron, copper, zinc or tin, which combine with dye molecules to produce aggregates, and prevents its removal from the dyed material (Moiz et al. 2009). Due to its toxic nature, chromium salts were practically banished, whereas tin and copper salts are used with restrictions; alum and ferrous sulphate are considered the safest among inorganic mordants and their application can lead to excellent results in dyeing.

Mordants in general are also used to increase colour yield, or even to obtain a different hue, depending on mordant selected.

Natural Mordants

Natural dyes in the textile dyeing are often used to emphasize the ecological and sustainable nature of products made from natural fibres. The toxicity of some metallic salts encouraged the search for alternative organic mordants. Besides, the use of an inorganic mordant in dyeing with natural dyes may compromise a "natural product" designation. Enriched tannins compounds have been used as mordant on textile dyeing for centuries. As referred by Jansen and Cardon (2005), black dyes combining tannins, most probably obtained from acacia pods, have been identified in ancient Egyptian textiles dating from the 18th Dynasty (1542–1302 BC). Shahid et al. (2013) states that phenolic hydroxyl groups of tannins tend to form effective crosslinks with different types of fibres and dyes and help in colour fixation. Tannic acid is a natural mordant found in many plants, and tannins are the best natural mordants for cotton and other cellulosic fibres (Tull 2013).

More recently some eco-friendly mordants are receiving renewed attention, such as the extracted solution from *Acacia catechu,* since its alternative application may reduce the risk of environmental pollution caused by toxic metal salts (Islam et al. 2013).

The extract obtained from the heartwood of the *Acacia catechu* tree, was indeed used in the past to produce a fast bronze colour on cotton fabrics, as well as a preserving agent.

Acacia Dealbata

Acacia dealbata is a fast-growing species, native to south-eastern Australia and was introduced in Europe as an ornamental plant at the end of eighteenth century. The species quickly settled, later becoming invasive in Mediterranean climates from Portugal to Italy, since it quickly resprouts following cutting or fire. The Portuguese decree 565/99 of 21st December defined this species as invasive.

Currently, *Acacia dealbata's* end uses had potential on biomass pellets for industrial application, and also at paper industry, namely in writing and printing paper production, according to Santos et al. (2006). In this work a novel application on *Acacia dealbata* tree bark were exploited: its use as natural dyeing bath, alone and along with extract of pruning wastes, to dye cellulosic fibres, with acceptable colour-fastness properties.

Natural Dyes Benefits

Natural dyes contain colouring compounds that are neither harmful for human health nor hazardous to the environment and its dyed materials exhibit soothing colours, with pleasant earthy hues (Moiz et al. 2009).

As defended by some authors, high standards on colour-fastness may not to be a top priority, mainly for environmentally aware consumers (Hill 1997; Moiz et al. 2009). Hill wrote in 1997 that "articles can be represented as being unique, with slightly uneven dyeing, or with variations in shade from one batch to another, or fade characteristically in use", and in fact, this is a strategy adopted by several clothing brands as a marketing benefit, by adding the attribute of "being unique garments". Moreover, the author refers as the poor colour-fastness at rubbing of indigo on blue denim became a fashionable attribute that is now emulated by other synthetic dyes.

The purpose of this work intends to explore natural dyes application as a viable co-partner or alternative to synthetic dyes in specific contexts.

Environmental Sustainability

Wastes: Source of Natural Dyes

Agriculture and timber industry wastes have attracted the attention of researchers as a source of natural textile dyes due to its abundance and availability at minimal costs. Additionally, its use can promote the ideal of "zero emissions", based on concept that wastes from one industry can be converted to raw material for another one, along with the added economic value to those wastes and minimizing its inherent environmental impact.

In this work, olive tree pruning wastes and *Acacia dealbata* tree bark were used to obtain colouring extracts as mordant to be used in dyeing bath for cellulosic. As per available literature it is the first time that such an experiment was carried out and it is intended to value agricultural and timber wastes, being a new sustainable concept and an opportunity to lower costs involved in natural dyes application.

Olive Tree Culture

Olive tree's (*Olea europaea*) culture is one of the most important and oldest agricultural activities, and the consumption of olives and olive oil foments a healthy diet; ironically its by-products are a severe environmental issue, mainly in the Mediterranean area (Arvanitoyannis and Kassaveti 2008).

According to data of FAOSTAT (2012), about 95 % of olive world surface is concentrated in the Mediterranean basin, and the producer countries from south of CEE contribute with 69 % of the worldwide production.

In Portugal, olives trees are mainly located in the northeast, at Trás-os-Montes region, in small groves, and in large plantations in the centre and south of the country. The most important species in Portugal is the 'Galega', representing about 60 % of the olive trees

Production of olive oil is still one of the most polluting agro-industries, generating large amounts of solid wastes, besides olive oil mill wastewaters. However, only a small part of those by-products are recycled or reused (Arvanitoyannis and Kassaveti 2008).

Olive Tree Pruning Wastes

Olive trees have to be pruned to control their continuous growth, although slow, mainly to maximize productivity and to facilitate olive harvesting. The resulting by-products are considerable amounts of olive leaves and branches. A typical olive tree pruning lot includes 70 % thin branches (by weight, with approx. one-third of leaves) and 30 % of wood (thick branches, diameter >5 cm approx.), estimated at 3 ton/ha, as an annual average (Cara et al. 2007), varying between 12 and 30 kg per tree. The chemical composition of these wastes is depending on the pruning severity and age of tree (young, adult or old); however protein contents is invariably low, which result in low digestibility coefficient, not adequate for animal feeding. Disposal of these residues is necessary to keep fields clean and to prevent propagation of vegetable diseases, often by burning or grinding and scattering on fields, and these activities add economic costs and environmental concerns.

The purpose of this work envisages developing a more eco-friendly technology by the application of a renewable bio-resource as a source of natural colouring material for dyeing cellulose substrates.

Materials and Methods

Cotton Pre-treatment

Before all dyeing processes, cotton woven was pre-treated to remove sizing agents as well as intrinsic fats and impurities, in order to allow dye molecules adsorption.

In this work, cotton desizing was carried out with a commercial enzymes solution (Rapidase L40), at liquor rate of 1:30 and pH 8.38, at 70 °C for 60 min, with agitation. The product's sustainability was enhanced with an enzyme's pre-treatment at a low temperature and minimal consumption of chemicals compounds.

Dyeing Bath Obtainment

Several extractions on olive pruning wastes were carried out at different conditions in order to optimize colouring extracted solutions to be used as dyeing baths. All experiments were performed in a closed system, in laboratory equipment Ahiba IR^{TM}, which allows higher pressures. The solvent was deionized water at a liquor ratio of 1:10.

The smaller liquor rate used in this work intended to obtain a concentrated bath with less energy consumption, small amount of water, and reduce the costs involved in the extraction of the colouring material at the dye-house.

This work comprised the following studies:

- Temperature of extraction: 95 and 120 °C;
- Time of extraction: 60 and 30 min;
- Pruning wastes pre-treatment: dry raw material (whole leaves and small twigs) and grinded material;
- Pruning wastes from young and adult trees.

Dyeing Procedure

Cotton dyeing was carried out in the laboratory dyeing machine Ahiba IR^{TM}, by the exhaustion method, at 95 °C, for 60 min, at a liquor ratio of 1:10, by using the extracted colouring solutions alone, as well as by addition of mordants and electrolyte to those solutions. After dyeing, the unfixed dyestuff was removed by rinsing with cold water.

The inorganic mordants used were $FeSO_4 \cdot 7H_2O$ and $KAl(SO_4)_2 \cdot 12H_2O$. An organic mordant was also applied, a solution extracted from *Acacia dealbata* tree bark. The electrolyte effect was studied by the addition of NaCl.

Colour Assessment

Dyed samples colour was assessed in terms of CIELab and LCh colour space under standard illuminant D65. K/S (colour strength) was also recorded.

Colour-Fastness Evaluation

The dyed samples were tested for colour-fastness to wet rubbing according ISO Standard NP EN ISO-X12: 2003. Washing tests were also carried out to assess colour changing degree and staining on multifibre fabric, as per ISO Standard NP EN ISO 105-C06: 1999.

Results and Conclusions

Extraction of Different Materials Studied

Extracted solutions were analysed in terms of pH, conductivity, total solid contents and volatile solid contents. The experimental results are given in Table 1.

At 95 °C, grinded materials yielded a higher rate of solid contents when compared with raw whole materials, i.e., 15.34 g/L versus 6.02 g/L.

It was realized that, by increasing temperature to 120 °C, total solid contents of raw material extraction (whole leaves and small twigs) was similar to the values of extraction with grinded material at 95 °C.

These results suggested that grinding pre-treatment of pruning wastes can consequently be eliminated, which minimizes inherent costs and time processing, as well as drastically improving the filtration rate and reducing clogging in the filtration

Table 1 Solutions extraction conditions and results of studied parameters

Studied parameters	Grinded		Raw (whole leaves and small twigs)			
	Adult		Adult			Young
	Extraction conditions					
	95 °C 60 min	120 °C 60 min	95 °C 60 min	120 °C 60 min	120 °C 30 min	120 °C 60 min
pH	4.63	4.53	4.57	4.25	4.35	4.32
Conductivity (mS/cm)	4.34	4.47	2.64	3.35	2.87	1.77
Total solid contents (g/L)	15.34	18.44	6.02	19.22	7.72	22.56
Volatile solid contents (g/L)	12.44	15.62	4.64	16.12	6.34	20.7

system. In addition, the extraction being performed in a sealed system prevents water evaporation and might contribute to improved process reproducibility by preventing the pH variation and oxidation of the compounds during extraction.

An additional extraction on raw materials was performed at 120 °C, but reducing the time for 30 min; however, solid contents were significantly smaller—only 7.72 g/L, and this option was, therefore, rejected.

Whole leaves and small twigs from an young tree were also used to carry out the last extraction at 120 °C to assess the tree age effect—in spite of different values in solution's conductivity, the results on total solid contents were quite similar, which suggest no influence of the tree age variable. These results allow us to conclude that the mixture of pruning wastes from trees with different ages does not cause problems and may even contribute to improve colour's reproducibility. Based on these results and on the colours strength of the dyed samples it seems possible to conclude that optimal extraction conditions of tree pruning wastes are at 120 °C for 60 min.

Dyeing Results

Extracted Temperature Solution and Material Conditions

Preliminary dyeing trials were carried out to assess the colour strength (K/S) due to different extraction conditions. Cotton woven samples were dyed with solutions obtained from grinded and whole pruning adult tree wastes (leaves and small twigs) extracted at 95 °C and from whole pruning adult tree wastes extracted at 120 °C. Results in K/S are given in Fig. 1. The colour strength is significantly increased in the dyed samples with extracted solution from whole material, at 120 °C, which is in agreement with previous results on total solid contents determination.

pH Effect of Dyeing Solutions in Colour Strength

The pH variations and its influence in colour strength of dyed cotton were assessed on both young and adult tree pruning wastes extractions, as shown in Fig. 2. Both extractions were carried out at 120 °C. K/S tends to increase with alkalinity's solution until pH $\approx$ 8.5.

Effect of Electrolyte and Mordant (Inorganic and Organic) on K/S

The effect of electrolyte and inorganic mordants were assessed in terms of colour strength on both extracted solutions from young and adult tree pruning wastes. NaCl salt, $KAl(SO_4)_2 \cdot 12H_2O$ and $FeSO_4 \cdot 7H_2O$ mordants were used at a 5 g/L concentration. A combination of both inorganic mordants, each one at a concentration of 2.5 g/L was also used.

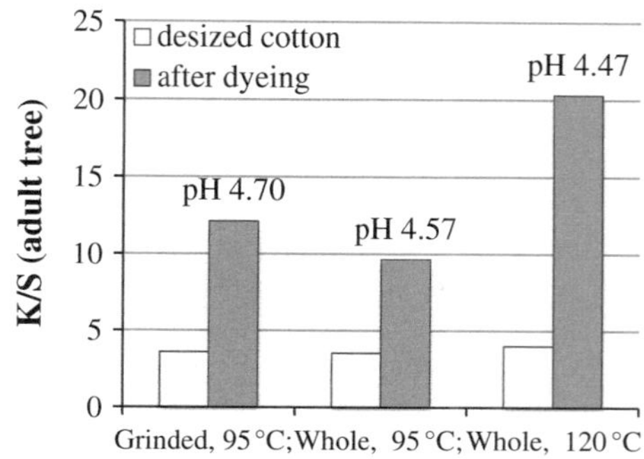

Fig. 1 K/S results of dyed cotton with different extraction conditions, and comparison with desized cotton (DSZ CO)

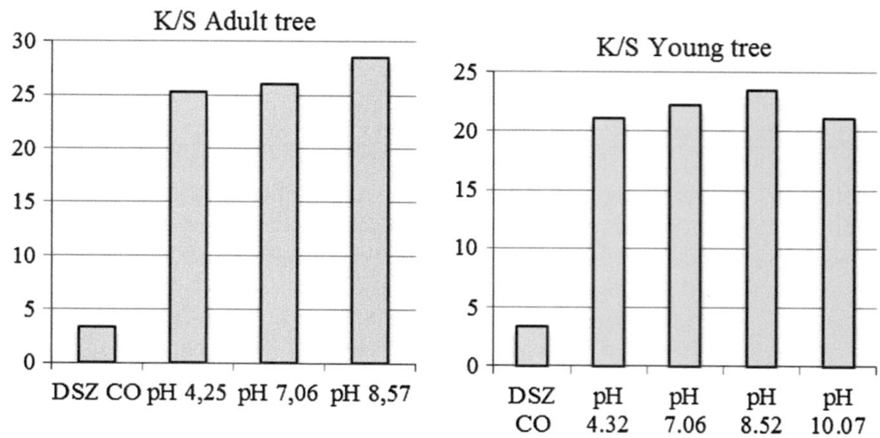

Fig. 2 pH influence on colour strength of dyed cotton (both extractions performed at 120 °C) and comparison with desized cotton (DSZ CO)

Two aqueous solutions were extracted from *Acacia dealbata* tree bark, at 95 °C, varying liquor rate at 1:5 and 1:10. By expecting they would act as natural mordant (NM), both solutions were used as a dyeing bath for cotton fibres, alone and along with extracted solutions from tree pruning wastes (OS). The mordanting method adopted was *meta*-mordanting; the mordant was applied simultaneously with the dye on the same bath. Results on K/S of dyed samples of salt and mordant solutions, as well as in the original solution (OS) can be seen on Fig. 3.

The NaCl addition had practically no effect on colour appearance whereas KAl and mainly iron sulphate, showed significant effect on dyeing result, yielding a different colour hue, apart from increased K/S. The combination of both inorganic mordants with OS yielded a deep colour, whereas extracted solution from *Acacia dealbata* along with OS dyed cotton at a bronze shade with a high value on K/S.

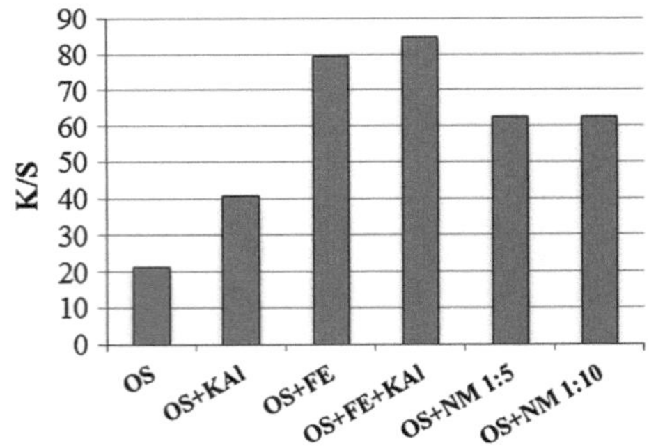

Fig. 3 K/S Results on cotton dyed samples with electrolyte and mordant addition to the original solution

Colour Coordinates and Colour-Fastness Results

Colour coordinates results of dyed samples can be seen in Table 2. Shade variations from desized cotton and between dyed colours are given in the grayscale column (1 corresponding to white and 100 corresponding to black). The colour-fastness results on wet rubbing are presented as well.

In general, colour-fastness results of wet rubbing on all dyed samples are quite acceptable, since the lowest degree achieved was 3–4 on dyeing of *Acacia dealbata* solution, in extraction rate 1:5 (this result can eventually be improved by NaCl addition). In spite of NaCl addition had no significant influence on colour strength, it was noticed some improvements on wet rubbing results.

Colour-fastness to washing has been assessed on selected samples for the changing in colour and staining on multifibre fabric, according to B1S standard (at 50 °C for 30 min). Results on staining were very good, since even at 50 °C washing, there was no visible contamination on tested fibres, and a degree 5 was attributed to all fibres, on all tests carried out (scale ranging from 1 to 5, and 5 is the best result). The extracted solutions used as dyeing bath on tested samples and results on degree colour changing are presented in Table 3.

Colour changing results on samples C and D are poor, classifieds at the lower degree. As previously mentioned these tests were performed at 50 °C and not at 40 °C, the most common temperature of domestic laundry washing.

Conclusions

This work intended to explore natural dyes application as an alternative to synthetic dyes in specific contexts.

Olive tree pruning wastes, such as leaves and small twigs, were used to obtain a colouring extract as dyeing bath for cotton. Based on the achieved results, the optimal extraction conditions are at a closed system (in a vessel under pressure), liquor ratio of 1:10, at 120 °C, for 60 min. The smaller liquor rate carried out in this work allows to obtain a concentrated bath with less energy consumption due to the

Table 2 Dyeing solutions and sample results (pruning wastes from young tree, whole leaves, 120 °C, 60 min)

Dyeing solutions	pH	L*	a*	b*	C*	h	K/S (sum)	Grayscale	Wet rubbing
Original solution	4.32	75.92	2.73	15.87	16.10	80.23	21.13		4–5
	7.04	72.89	3.61	16.58	16.97	77.81	22.88		4–5
	8.52	72.32	4.59	15.74	16.40	73.76	23.45		4–5
	10.07	73.66	5.01	15.66	16.44	72.26	21.15		5
Original solution + NaCl	4.32	73.93	2.38	16.29	16.46	81.67	24.00		5
	7.04	73.33	3.68	15.36	15.79	76.54	24.66		4–5
Original solution + $KAl(SO_4)2$	4.32	71.97	1.82	30.20	30.26	86.56	40.56		4–5
	7.04	71.44	1.71	31.49	31.57	86.89	41.00		4–5
Original solution + $FeSO_4$	4.32	51.24	0.69	11.79	11.81	86.67	79.50		4–5
	7.04	51.88	0.72	11.17	11.2	86.31	75.16		4
Original solution + Fe + KAl	4.32	50.88	0.78	13.16	13.18	86.59	84.92		4
	7.04	52.88	0.50	11.78	11.79	87.55	70.56		4
Original solution + nat.mord.1:5	4.32	56.48	10.16	16.86	16.98	58.93	62.36		3–4
Original solution + nat.mord.1:10	4.32	57.04	8.83	16.73	18.92	62.16	62.20		3–4
Desized cotton	–	88.78	1.25	10.33	10.41	83.11	3.40		

CIEL*a*b* refers to the colour space with a 10° observer angle, specified by International Commission of Illumination

Table 3 Dyeing bath solutions, K/S of dyed samples and colour change degree results of washing tests

Samples	Dyeing solution	K/S	Colour change degree
A	Adult tree—pH 8.57	28.48	4
B	Young tree—pH 7.04 + NaCl	24.66	4
C	Young tree—pH 4.32 + KAl	40.56	2
D	Adult tree—pH 7.06 + $FeSO_4$	64.47	1

small amount of water, since it is essential to reduce the costs involved in the extraction of the colouring material at the dye-house.

From this study one can conclude that olive tree pruning wastes extract can be used as dyeing bath with a low environmental impact. The dye extract is obtained without any chemicals additives and different shades were achieved only by pH changing.

Acacia dealbata, a fast-growing invasive plant in mediterranean-climate regions was also used as natural mordant. The dyeing experiments with the extract of pruning wastes from olive tree and *Acacia dealbata* as a mordant exhibited promising results on colour strength and colour-fastness on cotton dyed samples.

The sustainability of the dyeing process and valorisation of wastes are based on saving of resources such as water, energy and chemicals together with an ecological impact minimisation of the dyeing process in terms of effluents emission.

References

Arvanitoyannis IS, Kassaveti A (2008) Olive Oil Waste Management: Treatment Methods and Potential Uses of Treated Waste. In: Arvanitoyannis IS (ed) Waste Management for the Food Industries. Elsevier, New York, p 453-568

Cara C, Romero I, Oliva JM, Sáez F, Castro E (2007) Liquid Hot Water Pretreatment of Olive Tree Pruning Residues. In: Mulchandani A (ed) Applied Biochemistry and Biotechnology, vol 136-140. Humana Press, New Jersey, p 379-394.

Hill DJ (1997) Is there a future for natural dyes?. Review of Progress in Coloration and Related Topics 27(1):18-25.

Islam S, Shahid M, Mohammad F (2013) Perspectives for natural product based agents derived from industrial plants in textile applications - a review. J Clean Prod 57: 2-18. http://dx.doi.org/10.1016/j.jclepro.2013.06.004

Jansen PCM, Cardon D (eds) (2005) Plant Resources of Tropical Africa 3-Dyes and tannins. Prota Foundation, Wageningen.

Moiz A, Ahmed MA, Kausar N, Ahmed K, Sohail M (2009) Study the effect of metal ion on wool fabric dyeing with tea as natural dye. J Saudi Chem Soc 14:69-76. doi:10.1016/j.jscs.2009.12.011

Pereira L, Alves M (2012) Environmental Impact and Remediation. In: Malik A, Grohmann E (eds) Environmental Protection Strategies for sustainable Development. Springer, Heidelberg, p 111-162

Santos AJA, Anjos OMS, Simões RMS (2006) Papermaking potential of *Acacia dealbata* and *Acacia melanoxylon*. Appita Journal 59(1):58-64

Shahid-ul-Islam, Mohammad F (2014) Emerging Green Technologies and Environment Friendly Products for Sustainable Textiles. In SS Muthu (ed) Roadmap to Sustainable Textiles and Clothing - Environmental and Social Aspects of Textiles and Clothing Supply Chain. Springer, Singapore, p 63-82

Shahid M, Islam S, Mohammad F (2013) Recent advancements in natural dye applications: a review. J Clean Prod 53:310-331. http://dx.doi.org/10.1016/j.jclepro.2013.03.031

Tull D (2013) Edible and Useful Plants of the Southwest: Texas, New Mexico, and Arizona. University of Texas Press, Austin

Erratum to: Review of Wool Recycling and Reuse

Stephen Russell, Paul Swan, Mariell Trebowicz and Angus Ireland

Erratum to: 'Review of Wool Recycling and Reuse' R. Fangueiro and S. Rana (eds.), *Natural Fibres: Advances in Science and Technology Towards Industrial Applications*, RILEM Bookseries 12, DOI 10.1007/978-94-017-7515-1_33

The book was inadvertently published without the co-author names "Paul Swan and Mariell Trebowicz" of Chapter 33. The complete author names are given below:

Stephen Russell
Paul Swan
Mariell Trebowicz
Angus Ireland

The updated original online version for this chapter can be found at DOI 10.1007/978-94-017-7515-1_33

S. Russell (✉) · M. Trebowicz
Nonwovens Innovation and Research Institute, Leeds, UK
e-mail: s.j.russell@leeds.ac.uk

P. Swan · A. Ireland
Australian Wool Innovation Limited (AWI), Sydney, Australia

P. Swan · A. Ireland
International Wool Textile Organisation (IWTO), Brussels, Belgium

R. Fangueiro and S. Rana (eds.), *Natural Fibres: Advances in Science and Technology Towards Industrial Applications*, RILEM Bookseries 12,
DOI 10.1007/978-94-017-7515-1_36

Index

R. Fangueiro and S. Rana (eds.), *Natural Fibres: Advances in Science and Technology Towards Industrial Applications*, RILEM Bookseries 12,
DOI 10.1007/978-94-017-7515-1

MIX
Papier aus verantwortungsvollen Quellen
Paper from responsible sources
FSC® C105338

If you have any concerns about our products,
you can contact us on
ProductSafety@springernature.com

In case Publisher is established outside the EU,
the EU authorized representative is:
Springer Nature Customer Service Center GmbH
Europaplatz 3, 69115 Heidelberg, Germany

Printed by Libri Plureos GmbH
in Hamburg, Germany